**The "Postage Stamp" Tourmaline**
**Photo. Harold & Erica Van Pelt**

SANCO PUBLISHING
DIVISION OF SANCO LTD.
P.O. Box 177
Appleton, WI 54912-0177


Printed and bound in Singapore through Asiaprint Ltd. USA

Library of Congress Catalog Card No. 96-070150
ISBN 0-9653510-0-9

Designed by F. John Barlow

Dust jacket design
by
Harold and Erica Van Pelt

# The F. John Barlow Mineral Collection

F. John Barlow
Managing Editor

Robert W. Jones and Gene L. LaBerge
Associate Editors

SANCO PUBLISHING
APPLETON, WISCONSIN

F. JOHN BARLOW

Charcoal Drawing by World-Famous Artist George Pollard

# Contents

## The F. John Barlow Mineral Collection

# Dedication

*This book is dedicated to my beloved wife and partner for 58 years, Dorothy M. (Marx) Barlow (August 22, 1915-January 30, 1994).*

*"All of her interests centered around her family, and her life was devoted to the well-being of her husband and eight daughters. Her radiant smile and gentle kindness will be remembered by all who were privileged to know her."*

*Without Dorothy's constant backing and support, whatever successes I have achieved would not have been possible.*

*F. John Barlow*

# Acknowledgments

This book could never have materialized without the constant prodding and encouragement by my good friend, Professor Gene LaBerge, and the confidence and hard work of Bob Jones. My highest regard and appreciation to the great team of authors who contributed to this book: Bob Jones, Scottsdale, Arizona; Gene LaBerge, Oshkosh, Wisconsin; David Wilber, Golden, Colorado; Joel Bartsch, Houston, Texas; Charles Key, Portland, Maine; Richard Thomssen, Reno, Nevada; Terry Wallace, Tucson, Arizona; Bill Smith, Broomfield, Colorado; Marc Wilson, Pittsburgh, Pennsylvania; Peter Megaw, Tucson, Arizona; Peter Juneau, Mukwonago, Wisconsin; Vandall King, Rochester, New York; and Martin Jensen, Reno, Nevada.

A special thank you to artist George Pollard, Kenosha, Wisconsin; writer John Sinkankas, San Diego, California; Wendell Wilson, *Mineralogical Record*; Marie Huizing, *Rocks and Minerals*; Merle White, *Lapidary Journal*; and John Almquist for their permission to use their copyrighted material.

My appreciation to all the photographers for their professional expertise in helping to make this book what it is — Earl Lewis (EL) of Los Angeles, California; John Lewis (JL) of Kaukauna, Wisconsin; Mike Gray (MG) of Montana; Jeff Scovil (JS) of Phoenix, Arizona; Dr. Wendell Wilson (WW) of Tucson, Arizona; Nelly Bariand (NB) of Paris, France; Dr. Olaf Medenbach (OM) of Germany; Erica and Harold Van Pelt (VP) of Los Angeles, California; and Malcolm Hjerstedt (MH) of Munroe Studios, Neenah, Wisconsin, who provided most of the photography. (Each photo plate identifies the photographer by initials; UK indicates that the photographer is unknown.)

Thank you to all the collectors for their encouragement, assistance, and camaraderie through my collection years: John Almquist, Pauline Armstrong, Dr. Eric Asselborn (France), Dr. Peter Bancroft, Joel Bartsch, Leonard Bedale, Glenn Bolick, Ralph Clark, Carlton Davis, Dave DeBruin, Dr. Don Doell (Canada), Dr. Dale Dubin, J. Folch, Spain, Lance Hampel, Evan Jones, Dr. Gene LaBerge, Dr. Miguel Romero (Mexico), James Minette, Dr. Steve Neely, Joe Ondraka, Norman Pellman, Bill Pinch, Keith Proctor, I. Philip Scalisi, Art Sexauer, Harvey Siegel, Dr. Steve Smale, Marion Stuart, Ed Swoboda, Dave Wilber, Barry Yampol, and Martin Zinn.

Thank you to William R. Thompson, Neenah, Wisconsin for his guidance in printing; to Don and Kasnea Martin, Palm Desert, California for their helpful guidance during the early layout; and to Mary Dillmann for professional layout and design service.

In climbing the learning curve, my gratitude to Paul Desautels, John White, Dr. Pete Dunn, Paul Pohwat, and Dr. Jeff Post, Smithsonian; Dr. Carl Francis and Bill Metropolis, Harvard Mineralogical Museum, Cambridge, Massachusetts; Dr. Tony Kampf, Los Angeles County Museum; Stan Dyl II and Octave DuTemple, A. E. Seaman Mineralogical Museum, Houghton, Michigan; Klaus Westphal, Museum of Natural

Science, Madison, Wisconsin; Peter Embrey, retired, British Museum of Natural History, London, England; Dr. Leo Bulgak and Dr. Alexander A. Godovikov, Fersman Museum of Natural History, Moscow, Russia; Bill Roberts, South Dakota School of Mines, Custer, South Dakota; Natasha Pitomtseva of the St. Petersburg Museum; Dr. Vincent D. Manson, Gemological Institute of America; Dr. Pierre Bariand, Sorbonne Mineralogical Museum, Paris, France; Dr. Richard Gaines; Dr. Fred Pough; and Dr. Robert B. Cook, Auburn University, Auburn, Alabama.

In appreciation for the friendship and help in acquiring minerals of the finest quality, thank you to Jack Amsbury, Carlos Barbosa, Allen Bassett, Gerhard Becker, Udo J. A. Behner, Mike Bergmann, Dudley Blauwet, John Brottlund, Helmut Bruckner, Dave Bunk, Uli Burchard, Ernest Butler, Jerry Call, Larry Conklin, Ed Coogan, Rock Currier, Kevin Davy, Si Frazier, Gilbert Gauthier, Dr. Georg Gebhard, Harvey Gordon, Cal Graeber, Lance Hampel, Dr. Gary Hansen, Ed Harris, Rex Harris, Richard Hauck, Gonzalo Jara, Casey Jones, Michel Jouty, Tom McKee, Charles Key, W. J. Knoblock, Richard Kosnar, Dr. Rustam Z. Kothavala, Bill Larson, Gary Ledford, Bryan Lees, Wayne Leicht, Manak Lunia, Raz Lunia, Theo Manos, James G. McGinley, Al McGuinness, John McLean, Dr. John Medici, Frank Melanson, Herb Obodda, Don Olson, D.J. "Doug" Parsons, Don Pearce, G. E. Penekis, Lee Petretti, Julius Petsch, Jr., Sid Pieters, Clive Queit, John and Laura Ramsey, Gary Richards, Mike Ridding, Ken Roberts, Willard L. Roberts, Dr. Abe Rosenzweig, John Saul, Gene Schlepp, Manfred Schwarz, Robert W. Seasor, Kirby Sieber, Milton and Hilda Sklar, Vern Stratton, Abe K. Suleman, Bob Sullivan, Terry Szenics, Wayne Thompson, Roger Titeaux, Brad and Star Van Scriver, Mario Vizcarra, Jim Walker, Andreas Weerth, Richard Whiteman, John and Rosa Whitmire, Prosper Williams, Chris Wright, Victor Yount, and Julius Zweibel.

My grateful thanks to Deborah A. Schueller, Houghton, Michigan for copy editing the entire text. A special thank you to Richard A. Bideaux of Oro Valley, Arizona for his technical review of this entire text.

A very special appreciation to my family, daughters Alice Barlow, Bonnie Dignam, Grace Cohn, Jacqueline Barlow, Joan Barlow, Joyce Helein, Terri Barlow Brown, and Wendy Barlow-Juneau, and to my wife, Dorothy, for 58 years of 100% support.

I am deeply indebted to my hardworking helpers: my former secretary, Betty Erdmann; Holly Paulson; Carrie Johnson; Chris Muller; my grandson, Greg Helein; Ruth Leichtfuss; Sally LaBerge; Mark Loper; and pilot/bodyguard, Don Hoyman (deceased). And a very special thank you to my secretary, Mary Lou Grapentin.

# The F. John Barlow Mineral Collection

Managing Editor: F. John Barlow

Associate Editors: Bob W. Jones and Gene L. LaBerge

Authors:

**F. JOHN BARLOW**, Appleton, Wisconsin. B.S. Mechanical Engineering, University of Wisconsin, Madison (1937); Doctor of Science, Honoris Causa, University of Wisconsin (1994); Owner, AZCO INC., SANCO INC., and former owner of Earth Resources. Corporate board member of several corporations, SANCO Publishing, SANCO Minerals.

**JOEL A. BARTSCH**, The Woodlands, Texas. B.A. Seminary Studies, Concordia Lutheran College (1986); Colorado School of Mines, Mining Engineering (1980-84); Curatorial Assistant, Colorado School of Mines; Curator, Lyman House Memorial Museum, Hilo, Hawaii (1986-88); Curator/Director, California State Mining and Mineral Museum (1988-91); Curator of Gems, Houston Museum of Natural History (1991-present).

**MARTIN V. JENSEN**, Santa Monica, California. B.S. University of Nevada, Reno (1981); M.S. South Dakota School of Mines and Technology (1984); Research chemist (1985-86); Manager, Scanning Electron Microscope Laboratory, University of Nevada, Reno (1985-91). Presently mining.

**ROBERT W. ("BOB") JONES**, Scottsdale, Arizona. B.S. (1956), M.S. (1959) Education, Southern Connecticut State University; retired science teacher; member Rockhound Hall of Fame; lecturer and free lance writer of over 500 popular mineral articles; author of *Luminescent Minerals of Connecticut* and *Nature's Hidden Rainbows*; actively interested in minerals for over 50 years.

**PETER J. JUNEAU**, Mukwonago, Wisconsin. B.S. University of Wisconsin, Green Bay (1971); M.S. University of Minnesota, Duluth (1979). Exploration geologist (WGM, Inc., 1978; U.S. Steel, 1979-82). Own consulting business (1982-84). Science tutor/instructor (1984-86). Wisconsin teaching license (1987). Operation/database manager of Environmental Field Services group at Du Pont (1988-91). Geologist at consulting company in Milwaukee (1991-present).

**CHARLES L. KEY**, Cape Elizabeth, Maine. *Who's Who in Mineral Names* re: KEYite and LudLOCKite, both from Tsumeb, Namibia. Member Rochester Academy of Science. Over 35 years full-time involvement in the mineral and gem field with worldwide efforts in mining, prospecting, and collecting, especially in the Republic of South Africa and Namibia, which continues today.

**VANDALL T. KING**, Rochester, New York. B.S. University of Maine, M.S. State University of New York. Former staff mineralogist at Ward's Natural Science Establishment; author of *Mineralogy of Maine*; coauthor of eight other mineral related books. A director of Friends of Mineralogy and active in organizing Rochester Symposium. Recipient of the Distinguished Achievement Award, Eastern Federation of Mineralogical Societies (1993).

**GENE L. LaBERGE**, Oshkosh, Wisconsin. B.S. (1958), M.S. (1959), Ph.D. (1963), University of Wisconsin, Madison. Postdoctoral research fellow, Adelaide, Australia (1963-64) and Geological Survey of Canada (1964-65). Professor of Geology, University of Wisconsin, Oshkosh (1965-present). Awarded the F. John Barlow Endowed Professorship, University of Wisconsin, Oshkosh (1993). An active collector of minerals.

**PETER K.M. MEGAW**, Tucson, Arizona. B.S. (1976), M.A. (1979) University of Texas, Austin; Ph.D. University of Arizona (1990). Teaching assistant (1976 and 1984-86); Geologist (1978-88); President IMDEX, Inc., Tucson, Arizona; Past President Tucson Gem and Mineral Society (1984, 1990, 1991); Exhibit Chairman, TGMS Show (1987-93). Current owner of Megaw Minerals. Invited speaker to many mineral symposia and shows.

**WILLIAM R. ("BILL") SMITH**, Broomfield, Colorado. U.S. Marines (1941-45), served in the Pacific theater. University of Chicago, A.B., S.B. (1948-49). M.S. Math, University of Wisconsin (1951); University of California, Berkeley (1957-58). Retired National Security Agency (1980). Retired Cray Research Company (1987). Mineral collector since age 13.

**RICHARD W. THOMSSEN**, Carson City, Nevada. University of California, Berkeley and University of Arizona, Tucson. Graduate professional geologist with over 40 years experience in the field. A mineral collector for 50 years, specializing in microminerals with a computer-indexed collection of over 8,000 specimens. Past member Board of Directors and Past President/Executive Director of *Mineralogical Record*. Elected to the Micromounter's Hall of Fame, 1987.

**TERRY C. WALLACE, JR.**, Tucson, Arizona. B.S. Geophysics and Mathematics, New Mexico Institute of Mining and Technology (1978); M.S. Geophysics, California Institute of Technology (1980); Ph.D. Geophysics, California Institute of Technology (1983); Associate Professor (1988-92) and Professor (1992-present), University of Arizona. Curator, University of Arizona Mineral Museum (1984-present).

**DAVID P. WILBER**, Golden, Colorado. Systems analyst supervisor for Prudential Insurance prior to working for Commerical Minerals Company, New York (1968). Started own mineral business within two years. While assembling one of the great mineral collections, gained recognition as a premier mineral collector/dealer. Played a major role in lifting the mineral hobby to big business status.

**MARC L. WILSON**, Cheswick, Pennsylvania. B.S. (1975), M.S. (1979) Geology/Mineralogy, Michigan Technological University, Houghton. Exploration and remote sensing geologist. Mineralogist; Curatorial Assistant, MTU; Mineralogist/Curator, New Mexico Bureau of Mines and Mineral Resources, Socorro; Recipient of the Friends of Mineralogy Award of Merit, 1993; collector since age 8; currently Collections Manager and Head of Section for Minerals, Carnegie Museum of Natural History, Pittsburgh.

# PART I

## *THE COLLECTOR*

### *Introduction*

The F. John Barlow mineral collection is internationally known for its superb minerals, rarities, and exciting mineral suites from important localities. The collection is also a reflection of the man who assembled it.

Part I offers some insights into all of us as collectors. It also provides the reader with an abbreviated biography of John Barlow, the collector. This biography helps the reader gain a greater appreciation of John and the circumstances that brought him to assemble one of the great collections of our time.

> *To the enlightened amateur,*
> *preserver of so many good things of the earth,*
> *whose contributions to society, although great,*
> *have been little acknowledged.*
>
> *- Douglas and Elizabeth Rigby*
> Lock, Stock and Barrel
> The Story of Collecting.
> *J.B. Lippencott, N.Y. 1944.*

CHAPTER 1

# The Collector Instinct

*by Bob Jones*

**The story of John Barlow's collecting is really every collector's story.**

F. John Barlow's mineral and gem collection epitomizes the ultimate goal of a collector. It encompasses the rare, the scientifically interesting, the beautiful, the hard-to-obtain, the one-of-a-kind, and those specimens that simply appeal to the individual collector's taste. The collection holds the very best specimens John has been able to obtain. Simultaneously, the collection provides insight into the personality of the collector.

While working on this project, I was struck by characteristics in John's personality that I have seen so often in other collectors, including myself. A desire to accumulate something of interest, the effort to obtain the very best, constant study and learning that goes with the art of collecting, the joy of competition with other collectors as each seeks to outdo the others in obtaining the most beautiful or rarest, the thrill of the hunt, and, as mineral collectors, the expression of love associated with admiring nature's mineral art forms. All these aims are rooted within us to varying degrees. Collectively, they comprise what I call the "collector instinct."

*Plate #1-1*

When we collect, it is as if we are staking out, and adding to, our own territory much as animals do. To have a collection, be it marbles, baseball cards, minerals, animals, or what have you, becomes a very real territorial expression. The more significant the collection becomes, regardless of size, the greater is our sense of achievement knowing we control something that is tangible and real.

From our earliest age, we instinctively collect. Ask mineral collectors and I will bet you will find that as kids, they collected stamps, popsicle sticks, a certain type toy, colorful pebbles, a host of things. And, as we mature, so does our taste. Collection emphasis shifts to more

> *I was struck by characteristics in John's personality that I have seen so often in other collectors.*

complex, interesting, valuable, or aesthetically appealing "things," whatever they are.

Vandall King, in writing his chapter on tourmalines, recognized certain phases of collecting. "The initial stages of collecting are concerned with inward perception. Acquisition for its own rewards is the goal. The normal evolution of a collector involves maturation where one's place among his peers takes shape. The 'outward' phase of a collector's maturation involves awareness that allows a collection to blossom."

So, the true collector moves inexorably through a whole spectrum of collecting activities beginning with something quite simple, both in value and understanding, and moving toward the more valuable and complex as the art of collecting becomes a serious goal in the collector's life.

Mineral collecting is, in a very real sense, a recognition and expression of an art. It is the recognition of minerals as an art form. Mineralogist Frank C. Hawthorne, writing in the *Canadian Mineralogist* on the history of mineralogy, recognized that art is a part of the nature of man, saying, "Fascination with art seems an intrinsic property of the human psyche." And, since crystals have long been recognized as one of nature's great art forms, it is only natural that we are attracted to them and may ultimately want to possess them. Hawthorne takes the idea of minerals being intrinsic to man one step further as he says, "... the early development of humanity is intrinsically associated with the recognition and use of minerals." Thus, he makes not only a direct connection with man's regard for the arts, but also applies it to the development of humanity as we know it. Without saying so, he implies the strength and importance of man's innate curiosity, a curiosity that triggered much of humanity's development and progress. Little wonder, then, that people of all ages and backgrounds collect minerals! Even for noncollectors, minerals hold great fascination; this is easily revealed if you spend time in a museum observing the public as they view nature's inorganic flowers.

> *"...the early development of humanity is intrinsically associated with the recognition and use of minerals"*

## *Crystals of Gold*

*An oldtimer found sharp crystals of gold,*
*They were large, on matrix, that I did hold;*
*He quickly recovered*
*What he had discovered*
*So I stopped drooling; that piece*
*won't be sold.*

- John Almquist

John Almquist is a collector friend of Barlow. Coincidentally, Almquist was born on Barlow's 40th birthday.

While acquisition and recognition of beauty play a role in any collecting drive, it is not enough just to accumulate. Man is also very inquisitive, surely the most curious of nature's creatures. Curiosity relentlessly moves him to ask the obvious next question, to seek the next answer, to investigate that which awakens curiosity. No other creature possesses such a highly developed curiosity coupled with the ability to satisfy it. The outward expression of this curiosity and desire to know can be considered a collecting and learning drive. True, some birds and animals collect glittery and seemingly useless oddities, and they may be expressing an inborn collecting instinct. But that collecting instinct is most advanced in man: we don't just collect, we collect to learn.

### Odyssey of a Mineral Specimen

Serious collectors are as fascinated by the history of a good specimen as they are about the piece itself. To include the history of every Barlow specimen in this book would be impossible, even if the information were known. However, wherever possible, some information has been given to highlight a specimen's past.

It may prove informative and instructive to follow the trail of one specimen. For this, the writer chose specimen #1552, a stibnite from Japan (Plate #1-2). Though quite ordinary by collector standards, the specimen deserves attention for a number of reasons. It is historically interesting. It has been a "lively" specimen, going from collector to dealer to collector, never gathering dust in the drawer of some recluse or institution. It is also a prime example of John Barlow's dedication to searching out a specimen's provenance for his own satisfaction, the mark of a dedicated collector. Following is a description of the specimen and its provenance.

**Plate #1-2, Specimen #1552 STIBNITE – Ichinokawa Mine, Saijo, Ehim-ne-Ken, Iyo Province, Shikoku, Japan. This stibnite has a remarkable provenance. (MH)**

## STIBNITE (ANTIMONY SULFIDE)

**#1552 13.0×5.5×3.3cm Plate #1-2**

**Ichinokawa Mine, Saijo, Ehim-ne-Ken, Iyo Province, Shikoku, Japan**

Stibnite from this locality has long been considered the world's finest; it occurs as amazing blades up to 60cm and in groups of metallic-looking prisms weighing many kilograms. All advanced collectors desire a select specimen of Japanese stibnite. This particular specimen is relatively ordinary as Japanese stibnites go. It is composed of a tight cluster of six or so blades more or less parallel to each other and up to 10cm in length. The crystals show typical striations, fine pyramidal terminations, and good luster. In one crystal, the classic stibnite bending, due to movement along the gliding planes, can be seen. A small cluster of 2mm quartz crystals crowns the piece.

Fortunately, the history of the specimen can be traced from the first owner to the present day, a period of several decades. The specimen originated in the collection of Dr. Daisuke Tanaka, who obtained the specimen from the mine. Tanaka was a major force in collecting and preserving these famous stibnites. Tanaka died in 1954, and the specimen fell into the hands of Phoenix, Arizona dealer Dr. Ralph Mueller. Mueller, in turn, sold the specimen to Arch Oboler of California. Oboler started his career as a writer, and was involved in the radio program "The Shadow" when Orson Welles played the lead. Later, Oboler gained fame as a director of 3-D movies, including "Bwana Devil." Oddly, one person who attended the premiere showing of the movie was Harry Roberson, who acquired the stibnite years later! Oboler, a connoisseur of fine minerals, had a notable collection. During the construction of an addition to his Hollywood home for his mineral collection, a heavy rain filled the excavation and his small child fell in and drowned. This was a sad day for Arch Oboler. It ended his collecting days and he started selling his collection. Paul Desautels and Martin Ehrmann outmaneuvered Ed Swoboda in securing most of the collection. But, some of the pieces Oboler had obtained from Mueller, usually when he was in Arizona making films, were returned to Mueller's shop in Phoenix. By that time, Ralph Mueller had died, and his son, Jim, had taken over the business. The specimen went up for sale and was bought by Harry Roberson, noted collector, who later won a host of local and national awards with his excellent specimens. The stibnite, of course, was often displayed by him at Tucson and elsewhere.

*John could not resist helping out Father Raymond Lassuy financially when the priest told him the specimen came from Dr. Tanaka's collection and that it was in the collection of Arch Oboler, the writer of "Lights Out, the Shadow."*

When Roberson's interest in large mineral specimens waned, he placed the specimen back with Mueller on consignment. Here we find the only small gap in the specimen's history. Somehow, either by purchase or gift, or both, it ended up in the hands of a retired priest, Father Raymond Lassuy, who lived in Phoenix. Enter John Barlow.

Like all advanced collectors, John always attended the great Tucson Gem and Mineral

Society Show in hopes of finding new and classic additions for his growing collection. And he had been badgering dealer Ron Vance, in particular, to locate a good Japanese stibnite for him. By a remarkable coincidence, Father Lassuy was also in Tucson and overheard John talking to Vance about Japanese stibnites. As John left the dealer's hotel room, the priest spoke up, telling John he had a stibnite to sell. John immediately entered into negotiations for the specimen, which the priest had said was given to him by a parishioner. Remarkably, the priest named a price John felt was too low, so John insisted on paying the going rate for the specimen. A deal was struck. John flew his plane to Phoenix, met the priest after the Tucson Show, and the two made the exchange. John could not resist helping out Father Raymond Lassuy financially when the priest told him the specimen came from Dr. Tanaka's collection and that it was in the collection of Arch Oboler, the writer of "Lights Out, the Shadow."

*Collecting is the epitome of Man's curiosity...*

When I began working on this book, I visited John's collection in Appleton and immediately recognized the stibnite as having once resided in the Roberson Collection. John, in turn, had already begun a search for more information about the specimen. Learning that the piece had been in the hands of Arch Oboler, he wrote to Oboler's sister-in-law requesting more information. This gave John the stibnite's history while it was in California. With what I could add, we developed a nearly complete provenance on a fine specimen from a historically famous and mineralogically classical locality.

All this serves to demonstrate the lengths to which Barlow will go to learn about his minerals. He is a true collector.

Again, Vandall King touches on the subject in his tourmaline writings. "The search for a satisfying collection, scientifically, aesthetically, and perceptively, requires a depth of specimen material.... While some collectors are satisfied with having a single representative specimen, other collectors are curious about the objects they seek to the point that 'all is not enough.' Between the extremes, there is a large field of collectors [for whom] the study of minerals transcends the 'catalog' approach to collecting. Mineral collectors typically specialize early in their pursuit and strive to understand a world large enough to provide a lifetime of joy.... Each acquisition provides a basis for learning."

We express our collecting instinct in many ways, including collecting the earth's mineral treasures. It goes far beyond the natural instinct expressed by animals that collect food for survival or to satisfy a "pack rat" instinct. Collecting is the epitome of Man's curiosity coupled with a territorial need, a recognition and respect for the artful beauty of nature, and an instinctive desire to compete, explore, learn, and possess.

Clearly, the traits of a collector appear early in life. After all, they are an integral part of man's make-up. They are reinforced and nurtured through the years and, if all goes well, culminate in the assembling of a truly interesting and valuable collection, be it stamps, coins, minerals, or gems.

In reading John Barlow's profile and the stories of his collecting activities, you will see all the traits described above: his innate curiosity, his love of mineral beauty, his desire to compete and achieve, and his love of learning. The man is revealed in his mineral collection.

Yet, John's profile and the stories of his mineral collection are a two-way mirror. You'll see John, but you'll also see yourself and the traits that brought you to this great hobby and science of minerals. The story of John Barlow's collecting is really every collector's story.

## Crystals

*Crystals of gems and minerals I know;*
*They are flowers of minerals I like to show.*
*I don't like crystals chipped or flawed, out they go;*
*When I spot a specimen I want, I start to glow—*
*Like the one on matrix I see in that row.*
*I'd be buying more crystals if I had more dough,*
*But I used all the money I had to blow.*
*I hate using credit, but my gas tank is low,*
*So I'll pack what I bought, no gift wrap or bow.*
*Now I'll be eating scraps, just like that crow.*
*I better not drop this box of crystals on my toe,*
*Because damaged specimens bring tears of woe.*

- John Almquist

# CHAPTER 2

# Biography of a Collector

## F. John Barlow and His Mineral Collection

*by Gene L. LaBerge*

### Introduction

F. John Barlow has assembled one of the premier private mineral collections in North America, and perhaps in the world. Although his collection contains over 4,400 specimens, Barlow still recalls his first purchase — an attractive pyrite specimen acquired in a rock shop in Waukesha, Wisconsin in 1969. That specimen, no longer in the collection, along with a geode from his daughter Grace, was the beginning of a worldwide search for superb mineral specimens, now spanning more than 25 years.

As is true for most other aspects of John's life, once he undertook a project, he would be satisfied with nothing less than the very best. In effect, once John had decided to collect minerals, it was completely in keeping with his character that his collection would be outstanding. The basic traits that contributed to the formation of his important mineral collection were set early in his life. John Barlow, however, is a man of many dimensions. In addition to his mineral collection, he is part-owner of six corporations that do some $50 million in business annually, he holds an honorary doctorate in science, and he has long been a major force in civic affairs.

*Once John decided to collect minerals, it was completely in keeping with his character that his collection would be outstanding.*

### Early Life

Born to middle class immigrant parents in Milwaukee, Wisconsin in 1914, John Barlow has had a career that almost serves as a model of the American ideal. His father, Ernest, emigrated from Bristol, on the southwest coast of England, not far from the famous pirate lair of Penzance. Barlow counts some of the "Pirates of Penzance" as his ancestors. His mother, Alice Norton, grew up in Sedalia, Missouri, and, as a child, she and her family knew the James family, including the famous robbers, Jesse and Frank. The pioneering spirit of John's parents and their willingness to venture into new territory developed an important facet of Barlow's character.

His parents were devout members of the Episcopal Church, and from an early age John attended church every Sunday. His father and his uncle Harold were acolytes. As John grew older, he also served as an acolyte in the church. This early religious training had an important influence on John and was the basis for his long-standing relationship with the church. For example, after graduating from the University of Wisconsin and moving to Buffalo, New York with his new wife Dorothy, John joined the church and served as a

Sunday School teacher. When he moved to Appleton, Wisconsin in 1944, he joined the All Saints Episcopal Church. When the church was destroyed by fire in 1949, Barlow immediately became the project manager to oversee its reconstruction. During this period of rebuilding, a vacancy developed on the vestry of the church when Nathan Pusey, President of Lawrence University, resigned to accept the Presidency of Harvard University. John replaced Pusey on the vestry, remaining on that governing body for a number of years, rising to the position of Senior Warden, second only to the rector at the time. Dorothy was also very active in church affairs. Although on a reduced scale, Barlow's participation in church affairs continues to the present.

Early in his life, Barlow began to develop his independence and self-reliance. By age seven, he had begun taking the street car alone from his home on the west side of Milwaukee across town to the east side to go swimming at the Milwaukee Athletic Club, where his father was a member. This self-reliance continued to develop, and by age nine he was already a leader among his peers. In an era when playing marbles was a major pastime for youngsters, Barlow's competitive nature emerged as he won all the marbles from his friends. This early kindling of the spirit to do more and better than the competition became a major personal challenge throughout his life.

About this same time, he joined the midget softball league as a pitcher and also organized the Tiger Athletic Club, a youth softball club consisting mainly of boys several years older than himself. As his leadership skills developed, Barlow organized an American Legion baseball team that moved up to Class A in the Milwaukee Sandlot League.

Barlow's competitiveness is perhaps exemplified by an experience he had when he was 12 years old. He entered the Junior Olympics for 13- to 16-year-olds, though he was not yet 13. Contemporaries told him he was foolish to try because he had little chance against one 16-year-old "super" athlete. Although the 16-year-old won the overall competition, Barlow placed second overall and bested him in the baseball throw. The important thing to Barlow was the challenge.

*He formed the "Sunset Stamp" Company at age 12 and placed advertisements, which cost him $.25 per month, in* Popular Mechanics.

He participated in softball, football, boxing, and track in high school. He was named All-City Fullback using the old double wing back system in his senior year of 1931. Again, Barlow's competitive nature was developing into a major driving force that continues today.

***Plate #2-1, L to R*** **Hugo Marx, Barlow, Harry Marx, and Ken Beckman. (UK)**

The interval between high school and college also provided Barlow with some memorable experiences. The Olympic Games were held in Los Angeles in the summer of 1932. John, his two future brothers-in-law, Hugo and Harry Marx, and the high school quarterback, Ken Beckman, decided to go see the games. With approximately $50.00 between them, they left Milwaukee in early July in a 1922 Chalmers, bought for $50.00. It was the trip of a lifetime for four young men just out of high school, and the experiences had a great influence on their future lives. There were no interstate highways in those days. In fact, there was only one major transcontinental highway, the Lincoln Highway (Highway 20). The pavement was very narrow, and in the mountains the road was washed out in places so it was necessary to detour along dry river beds to get back to the main highway. After several weeks of travel, they arrived in Los Angeles and did manage to see some of the Olympics. The trip back took them to several National Parks and provided more experiences they would recall for the rest of their lives.

Early on, John coupled his drive with an emerging sense of business. Typical of kids during the 1920s and 1930s, John sought ways to earn money. He had developed an interest in collecting stamps and soon turned that collecting effort into a business. He formed the "Sunset Stamp Company" at age 12 and placed advertisements, which cost him 25¢ per month, in *Popular Mechanics*. Using the mails to send approval books and cancelled stamps, John dealt in world-wide stamps, which he had accumulated, sorted, and prepared for sale. The asking price on stamps was ½ to 1¢ each and he made individual sales averaging 48¢ monthly through his ads.

***Plate #2-2*** **St. Johns Military Academy taught Barlow a major lesson in discipline. (UK)**

At the same time, he worked door-to-door selling the *Ladies Home Journal* and the *Saturday Evening Post* for 5¢ per copy, and he held a license as a newspaper carrier as well. He well remembers 1927 as he went down the streets of his route calling, "Extra! Extra! Lindy Crosses the Atlantic!" and later, "Tunney Knocks Out Dempsey!" His success in earning money on his own initiative caused him to decide early on that he would pursue a career in business.

By the time he graduated from Washington High School in 1932, Barlow had narrowed his choice of colleges to several schools where he could pursue his goal of studying engineering. M.I.T. and Purdue University had excellent reputations, but the world was in the depths of the Great Depression and financing an education at those schools was out of the question. However, an important incident in Barlow's life occurred the year after he graduated from high school. His high school football coach, Lisle Blackburn (who became football coach at Marquette University in Milwaukee and later coached for the Green Bay Packers), recognized Barlow's speed and tenacity as a football player. He suggested that John increase his weight and strength in order to compete at a major university. Blackburn offered to help Barlow get a scholarship to play football at St. John's Military Academy in Delafield, Wisconsin.

Like other private schools, St. John's enrollment was down drastically owing to the Depression, and the administration felt that fielding an outstanding football team might improve their enrollment. Barlow was offered a scholarship and became a member of the most remarkable football team of any private school in the midwest that year. Several of the players later made All Big Ten teams, including guard Mike Calvano, halfback Steve Swisher, and George Wilson, who later went on to coach the Detroit Lions of the National Football League. Shortly before the season started, Barlow underwent an emergency appendectomy, greatly curtailing his participation in football. After years of being the star athlete in high school, Barlow learned a major lesson in discipline. Although it was difficult to accept the discipline at the time (including many hours walking guard duty), it was to pay great dividends later in his life.

During the year at St. John's Military Academy, John had decided to pursue his education at the University of Wisconsin. He spent the summer of 1933 working as a laborer on the beer barrel line at A. O. Smith Corporation in Milwaukee. Taking the job developed his strength in preparation for playing football and helped raise the $27.50 tuition for the university, plus money for living expenses. That fall he tried out for the university football team and made the freshman team. He also got a job washing pots and pans at the Student Union to augment the money his father was able to provide. He made the varsity team in his second year, but he realized his future was in engineering and business, not football. So he concentrated on his academics, pursuing his professional goals with a major in Mechanical Engineering, and elective courses mainly in the School of Business.

Barlow was pledged a Sigma Chi, and in 1934 he became a member of the fraternity. Like all fraternities, Sigma Chi had been hit by the Depression. In his junior year, Barlow's business acumen was tested when he was selected by two

Milwaukee alumni, Hal Gaffin and Ed Bartlett, to put the Sigma Chi house back on its feet financially. His business savvy learned in his younger years served him well, and within a year he had resolved most of the fraternity's financial problems. He formed a corporation called Alpha Lambda House Corporation. He then settled the outstanding accounts at 10¢ on the dollar to reestablish the financial stability of the fraternity. The next year he became President (Consul) of Sigma Chi house and began making money for the fraternity. Again, he had accepted the challenge and won. In his final year at the university (1937), he was selected as *Saint Patrick*, the outstanding engineering student in his class.

***Plate #2-3*** **Dorothy's life was devoted to the happiness and well-being of her husband and their eight daughters. (UK)**

Clearly, the traits of leadership, organizational ability, and competitiveness that were to become the backbone of Barlow's future success in business and as a mineral collector were evident early in his life. One other event was to have a profound influence on his life—his marriage to Dorothy Marx, his high school sweetheart.

John first met Dorothy through her brothers Harry and Hugo Marx in 1927, when John was thirteen years old. He and Harry were avid ice skaters, and they spent many evenings skating on a pond near their home. One evening, these teenage boys went to their favorite skating rink, where Barlow noticed a young girl who always seemed to be in his way. He told her to go home, he didn't need her around their skating rink. Although John was quite unimpressed with Dorothy at this first meeting (typical of a 13-year-old), his friendship with her brothers resulted in numerous contacts with her in succeeding years. Neither Dorothy nor John indicated any inclination to have much to do with one another. But, by the time John started college, he realized that he cared a great deal for the beautiful young lady called Dorothy. They were married in 1935, and shared 58 wonderful years together. Dorothy's life was devoted to the happiness and well-being of her husband and their eight daughters, Joyce, Bonnie, Joan, Grace, Jacqueline, Wendy, Terri, and Alice. Their marriage proved to be a great source of comfort and stability for John during the very hectic years of his business life.

*The traits of leadership, organizational ability, and competitiveness that were to become the backbone of Barlow's future success in business and as a mineral collector were evident early in his life.*

John and Dorothy's eight daughters provided their parents with a wide variety of experiences. Despite his very busy business schedule, John made a real effort to spend time with his family. To transport his children on family vacations, he purchased a camper trailer, which he towed behind his truck. The trailer was soon replaced with a motor-driven recreational vehicle. Some of his mineral-collector friends may find it interesting to picture John Barlow driving a recreational vehicle with his wife and a bevy of daughters on family vacation trips to various campgrounds in the midwest and elsewhere. Their daughters also accompanied John and Dorothy on a number of trips to far-flung places in the Barlow airplane. The Barlow family also traveled via railroad and boat to a cabin in the wilds of Canada. The cabin would become a favorite summer destination for almost 20 years. These trips generated many lasting memories for the Barlow clan.

The Barlow household not only had eight daughters, it also hosted AFS foreign students. One in particular, Maria del Bosque from Aguascalientes, Mexico, became almost a member of the Barlow family. She spent the 1969–70 academic year with the Barlows in Appleton. The contact has been maintained over the years, and they have visited back and forth. These experiences have been very special for John and Dorothy.

Recollections from the Barlow daughters provide glimpses into some aspects of John's character as a father. For example, Grace recalls a memorable incident when she was about five years old that illustrates John's boundless enthusiasm and optimism. She writes:

*I have a memory of my dad when I was about five years old that comes to mind during idle moments. I don't believe I've ever told him about it, although I've intended to do so for a number of years.*

*As I recall, it was evening and Dad had come home from a long day's work. Clearly he had no intention of plopping into his easy chair to read the newspaper this night. I could tell that something was up, but he was clearly too preoccupied to ask him what it might be. He was kind of giddy with excitement, and I was hooked! I watched him with a fascination I can almost feel to this day.*

*Dad sat down and took a little notepad, pencil and slide rule out from his pocket. He proceeded to make hurried calculations — scribbles to me! — working furiously for maybe almost an hour. All the while a gang of us girls were causing our usual evening ruckus, running about trying to one-up each other, hitting, tripping, screaming, jumping on the furniture, giggling our heads off! Think about it; it could have been much worse. Think of the mayhem eight sons might have been capable of committing!*

*Dad was sitting just a few feet away, oblivious to all the commotion, in total concentration on his goal. At an opportune moment I asked what he was doing. He said he was trying to win a contest he heard on the radio, which involved finding the solution to a complex mathematical puzzle. The first person to bring in the correct answer would win $100 and the deadline was twelve o'clock that night. No wonder the excitement and the frequent glances at the wrist watch.*

*I remember feeling exhilarated just watching him go at it with such enthusiasm and confidence. I truly felt in awe of him at that moment. I wanted my dad to win so bad that what he did next threw me for a loop (or, depending on your generation, "blew me away"). Dad laid down his work, and said he thought he had it right. He looked at his watch and dashed upstairs to put on a clean shirt and to shave before taking his solution down to the radio station. I couldn't understand why he would waste time doing that, and risk not getting there first. His shirt was perfectly clean and he shaved every morning!*

*The sheer output of boundless energy, enthusiastic delight, and seriousness of purpose, so characteristic of all my dad's endeavors, was captured in that early memory emblazoned in my mind. I honestly don't remember if he won or not — so obviously that wasn't the important part of this story — to me, at least.*

Jackie Barlow remembers a compassionate, supportive father. She writes:

*My father and mother were very influential in my music studies and career. Dad bought me my first piano at age 16, and my parents encouraged me in my wish to begin singing lessons at Lawrence Conservatory. Through high school and college, at the University of Wisconsin in Madison, my parents were supportive in my musical and dramatic performances, coming to see me frequently. Then, when I studied music in Germany, and got married there in Heidelberg, my parents came to visit me. That was 1969, and I remember that was the time that my dad was beginning his interest in minerals. When I was going through a divorce in 1976, my parents were very supportive. I moved in with them to return to college at Lawrence University. They took in my son, Christopher, who was then three years old, as if he was their child. We lived with them for two years. When I moved to Columbus, Ohio to pursue graduate school, my father continued to offer financial support. My parents came to performances of operas that I was in at Ohio State University and*

*Opera Columbus. I now teach at Capitol University and my son is now an OSU student. My father continues to generously offer him financial help to finish his studies and continues to be a solid role model for him.*

Terri recalls a trip to Europe with John in 1977, where he displayed his enthusiasm for "shopping" (an important trait for a mineral collector) and his sense of humor. She writes:

*After six months in Miami working as a nurse, I decided the cockroaches could have the place and moved back to Appleton. While I was at home my dad asked if I would like to go to Europe for a gem show. That was in 1977.*

*We flew into London to take in the sights. I remember my dad showing me the beautiful Westminster Abbey where we tried "brass rubbing." We went to the Tower of London where so much history has taken place.*

*My dad rivals most of us with his love of shopping! Harrod's of London definitely was an experience.*

*Before we left London, we had to stop at the Natural History Museum. I can remember my dad disappearing into the mineral hall not to be seen for hours.*

*Then we traveled to Idar-Oberstein, Germany to visit some of Dad's friends. I remember quite a few gem and mineral stores. The twin cities are very quaint.*

*The last stop was the gem show in Zurich, Switzerland. What a great show it was. I remember buying a cut emerald and my unusual lapis ring.*

*We spent some time with a lot of other gem dealers in Zurich. Dad and I were able to laugh when everyone thought I was his girlfriend rather than his daughter. So we played along for a while, at least through breakfast.*

*My dad's a seasoned traveler who still gets excited showing others around. And that's what he did.*

Joyce also recalls a trip to Europe where John's enthusiasm for sharing the good times with others was obvious. She recalls:

*I remember, fondly, a trip my dad arranged for my sister, Grace, and me during the early 1970s. Grace and I traveled to Rome where we met Dad and the three of us had a wonderful time touring all the historical sites — the Vatican, the Sistine Chapel, the Coliseum and, of course, we hit all the famous Italian restaurants. Dad went on to India on a hunting expedition, but Grace and I, so young, so naive and giggling most all the way, continued on to Florence, Milan, Frankfurt and then to Heidelberg, Germany, to visit Jackie who was living there at the time. She and her husband got us passes to get into East Berlin. What a great experience that was — truly history today, to have been beyond "the Wall." I flew home alone from there, but Grace stayed in Europe for many months. It was certainly a trip I will remember my whole life.*

## Business Career

Graduating from the University in 1937, Barlow had planned to work as an engineer for the A. O. Smith Corporation in Milwaukee, where his father worked as manager of the sales department. But job opportunities were meager during the Depression. No company had come to the university to interview prospective employees since the stock market crash in 1929. There were eight jobs known to be available, however, and Barlow's professors insisted that he interview with several of those companies.

He was offered two jobs, one with Buffalo Forge Company for $25.00 per week, the other with Air Reduction Sales, a division of the large Krupp syndicate from Germany, for $100.00 per month. He chose to work for Buffalo Forge, and, with his wife, Dorothy, moved to Buffalo, New York. He took an intensive two-year training program concerned with design and application of fans and heat-transfer equipment. Although he was hired as an engineer, Barlow's desire for advancement soon resulted in his transfer to the sales division. His technical background and business acumen served him well. One of the first persons to go out into the field for Buffalo Forge Company, he was sent to the Chicago and Milwaukee industrial markets.

In 1940, the business community became an important part of the war effort, and John was recruited by A. O. Smith Corporation to work on military projects. His expertise in designing specialized piping and heating and air-conditioning systems was in great demand. He resigned his position with Buffalo Forge to work for A. O. Smith, with the proviso that he return to Buffalo Forge after the war.

*At A. O. Smith, Barlow's expertise was re-directed toward the design of various wartime products such as incendiary bombs, "block-busters," torpedoes, landing gear for B-25, B-17, and B-29 bombers, and high-performance hollow-bladed welded chrome-molybdenum propellers for airplanes.*

At A. O. Smith, Barlow's expertise was re-directed toward the design of various wartime products such as incendiary bombs, "block-busters," torpedoes, landing gear for B-25, B-17, and B-29 bombers, and high-performance, hollow-bladed, welded chrome-molybdenum propellers for airplanes.

From his ROTC days at the university, John held a commission in the U.S. Army. In 1941 he was given 10 days to decide whether to serve in the Army or resign his commission to continue his engineering activities. It was a very difficult decision, but officials at A. O. Smith finally convinced him that he could help the war effort more by resigning his commission and working in industry. So Barlow worked 60–70 hours a week (at $1.35 an hour) for A. O. Smith. He was also asked to work as an inspector at the Bailey Blower Company in Milwaukee. He took this job in addition to his position at A. O. Smith. In 1942, he received his certification as a registered professional engineer.

In 1944, Barlow's career took another major turn when he accepted a position with Western Condensing Company in Appleton, Wisconsin. The company bought the waste (whey) from the many cheese factories in eastern Wisconsin and processed the whey to produce lactose. A few years earlier, several British scientists had discovered that lactose increased the production of penicillin thousands of times over sucrose, fructose, or other products that served as nutrients for penicillin. With the high demand for penicillin during the war, maximizing production of this life-saving drug became a major priority. Barlow's expertise in design engineering and heat transfer in the production of evaporators and dryers was essential to this effort. So, with a significant increase in salary, Barlow moved to Appleton with his family to begin a new phase of his life.

At Western Condensing a problem with John's name arose. It provides an interesting anecdote to the earlier part of his life. His given name is Frank John Barlow. Frank Barlow was his grandfather's name. His family and close friends called him Jack, and through much of his schooling he was known as Jack Barlow. He chose the name "F. John Barlow" to go on his diploma from the University of Wisconsin, but he continued to use "Jack Barlow" until he moved to Appleton in 1944. The chief chemical engineer of Western Condensing was Jack Crews. Because both chief engineers were named "Jack," Barlow thought confusion would result. Therefore, he decided to begin his new job with the name "John Barlow." It took a bit of getting used to (he sometimes didn't respond to "John"), but from the time he began his career in Appleton, he was known as "John Barlow" — although his family continued to call him "Jack."

Near the end of the war, Barlow again branched out from his engineering duties to manufacturing and sales. He was given responsibility for six plants in Wisconsin, but he soon expanded his territory throughout the midwest, and within six to eight months, had responsibility for all 28 Western Condensing plants in the United States. In the years following World War II, Barlow was the number-three man in the company, doing very well financially. This employment provided important engineering and management experience, and, although the promotions were rewarding, Barlow decided that he wanted more of a challenge.

The years with Western Condensing were also an important time in the life of John Barlow the family man. As the letter from his daughter Bonnie illustrates, the Barlow family was a close-knit group.

> *I remember special times with my dad in the '50s when life seemed simpler and leisurely drives on Sunday afternoon were*

*much to be treasured. We would all pile into the car, sing and play word games. In the summer we would drive to Lake Michigan to swim and sun; in the fall we would go to the marshes to watch the ducks; and, in the winter to see the Christmas decorations in the Valley. One summer when Dad was traveling a lot, checking cheese factories for Western Condensing, my mother said I could go along and chatter and sing and keep my dad awake. We traveled to many plants, slept in a motel, and then we drove into the big city of Chicago. He took me on a tour of the downtown. We ate tamales on the street corner and had real Chinese food in a restaurant. As we crossed the bridge at night, Chicago was all shimmering lights. On the long drive home we stopped at a custard stand, but we only had fifteen cents, so we split an Eskimo Pie and listened to "The Shadow." Dad taught me about the constellations and the moon, never realizing that in our lifetimes man would land on that shining, smiling face.*

*I fell asleep knowing that if a bear did come, that Dad would yell, "Get out of here..." chasing the bear away, and we would be safe.*

John was very much an outdoorsman and, through his hunting experiences, had developed an expertise in tracking wild animals. He enjoyed sharing that expertise with his family, and his daughter Wendy reminisces:

*As I gazed out my window at the different animal tracks in the snow, I was reminded of a thrilling early morning tracking experience some years ago at my parents' home. My dad rushed into the house yelling, "Wendy, get dressed. Come outside. I've discovered something so exciting! I think you'd love to see it!" Throwing a heavy coat over my pajamas, boots over bare feet, I hurried out the door. From the sound and expression on his face, I knew this discovery must be exceptional. Entering the snowy outdoor stage where a drama of the reality of predator and prey left only their impressions, once again my dad pipes up, "You have to see this incredible scene!" He walked me through the drama in the snow where a rodent had traversed from one corner of the large front yard to the opposite corner, and then winding its way around the sugar maples and tall ash trees, it had come toward the front door. Once again the tracks crossed over the old route, traveling back toward a maple tree. The tracks suddenly stopped. At our feet, in the clean white snow, was a perfect imprint of two large wings of an owl. We paused and reflected with wide open eyes...*

*This memory of my dad in his excitement and pleasure of discovery, and his desire to share that discovery, transports me back to a particular camping experience, an experience that also showed my dad's great ability on vacations for mischief and play. This was truly an exciting vacation, my first vacation up in Canada, with my mom, my dad and four of us girls.*

*The August adventure began by traveling by train up to the river and then motoring up the river to the camp. I remember our excitement in spying the cabin as we rounded the bend into the bay. The camp was fondly called "Wisont," a contraction for Wisconsin and Ontario, and resided in what many called "God's Country." We spent many summers at Wisont and surrounding territory, magnificent for its granite cliffs, wooded islands, quiet bays, waterfalls and hidden lakes. An exciting, adventurous, wild wonder of a place that my father, some years before set his mind to finding. By airplane and then by boat, he scouted out the special spot. A logger's cabin was purchased and hoisted up the steep bank and set into place. Perfect for its open space, but seasoned once again with nature's young green boughs and branches.*

*In back of the cabin were thick woods and old logging trails. It was here that my father took us tracking and told us stories, imagined and real.*

*My dad also enjoyed telling us stories at night when we were all tucked in our bunks, listening to the fire hissing and crackling. Dad had a wonderful way of telling stories. He would just be talking quietly, telling us what to do if ever we should meet a bear. We would begin to doze off when my dad would deepen his voice and yell out (to the bear), "GET OUT OF HERE. THIS IS MY HOUSE!" His yelling would startle us, and as we tried to fall asleep, we listened for any sound, however small, that would warn us of a bear outside, ready to break into the cabin. I fell asleep knowing that if a bear did come, that Dad would yell, "GET OUT OF HERE..." chasing the bear away, and we would be safe.*

*One afternoon my father took us on a long expedition into the woods, beginning on the logging trail. We hiked back to a falls and then leaving the trail, we trudged through the thick Canadian woods to a hidden lake. Being a hunter, Dad was very knowledgeable in tracking animals. He loved to interpret the animal signs for Mom and us girls. There were signs that were left in the mud, by broken branches, and as marks on trees. "These tracks are a deer, probably a doe; see how small the hooves are, and how close together? She probably walked in this direction. See how the hooves settled into the dirt on this side and disturbed the dirt on the opposite side when she picked up her feet?"*

*Leaving the trail to walk through the woods, Dad suddenly discovered a tree where a patch of bark was gone. "See where the bark has been worn off this tree? A BEAR was here! He used this tree to scratch his back and sharpen his claws." My heart was pounding as Dad continued to speak quietly, "and probably not too long ago!"*

*Back on the trail we discovered some animal dung. "This dung is full of grayish hair and tiny bones. It probably was a mouse eaten by an owl."*

*I will always treasure my father's sharing these nature discoveries, tracking experiences and the wild animal stories at night, as well as his relaxed sense of fun and his love of discovery at Wisont. I shall also remember the excitement, awe, and my dad's and my reflecting upon the wing impressions left in the snow on that cold winter morning so long ago.*

***Plate #2-4*** **Barlow at the office. (MH)**

In writing about the ten most important people in the Fox Cities, Bob Lowe, *Appleton Post-Crescent* staff writer, said of John:

> *The typical entrepreneur. Tenacious. A take-charge guy. Intimidating. Demanding. A workaholic. A self-made man. Dynamic. An activist in civil affairs.*
>
> *These were words used to describe F. (for Frank) John Barlow by friends, associates and people who know him or have worked with him.*
>
> *Even before he was elected president of the Appleton Development Council, the group that successfully spearheaded the development of The Avenue Mall in downtown Appleton, Barlow had racked up a list of accomplishments that few can match.*
>
> *His imprint on the Fox Valley is deep and widespread.*

In 1949, with three partners, Frank Jenkins, Harry Koller, and Al Bassett, and a $5,000 loan, Barlow organized an engineering contracting company he called AZCO, Inc. The name stood for the company's ability to undertake anything from A to Z in the field of engineering. After several years of working day and night to get the company established, Barlow bought out his partners and assumed full control of the corporation. He ran his business with the philosophy that it is essential to be honest and fair, especially when it was necessary to make "tough" decisions.

Under Barlow's leadership, AZCO grew into a conglomerate of six companies: THE AZCO GROUP LTD., AZCO HENNES INC., B & B LEASING INC., SANCO LTD., EARTH RESOURCES, and PAFCO LTD., with an aggregate business of $50 million annually and employing over 1000 people. Needless to say, his business enterprises play a major role in the development and economy of the Fox River Valley in northeastern Wisconsin. Barlow served as Chairman of the Board of each of these corporations, though his sentimental favorite has to be EARTH RESOURCES, a division of SANCO LTD. Dealing in gems, minerals, and jewelry, Earth Resources was conceived and developed in 1975, paralleling his strong interest in natural resources, particularly minerals. John was also on the Board of Directors and Executive Committee of Air Wisconsin (now United Express), of which he was cofounder in 1965. He's also past director of the First National Bank of Appleton, Transpace Carriers, and Beta Color, Inc.

**Plate #2-5 (MH)**

*When he was awarded the Distinguished Service Award, he was introduced as the "uncommon man."*

Barlow was responsible for bringing cable TV to Appleton. On July 1, 1970, he secured the first ten-year franchise, which was later sold to Fox Cities Communications (now Cablevision of the Fox Cities).

John Barlow has received numerous prestigious recognitions in his profession as an engineer and businessman. He has served as State and National President of the Mechanical Contractors Association. He was chosen by President Gerald Ford to attend the 1974 Summit Conference on Inflation and was selected to represent the U.S.A. with other Common Market countries in an international group known as *Genie Climatique*, which focused on European construction problems. In 1983, Barlow was awarded the Distinguished Service Award, the highest award given annually to a member of the Mechanical Contractors Association of America. Thus he is nationally and internationally recognized for his professional accomplishments. When he was awarded the Distinguished Service Award, he was introduced as the "uncommon man."

One of the many things considered to be uncommon was the purchase by Barlow in 1964, at the age of 50, of a fast "V"-tail Bonanza aircraft when Barlow did not know how to fly any kind of airplane. Without anyone except his federal instructor knowing about it, John got his solo license in 12 hours. He received his instrument rating in 1969. But why?

On his hunting expeditions in the 1950s and early 1960s, Barlow often flew with his good friend, Karl Baldwin, who had learned to fly on the G.I. Bill after World War II. As they were flying to one of their hunting locations in Baldwin's airplane, Barlow thought, "If Karl can fly his own airplane, I can, too."

Then, in 1964, Baldwin traded in his Bonanza on a new, larger Cessna. Barlow decided to act, and he secretly purchased Baldwin's Bonanza. At first he was licensed to fly only under VFR (visual flight regulations), in which navigation is accomplished by following various geographic features. These rules require that one be able to see the ground. During their many flights to parts of the United States and Central America, John's wife, Dorothy, served as navigator, reading the charts and locating appropriate features on the ground. She soon became an accomplished navigator on the Barlow aircraft.

Perhaps the most harrowing experience John and Dorothy encountered occurred in 1965, while they were flying from Appleton to

Woodbine, New Jersey. Crossing the Appalachian Mountains in eastern Pennsylvania, they suddenly found themselves in the clouds, and they lost their bearings. Lacking the instruments necessary to guide them to an airport, John realized that he must somehow get below the clouds in order to see the ground — without crashing into a mountain! After searching for some time to find a hole in the clouds, John and Dorothy finally saw the most welcome sight of green fields and a river below them. John flew down through the hole to regain contact with the ground. Due to their circling above the clouds, however, they did not know where they were. While frequently talking to a control tower operator on the radio, John followed the river until they came to a town — but, which town? The radio operator asked if they could read the name on the water tower. No luck. So John flew at about 500 feet above the main street. He finally read the name of the First National Bank of Allentown, Pennsylvania. With this information, the radio operator guided them to the airport about eight miles away. After landing, John called the party in Woodbine, who instructed them on the route to fly for the remainder of their journey. Interestingly, throughout the entire ordeal, both John and Dorothy remained calm. Had they panicked, the situation could have been a disaster.

John received his instrument flight rating in 1969. This IFR rating provided more mobility and enabled John and Dorothy to make longer trips. His flying experiences then included "white knuckle" landings and takeoffs on small, remote airstrips in Baja California Sur, Mexico and transoceanic flights from Puerto Rico to Florida, where one had to fly well enough to "pick up the navigation beam" after flying for hundreds of miles over featureless ocean.

The airplane gave John great mobility for his business ventures. It also significantly expanded his hunting experiences, afforded many opportunities to obtain mineral specimens, and last but not least, provided him and Dorothy much pleasure on their vacation trips. In short, the airplane was a great boon to the many aspects of John's life. As the years went by John hired a pilot, Don Hoyman, to fly the plane on many of his trips. However, he never lost the thrill of flying his own airplane.

Recently Barlow was awarded an Honorary Doctorate of Science by the University of Wisconsin in Oshkosh. The Board of Regents of the University of Wisconsin System approved his nomination for this signal honor. The degree was conferred on May 14, 1994, in recognition of his numerous accomplishments in his profession of engineering, as well as his many contributions to his community, to the State of Wisconsin, and to the nation.

What type of person accomplishes the amount Barlow has in his lifetime? An insight comes from a story told by John's son-in-law, Peter Juneau:

*Most people think of John as the ultimate businessman — very goal oriented, very hard driven and very decisive — but never see his personal side. Being his son-in-law, I also had this impression when I first married into the family* [to John's daughter, Wendy]. *However, over the years, I have come to see John's personal side and how he relates to people. He has the uncanny ability to make those he meets feel that they are important and that what they say and do is important. I believe this is a major factor in his success in business enterprises. I now treasure the time I spend with John in his mineral room, doing the work that's "not too exciting but needs to get done" and listening to his stories.*

*One of the most touching scenes between friends that I've ever seen was when I accompanied John to the 1995 Tucson Gem and Mineral Show. John and I were making the rounds at the Executive Inn when John spied Carlos Barbosa sitting in one of the rooms. John walked into the room announcing, "Greetings, my old friend!" Carlos slowly stood up, walked over to John, and as John and Carlos gripped each other's forearms, Carlos stated, "Congratulations, John."*

*As the gripping of the forearms turned into a warm embrace, John had a quizzical look on his face. His eyes were darting about and I could tell he was searching his brain for some item or event of honor deserving of congratulations. After several seconds, John asked "For what?" As both men affectionately patted each other's back, Carlos continued, "We are both still alive, my friend!"*

*It was an extremely memorable scene.*

*Plate #2-6* **Fluorescent mine tunnel, Children's Museum, Appleton, Wisconsin. Barlow's identical twin granddaughters Ariel and Zoe Brown. (MH)**

## Community Service

John Barlow has been extensively involved in community organizations for many years, where his leadership and organizational abilities are sought out. He is currently a member of the Appleton Development Council, Appleton Elks (he was "Elk of the Year" in 1984), Rotary Club (he received the Rotary International "Paul Harris Fellow Award" in 1987), Masons, Shriners, Butte des Morts Country Club, and Riverview Country Club. He has been a member of the Chancellor's Advisory Council at the University of Wisconsin, Oshkosh since 1985, and he was awarded the Chancellor's Medallion in 1987. He has served as a Councilor for the Wisconsin Academy of Sciences, Arts and Letters, providing advice on the business affairs of the academy. A generous contribution from John in 1993 established the F. John Barlow Endowed Professorship in the College of Letters and Science at the University of Wisconsin, Oshkosh.

Barlow has also been a major contributor to, and currently serves as Chairman of, the Board of Directors of the Bergstrom-Mahler Museum in Neenah, Wisconsin, where his contribution of a major limited-edition paperweight collection represents a significant addition to their holdings. He was elected President of the Bergstrom Museum in 1995.

His most recent community project has been the Children's Museum in Appleton, of which he is a major benefactor. He was responsible for the construction of an impressive and realistic walk-through mine tunnel of fluorescent minerals from Franklin, New Jersey, the fluorescent mineral capital of the world (Plate #2-6), as well as a number of large, attractive mineral samples for children's hands-on experience. The museum is recognized as an outstanding learning experience for children of eastern Wisconsin.

## Big Game Hunting

Big game hunting was a major part of John Barlow's life for many years, and his success in the field was a prelude to his equally successful mineral collecting. But, how does a young man growing up in Milwaukee become a big game hunter or a leading mineral collector? For John Barlow, big game hunting started as a result of a trip to Colorado in 1935. John, Dorothy, and her brother Hugo Marx visited the Thomson's ranch. Rich Thomson and his son Bob were outfitters with 60 to 80 horses on their ranch. Rich was the 1904 and 1905 World Champion Bronco Buster as well as a former forest ranger who knew every trail and stream on the western slope. He loved to tell colorful stories about people and adventures in the high country. Bob took John and Hugo on a three-day pack trip with horses to go trout fishing in Grizzly Canyon near Glenwood Springs. This first experience in the mountains thrilled John and triggered a long love affair with the great outdoors.

After John graduated from the university in

*Plate #2-7* **Point Hore, Alaska, 38° below zero, March 1971. (UK)**

1937, he and Dorothy went back to Glenwood Springs for another two week fishing and camping trip before leaving for John's new job in Buffalo, New York. The horseback trip into the mountains was so moving that Barlow knew he would be returning again to experience more of the real wilderness. The trip also kindled a new interest: big game hunting, a hobby that Barlow would enjoy for many years.

The hectic schedule of starting a career, however, left little time for hobbies, and with John's salary of $25.00 a week, there were no hunting trips for several years. In 1940, Rich Thomson invited Barlow, Hugo Marx, and Harry Koller, a good friend of John's, to join him in a mountain hunt for mule deer on the flat tops above Glenwood Canyon. They camped in deep snow, and John got his first mule deer. This first of numerous hunting trips introduced John to the challenge of the wilderness hunt.

The next year, the same group went out again to Camp Defiance. They found an old Ute Indian trail up the sheer cliffs of the mountain and camped there for two weeks. John was so exhilarated and challenged that hunting trips became an annual event. These trips were a real diversion from the extremely busy daily life that John had during those years. On one of these hunting trips in Colorado in 1946, John saw his first bighorn ram from a great distance. He was complimented for even seeing it, for they are exceedingly

***Plate #2-8*** **African "record book" buffalo with one shot, 1967. (UK)**

## *Not Even the Best of Hunters Always Succeeds*

Fall in Wisconsin is a great time to go buck hunting. It was a weekend with no wind, temperature in the high 20s, and overcast sky — perfect for tracking deer in the snow — and John had the "itch." He called his hunting friend Harry Koller, and they quickly developed a plan. Harry would take one of his sons and John would take one of his daughters. That presented a problem for John since the girls hunted wild mushrooms and watercress with their father, never wild game.

Joan was the daughter who most appreciated wild animals, so John asked if she would like to go deer hunting with bow and arrow the following Sunday in the Necedah Wildlife Preserve. She surprised John by saying, "Yes, what do you have in mind?"

The season for deer was legally open, but only for bow and arrow hunting. Leaving early in the morning, they drove to Necedah, meeting Harry at a prearranged point. John and Harry decided that they would take different directions following deer tracks and that they would rendezvous before dark. Joan and her father had great fun tracking deer in the snow, but they never did see one.

About 4:00 PM, John decided not to track; instead, he found a high spot under a large pine tree and suggested they sit down there quietly and wait, because this was the time that the deer would be moving. John explained to his daughter the sportsmanship involved and the various aspects of hunting wild game. They had been there less than 15 minutes when John heard the movement of deer below the ridge where they sat.

John and Joan waited for the deer to come into view above the ridge — it seemed like an hour. Fewer than 35 yards away, coming up the ridge were two deer. The deer stopped and looked around, realizing there might be danger. John rose to his knee and pulled back the arrow on the bow. At the sound of the rustle, the deer moved quickly. John let go with the arrow, but neither animal stopped. In the commotion John couldn't understand what had happened and ran for the expected point of impact. The deer were gone. He looked up and saw the arrow embedded in a tree.

Then John and Joan had to move quickly to the rendezvous point, as it was getting dark. It was a normal day in a sportsman's life, tiring but exhilarating. John will never forget his best hunt with his daughter Joan.

*In 1966, Barlow became the first Wisconsin hunter to achieve a "Grand Slam."*

wary animals, inhabiting the highest peaks and possessing exceptional eyesight.

Barlow's competitive nature was challenged, and his quest to bag a bighorn ram began. The ram is the ultimate symbol of a great trophy hunter. He joined Reg Meade, an associate from Western Condensing Company, on a 30-day hunting trip across the Smoky River, north of Jasper National Park, Alberta, Canada in 1948. On this hunt John bagged his first bighorn ram, an elk, and a mountain goat.

In 1958, on a hunting trip to Alaska, he bagged a beautiful white Dall sheep. His group had flown into McGrath, then traveled to Dillinger Creek, west of Mt. McKinley in the Alaska Range where John got the ram. Four years later, on a hunting trip to British Columbia with Karl Baldwin and Harry Koller, he succeeded in bagging a Stone ram. This gave John three of what are known as the "Grand Slam" four species of bighorn sheep that inhabit the mountains of western North America. Only the elusive Desert ram in Baja California Sur, Mexico remained. The challenge was twofold: first to secure a permit, then to bag the ram.

Barlow accomplished his quest for a "Grand Slam" in 1966, when he became the first Wisconsin hunter to achieve this feat. The official list of "Grand Slammers" numbers about 600. John ranks 46th on the list, putting him in a select group of hunters. The stage was now set for a bigger challenge, the big game of Africa and India.

*He won virtually all of the major competitive mineral awards.*

In 1967, Barlow accomplished what African big game hunters strive for, the "Big Five." He was fortunate and skillful enough to bag a lion, an elephant, a leopard, a cape buffalo, and a white rhino on two hunting trips in eastern Africa. Again, we see him accept the challenge and come away a winner.

In 1970, Barlow went to India in search of the Bengal tiger, the dream of all big game hunters. This trip, however, was a major disappointment. Not only did he fail to bag a tiger, he was nearly killed by a panther. Thus, not even the best of hunters always succeeds.

## Mineral Collecting

All this success — in sports, in school, in business, and in hunting — reflects John's boundless energy and his commitment to excellence. The stage was set for a new challenge, a new adventure. In addition to his responsibilities in his own business ventures, his state and national offices in contract engineering organizations, and his big game hunting in the 1960s, Barlow began to develop an interest in fine mineral specimens. His business and hunting trips took him throughout the world, and he visited museums to learn more about his new hobby. Some of his trips afforded him an opportunity to acquire mineral specimens, the beginnings of his superb collection.

Yet Barlow really owes his introduction to mineral collecting to his daughter Grace, who gave him two fine amethyst geode halves that she had acquired on a trip to the Atlas Mountains of Morocco in 1969. He still has these attractive pieces in his collection.

In 1970, two other daughters, Wendy and Alice, and John's wife, Dorothy, took courses at a local vocational school to learn lapidary, the art of cutting and polishing gems. Their enthusiasm was infectious; John bought equipment and learned lapidary himself. He became a lapidarist for all his friends, making tie clasps, bolo ties, cuff links, brooches, etc., which he gave as gifts. Meanwhile, he began to purchase mineral specimens. Starting with that pyrite embryo in Waukesha (now aborted), his interests gradually changed from lapidary to mineral collecting.

The problem for an engineer and big game hunter, or anyone else, was how to gain the necessary knowledge to be successful in assembling a major mineral collection. Barlow attended mineral shows, studied exhibits and materials offered for sale, and talked to dealers and collectors. His natural competitiveness led to his entering a case of minerals in competition

*John participated directly in several mining operations, something few collectors have an opportunity to do.*

in a national show, where he soon learned some of the perils of competitive exhibiting. Although he failed to win in some of his first attempts, he became determined to show that he could compete in this arena as well. His later competitive exhibits showed how well he had learned the lessons of selecting top-quality mineral specimens. From 1975 to 1980 he won virtually all of the major competitive mineral awards including the Federation Trophies, the prestigious McDole Trophy in 1975 and 1978 at Tucson, the Detroit Challenge Cup in 1977 and 1978, and the Desautels Trophy in 1993 at Tucson.

Barlow has experienced many of the problems associated with assembling a great mineral collection. Mixed communications within his group at a show in Florida in 1979 resulted in his entire exhibit being left in the hotel room after Barlow had returned to Appleton. Quick action by Barlow, his mineral collector friend, Dr. Dale Dubin, and a "private eye" recovered the minerals the next day, avoiding a huge financial loss. (Refer to "Cops and Robbers" in Chapter 17, page 397.)

John was not as lucky with a mineral exhibit during a showing of a Fabergé egg collection at the Paine Art Center in Oshkosh, Wisconsin in 1981. Thieves made off with most of the exhibit, including thousands of dollars worth of

***Plate #2-9, Specimen #922*** **Self-collected tourmaline named "The 4th of July Barlow Buster." Tourmaline Queen Mine, Pala, California, 1974. (MG)**

***Plate #2-10*** **Mining out a gem pocket. Tourmaline Queen Mine, Pala, California, 1974. (MG)**

Barlow's cut gemstones. Although the Fabergé eggs were recovered, Barlow's gemstones never were. These are but two examples of the danger of showing a part of one's collection to help an exhibition.

But prize specimens can be "lost" in other ways as well. A collector may try to "protect" his top specimens by placing a high value on them. As Barlow can attest, however, other collectors and dealers may be willing to pay the price you have set on a particular specimen. A number of outstanding minerals, including a superb emerald, a spectacular orange-red imperial topaz, and a large red rhodochrosite rhombohedron are but a few of the specimens that "got away" from John over the years, much to John's regret.

Many collectors concentrate on a limited number of mineral species, but Barlow has assembled an eclectic, comprehensive, broad-based collection. Although admittedly biased, as any judgment of mineral specimens is, his collection contains 70–80 pieces that rank, pending new finds, as "Best in the World," 30–75 that rank as "Best in the U.S.," and around 1000 specimens that are considered "World-Class." These outstanding specimens include red beryl, red topaz, apatite, acanthite, tourmaline, hessite, jeremejevite, hydroxylherderite, native silver, bornite, chalcocite, libethenite, aquamarine, corundum (sapphire), mimetite, many golds, miargyrite, proustite, sperrylite, stephanite, betafite, euclase, rhodonite, red scheelite, and tanzanite, all described herein. Experts agree on the superb quality of these pieces.

In addition to numerous outstanding individual mineral specimens, the collection has several outstanding subcollections of mineral species, elements, and minerals from specific mines and certain countries. Included in this group of subcollections can be found the best suite of red beryls from the Harris Mine in the Wah Wah Mountains, Utah; probably the best private collection of silver-bearing minerals in the U.S.; an outstanding collection of crystallized gold; a comprehensive collection of minerals from the Montreal Iron Mine in northern Wisconsin; the finest collection to date of the world's best chalcocites from the Flambeau Mine in Ladysmith, Wisconsin; and the preservation of one of the finest assemblies of a suite of the phosphates of the Tip Top Mine in South Dakota. His Mexican minerals are a superb representative collection. His collection of minerals from Tsumeb, Namibia and elsewhere in Africa ranks among the best in the U.S., as does his tourmaline crystal collection, which includes a unique reconstructed gem tourmaline pocket, the "postage stamp tourmaline," and outstanding tourmalines from worldwide localities. Each of these subcollections is represented by a chapter in this book.

Such diversity requires its assembler to be knowledgeable on a wide variety of minerals. In addition to attending many mineral shows and visiting mineral museums around the world, John participated directly in several mining operations, something few collectors have an opportunity to do. John has excavated a gem tourmaline pocket at Pala, California, mined red beryls in the Wah Wah Mountains in Utah, and self-collected at the Amelia Mine, Boleo District, Santa Rosalia, Baja California Sur, Mexico. He has also provided financial assistance for several mining efforts in Arizona and Nepal.

**Plate #2-11 Professor Gene LaBerge working in the Barlow museum. (MH)**

John's quest for knowledge about minerals has taken him to many areas from which mineral specimens are recovered. Some areas are relatively nearby, such as the Flambeau Mine at Ladysmith, Wisconsin, or the pegmatites in the Black Hills, or the mines in Colorado. However, other areas are much more remote, such as South Africa, Namibia, and Brazil. As the following letter from Alice Barlow attests, some of these trips turned out to be real adventures.

*I learned a lot about my father on a trip to Brazil in 1979. I realized how much he truly loves people and finds everyone interesting (no matter who you are or what you do); how far he will go in search of a goal, including a total disregard for his own discomfort or conditions being less than desirable; and, just how much he loves life.*

***The trip:***

*Being the adventurer that my father is, we had no itinerary, no plans on where to stay nor exact plans as to where we were going. We just headed for Governador Valadares with a mission to explore.*

*We did many things, including visiting dozens of mineral dealers and collectors; visiting warehouses filled with unbelievable quantities of materials; and looking at hundreds of thousands of loose gemstones. (We made purchases of course!)*

*One vivid event took place after nearly a full day's drive into mining country in a four-wheel drive, with the last several hours of the trip being on the worst "roads" I've ever been on. Anything to visit a mine. And this was a crude mining operation. I took one look down this tunnel (straight down) and said, "No way." My father, of course, jumped at the chance to go down. It was narrow, no steps, one*

***Plate #2-12*** **Barlow in his museum. (MH)**

*rope to hold onto and, of course, no lighting. OSHA would have had heart failure. In fact, that was the year of my father's heart problems. He was terribly out of breath climbing up the mountainside just to get to the opening of the mine. He didn't tell me he had bad chest pains until after he went down the tunnel. Within a few months of that trip he had open heart surgery.*

*The trip back at nearly dusk was interrupted because our four-wheel got caught behind a large flatbed truck hauling quartz crystals, many weighing hundreds of pounds, from the mine. It got stuck in a creek and we couldn't get by. Much to my dismay it was well into the night before we got out. My dad was in the best of spirits even though it had been many hours and it was getting quite cool, and even after he was attacked by biting ants which crawled up his pant legs. He even laughed at himself with the others as he jumped around and finally took his pants off! Definitely a good sport.*

F. John Barlow is much more than a collector of minerals. He has gone to great lengths to learn from the minerals he collects. For example, he was fortunate enough to participate in the mining out of a large gem pocket of tourmaline in the Tourmaline Queen Mine at Pala, California in 1974. He obtained numerous pictures underground as the pocket was exposed and excavated. When he learned that a gem tourmaline pocket had never been reconstructed, he acquired the necessary tourmalines and morganites recovered from the pocket, loaded several barrels of matrix feldspar, quartz, lepidolite, and pocket clay, and shipped them to Appleton, Wisconsin, where the pocket was reconstructed over a period of several months with the help of Dr. Peter Bancroft. Anyone who has seen it acknowledges that it is a masterpiece of *authentic* reconstruction of a gem tourmaline pocket. (Refer to Chapter 4, Plate #4-1.) And the enthusiasm with which Barlow tells the story of the pocket is, to this day, an inspiration to hear. The complete story is in Chapter 4.

Barlow's penchant for sharing wonders with others, and his interest in teaching, is demonstrated in several ways. He has donated numerous mineral specimens and gemstones to a number of museums. His contributions to the Geology Museum at the University of Wisconsin, Madison, constitute a significant part of its holdings and greatly enhance its educational functions, where some 20,000 visitors (mainly school groups) annually view the exhibits. He has been named a life member of the James Smithson Society and was awarded the Silver Medal and the Gold Medal for his donations of gems and mineral specimens to the Smithsonian's National Museum of Natural History.

Barlow has also been awarded the Chancellor's Medallion by the University of Wisconsin, Oshkosh, in part for his contributions to the education of hundreds of geology students who have visited his collection. Barlow has encouraged and often hosted these visits, recounting some of his experiences and knowledge of minerals to the students. To encourage the students to consider various attributes of mineral specimens, Barlow commonly asks each of the students to choose the mineral they like best of the approximately 1,000 specimens on display. This exercise generally takes place after the students have had several hours to look at the minerals. Barlow then asks the students to explain what they especially like about the specimen of their choice and why they made their selection.

Then, he discusses with them the merits of their selection. The students always come away better informed on minerals and are genuinely impressed with Barlow's knowledge of minerals, his enthusiasm, and his friendliness.

In his never-ending hunt for fine mineral specimens, a hunt often patterned after his big game hunting days, F. John Barlow has traveled extensively to mining areas and to major museums around the world. He has donated and traded specimens with such prestigious museums as the Smithsonian in Washington, D.C., the Harvard Mineralogical Museum, the A.E. Seaman Mineralogical Museum, the Chicago Field Museum, the British and Austrian Museums of Natural History, the Sorbonne Museum in Paris, the Fersman Museum, Moscow, and the National Museum of China, Beijing. To Barlow, museum visits are especially rewarding in his quest for knowledge about fine minerals and gemstones.

John Barlow has lived his life with gusto and with class. He has gained much in experiences and in material things, including a superb mineral collection. And he has given a great deal to his community, his family, his profession, his chosen hobby, and the world. Perhaps the following poem would be an appropriate way to conclude this profile:

*'Tain't what we have,*
*But what we give.*
*'Tain't where we are,*
*But how we live.*
*'Tain't what we do,*
*But how we do it.*
*That makes this life*
*Worth going through it.*

*- Anonymous*

That is F. John Barlow, and this is his mineral collection.

**REFERENCE**

Hawthorne, Frank C., *Canadian Mineralogist.*

# PART II

## *THE CLASSICS*

### *Introduction*

Chapter 3 describes what experts recognize as mineral specimens that best exemplify John's collecting tastes, offering readers information on the world's significant minerals. Many of the 100 specimens described and illustrated here are included and briefly described in other chapters; with a few exceptions, their inclusion in Chapter 3 highlights their higher status with collectors.

*The collection is the visible embodiment,*
*as well as the extension,*
*of the aspiring, successful self.*

*-Douglas and Elizabeth Rigby*

# Worldwide Classics

*by David P. Wilber and Joel A. Bartsch*

## Introduction

Over the course of the last 25 years, F. John Barlow has succeeded in building one of the world's finest private collections of gems and minerals.

Indicative is the fact that Barlow has won virtually every award possible for his exhibitions of fine-quality minerals, including the McDole Trophy, the Desautels Trophy, the Detroit Challenge Cup, and numerous American Federation of Mineralogical Society national awards. (See Chapter 2.) However, this collection is much more than a private showpiece. While the depth and diversity of the collection rank it among the finest private collections in the world, the overall quality of the specimens, supplemented by Barlow's meticulous curation, make the collection the envy of every advanced collector and major museum.

Thanks to Barlow's generous hospitality, the authors had an opportunity to study this magnificent collection with the intention of selecting the most important pieces for inclusion in this section of the book. However, within the first few hours we realized the impossibility of this dream. The collection contains hundreds, perhaps thousands, of important crystallized mineral specimens and polished gemstones, some rare and some common, but impossible to list in a single chapter. Furthermore, the authors chose not to preempt information that will be found in other chapters (recognize the prejudices in Barlow's collecting: color red, rarities, gold, silver, chalcocite, and tourmaline). Only by studying the collection for several days, and by thoroughly reviewing the collection catalog, with more than 5,000 entries, can one truly appreciate this major collection.

*Only by studying the collection for several days, and by thoroughly reviewing the collection catalog, with more than 5,000 entries, can one truly appreciate this major collection.*

Collectors come in all shades. Many focus on special areas of interest. Some concentrate on acquiring only rare species, or perhaps collect minerals from a single mine or mining district. Others prefer minerals containing a common chemical element or anion group. Still others collect only gem crystals or perhaps only those minerals that are superbly crystallized and breathtakingly beautiful.

But how many collections have the breadth of this collection, which boasts not only hundreds of world-class specimens of exceedingly rare species, but also extensive suites of minerals from places like Tsumeb, Namibia; San Diego County, California; Wisconsin's Montreal and Flambeau Mines; Utah's Wah Wah Mountains; and Michigan's

Keweenaw Peninsula, as well as some of the finest gem crystals in existence and dozens of top-quality mineral specimens that for aesthetic beauty and crystal perfection from world-famous localities are considered among the world's best? The collection contains all this and more, making ours a formidable task indeed.

Of the hundreds of specimens worthy of inclusion in this chapter, our lengthy and often spirited discussions settled on a selection of 100 specimens to be photographed.

The decisions were based on the following highly subjective criteria: aesthetic beauty, crystal perfection, crystal size, rarity, and personal bias.

For this chapter the selections are alphabetized. Some specimens are described in subsequent chapters again as the several authors view them differently.

---

***Plate #3-1, Specimen #988*** **ACANTHITE – Segen Gottes Mine, Rossvein, Gersdorf, Saxony, Germany. An exceptional example of a European classic. (MH)**

## ACANTHITE (SILVER SULFIDE)

**#988 11.0×8.0×8.5cm**
**Plate #3-1**

**Segen Gottes Mine, Rossvein, Gersdorf, Saxony, Germany**

A small matrix, largely of calcite, is completely covered with modified cubes of acanthite to 8mm, showing the typical black color. This specimen is an exceptional example of a European classic. The specimen's most prominent feature is its high crystal relief, with two 5.0cm branches composed of intergrown stacks of acanthite crystals. The specimen was formerly in the British Museum of Natural History and the William Larson Collection.

## ANGLESITE (LEAD SULFATE)

**#2590 18.5×12.0×1.5cm**

**Tsumeb, Namibia**

For those who appreciate the spectacular crystallized minerals from Tsumeb, Namibia, this large, flat, single anglesite crystal is truly exceptional. It is a crystal of enormous size, its longest dimension exceeded only by some 60cm crystals found at Touissit, Morocco. It was acquired from Namibian dealer Sid Pieters and contains a large grayish "phantom" rimmed with a secondary overgrowth of yellowish anglesite.

*Plate #3-2, Specimen #4850* **ANHYDRITE – Kilometer point 9.498-9.499, North Side, Simplon Tunnel, Valais, Switzerland. A world-class specimen of a Swiss classic. (JS)**

## ANHYDRITE (CALCIUM SULFATE)

**#4850 6.0×3.0×1.5cm Plate #3-2**

**Kilometer point 9.498-9.499, North Side, Simplon Tunnel, Valais, Switzerland**

Anhydrite crystals, uniquely developed from this locality, are highly revered not only by collectors of Swiss minerals but also by collectors of worldwide classics. This large, world-class specimen consists of two lustrous, lilac-colored, intergrown crystals to 5.5cm lying flat, one against the other. The crystals are well terminated and are typically striated vertically; the lilac color is delicate yet strong. Acquired from the Eric Asselborn Collection.

## AZURITE AND MALACHITE (COPPER CARBONATE HYDROXIDES)

**#2344 6.0×5.0×2.0cm Plate #3-3**

**Czar Shaft, Bisbee, Cochise County, Arizona**

This aesthetic Bisbee rose with spheres of malachite is a "classic" Bisbee specimen, a gem to this collector. Azurite forms four overlapping mounds from 2.0 to 3.0cm across composed of vivid royal blue 3-4mm intergrown bladed crystals. The azurite is flanked by six 1.0cm smooth rounded masses of deep green malachite. An outstanding miniature specimen.

*Plate #3-3, Specimen #2344* **AZURITE and MALACHITE – Czar Shaft, Bisbee, Cochise County, Arizona. A classic Bisbee specimen. (VP)**

*Plate #3-4, Specimen #2442* **BERYL, var. aquamarine – Minas Gerais, Brazil. A 2,100g gem crystal known as the "Baseball Bat." (VP)**

*Plate #3-5, Specimen #1183* **BERYL, var. aquamarine – Minas Gerais, Brazil. A 242g textbook gem crystal. (MH)**

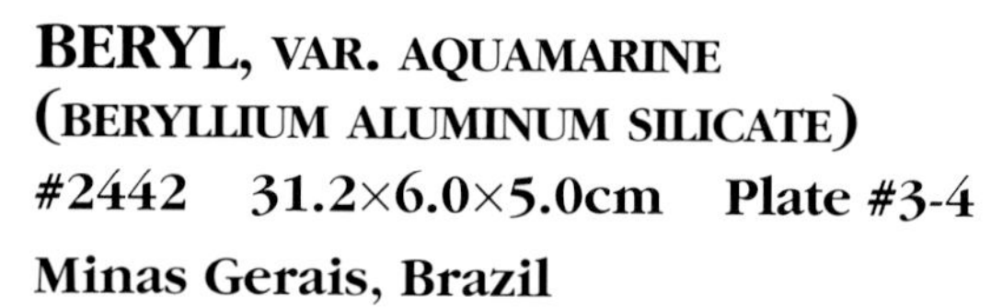

**BERYL, VAR. AQUAMARINE (BERYLLIUM ALUMINUM SILICATE)**
**#2442 31.2×6.0×5.0cm Plate #3-4**

**Minas Gerais, Brazil**

This magnificent aquamarine crystal, a beautifully terminated prism weighing 2,100g, was liberated from a private German collection in 1976 to pass through the hands of dealers Herb Obodda and Dave Wilber on its way to the Barlow Collection. Known as the "Baseball Bat," this slightly bluish, sea-green crystal has a remarkably glassy surface showing light etching that in no way detracts from its beautiful color and impressive size. (See Chapter 5, page 128.)

**BERYL, VAR. AQUAMARINE**
**#1183 13.2×3.8×3.2cm Plate #3-5**

**Minas Gerais, Brazil**

This gem, natural green crystal of aquamarine, which weighs 242g, has six facetlike terminations and was acquired by Barlow in 1977 at Dave Wilber's famous "Detroit Fire Sale," which was held to raise a huge sum of cash immediately. (See Chapter 5, page 128.)

*This superb, doubly terminated, slightly etched crystal is the ultimate aquamarine.*

**Plate #3-6, Specimen #5131** **BERYL, var. aquamarine – Near Jos Plateau, Nigeria. A huge, complex crystal among the finest found here. (VP)**

### BERYL, VAR. AQUAMARINE

**#5131 25.0×3.7×3.5cm Plate #3-6**

**Near Jos Plateau, Nigeria**

This superb, doubly terminated, slightly etched crystal with intense blue color is the epitome of an aquamarine.

Attached at one end of the main crystal and jutting out at an angle is a second crystal of comparable quality. It is described in detail in Chapter 5, page 129 and in Chapter 12, page 314. From Jurgen Henn and Mike Ridding.

**Plate #3-7, Specimen #5020** **BERYL, var. emerald – Near Jos Plateau, Nigeria. A phenomenal gem emerald of recent vintage. (VP)**

### BERYL, VAR. EMERALD

**#5020 9.5×1.0×1.3cm Plate #3-7**

**Near Jos Plateau, Nigeria**

The extraordinary, often doubly terminated, light blue-green beryl crystals from this locality are classified on the basis of their chromium content as emeralds by some gemologists. However, the color of these crystals is much lighter than that of most typical emeralds, and when viewed perpendicular to the *c*-axis, they exhibit a bluish green color similar to that of phosphophyllite. This single crystal is a virtually perfect, doubly terminated gem pencil. It is nearly flawless.

The prism faces show no etching and are planar and bright, resulting in a prism of exceptional clarity. It is terminated on one end by a pinacoid with a steep pyramidal termination opposite. This gives the effect of asymmetry unusual for beryl. Acquired from African dealer Mam Adi through dealer Bill Larson in 1992. (See Chapter 5, page 132 and Chapter 12, page 315.)

***Plate #3-8, Specimen #5160*** **BERYL, var. emerald – Cosquez Mine, Otanché, Boyaca, Colombia. A most unusual tabular gem. (MH)**

## BERYL, VAR. EMERALD

**#5160 7.0×4.5×4.0cm Plate #3-8**

**Cosquez Mine, Otanché, Boyaca, Colombia**

In the last few years, due to the efforts of Don Bachner and José Vesga of Delta Bravo Gems, several superb specimens of emerald crystals in matrix have been retrieved from the emerald mines of Colombia, particularly at Cosquez. Normally, matrix specimens get little care at the mine, but that is changing. This specimen consists of a group of unusually elongated, beautiful, lustrous, tabular crystals, slightly distorted yet facetable, lying across a calcite matrix and parallel to each other. The largest crystal is 2.5cm long. The crystals exhibit a saturated, deep green color of exceptional intensity, consistent with fine-quality gem-grade emeralds. On the dark carbonaceous shale matrix is the typical calcite of Cosquez — gray-white and smokier than the more stark white "cannizarite" calcite matrix seen at Muzo. (See Chapter 5, page 132.)

Known for nearly 1,000 years, the emerald deposits of Colombia are slowly giving up their secrets to scientists.

In a recent study (Cheilletz, A., et al., 1994), the age of the deposits at Coscuez and Muzo, now often referred to as the Quipama-Muzo district, was determined. In addition, a new theory on the origin of the emeralds was developed. By studying fluid and solid inclusions in the emeralds, the writers of the article determined the age of the Muzo (Quipama) deposit as 31.5 to 32.6 million years and that of Coscuez as 35 to 38 million years.

On the basis of their findings, they describe the deposit as being a moderate temperature epigenetic hydrothermal deposit in sediments. A number of researchers had already proposed the hydrothermal idea, and with good reason.

The host rocks are sedimentary carbonaceous shales, with minor limestone members. The formations have complex fractures through which hydrothermal solutions circulated. An intense hydrothermal fluid-rock interaction occurred. So, it is believed the host rocks provided the beryllium and the chromium in order for emeralds to form in a carbonate vein network. *R.W.J.*

***Plate #3-9, Specimen #4950*** **BERYL, var. emerald – Adams Property, Hiddenite-Emerald Mine, Hiddenite, Alexander County, North Carolina. One of the top three specimens found circa 1970. (VP)**

## BERYL, VAR. EMERALD

**#4950 9.0×5.5×2.7cm Plate #3-9**

**Adams Property, Hiddenite-Emerald Mine, Hiddenite, Alexander County, North Carolina**

The Adams property near Hiddenite, North Carolina has produced some of North America's finest emerald crystals. This exceptional specimen is one of three major pieces recovered during the late 1960s and early 1970s mining effort.

Found in 1971, it consists of two well-terminated, glassy to selectively etched emerald crystals of nearly equal length that diverge from the base in a "V" cluster.

The larger crystal is 6.0cm long. Smaller crystals cluster in subparallel form around the base. One of the main crystals shows some parallel growth development. Lying flat against the surface of several of the emerald crystals are numerous lustrous, reddish brown needles of rutile that offer no distraction. The color is considerably darker than most emeralds from this locality. The interior is mostly cloudy with minor gem areas. Barlow acquired this specimen from the collection of Glen Bollick after noted Tennessee collector Steve Neely brought the piece to his attention. It ranks in the top three from this site. The other two fine emerald specimens from this discovery were acquired by the American Museum of Natural History and the Smithsonian Institution. (See Chapter 5, page 133.)

***Plate #3-10, Specimen #5376*** **BERYL, var. emerald – Crabtree Mine, Spruce Pine, Mitchell County, North Carolina. Possibly the best emerald found at the Crabtree Mine. (MH)**

## BERYL, VAR. EMERALD

**#5376 6.7×4.3×2.3cm Plate #3-10**

**Crabtree Mine, Spruce Pine, Mitchell County, North Carolina**

This is a cluster of gemmy to cloudy emerald crystals with minor black biotite matrix from the Crabtree Mine in Spruce Pine. The emeralds, although relatively uniform in color, have a few barely perceptible yellowish emerald highlights. The crystals show simple hexagonal prisms with an interrupted steplike growth of the basal face. The specimen was collected by former miners Bill Collins and Ted Ledford, Sr. in 1968 and was obtained from current mine owner Gary Ledford. Collins and Ledford feel that this is the best from the Crabtree Mine and that it is an important North American emerald. (See Chapter 5, page 133.)

## BERYL, VAR. RED

**#5330 4.7×2.7cm**

**Refer to Chapter 14, Plate #14-1**

**Harris Mine, Wah Wah Mountains, Beaver County, Utah**

A violet-red color, this crystal reaches the awesome length of 3.6cm. Though not gemmy (due to rhyolite inclusions), this is the largest crystal on matrix known to date by mine owner Rex Harris and Barlow. Some of the prism faces and the terminations show contact marks from the enclosing rhyolite, since removed. (See Chapter 14, page 357.)

**BERYL, VAR. RED**

**#4771 6.0×2.7×2.6cm Plate #3-11**

**Harris Mine, Wah Wah Mountains, Beaver County, Utah**

The Barlow Collection undoubtedly contains the finest suite of red beryl specimens in any one collection. This specimen consists of an exceptionally large, doubly terminated crystal 2.7cm long, nearly fully exposed on the matrix, with smaller crystals of similar quality clustered around its base. The main crystal, a perfect, lustrous hexagonal prism with luscious violet-red to ruby-red color, is barely attached to a cream-colored matrix of granular rhyolite. The interior of the large crystal shows no significant gem areas though the smaller crystals do. As of this writing, the authors and Rex Harris, who operates the mine, consider this specimen the finest of its kind from any source. This specimen, when found by the Harrises, was specifically earmarked for the Barlow red beryl suite. (See Chapter 14, page 357.)

***Plate #3-11, Specimen #4771*** **BERYL, var. red – Harris Mine, Wah Wah Mountains, Beaver County, Utah. Considered the finest red beryl cluster ever found. (VP)**

***Plate #3-12, Specimen #1878*** **BERYL, var. red – Harris Mine, Wah Wah Mountains, Beaver County, Utah. Probably the most famous red beryl prism from this locality. (MH)**

**BERYL, VAR. RED**

**#1878 2.6×1.7×2.0cm Plate #3-12**

**Harris Mine, Wah Wah Mountains, Beaver County, Utah**

Of all the specimens in the Barlow red beryl suite, this single crystal is perhaps the most famous. It was illustrated by John Sinkankas and included as Figure 2 in his book *Emerald and Other Beryls*. The color of the crystal is slightly pinkish due to rhyolite inclusions on one face and just below the surface. This piece is an absolutely world-class thumbnail-sized specimen and, when dug in 1980, was the best doubly terminated crystal known. It is perfectly terminated on both ends with the typical flat pedion faces. The prism faces are brilliant and, as is typical of the red beryls from this locality, unetched but with minor growth patterns. (See Chapter 14, pages 359-361.)

## BETAFITE (CALCIUM TITANIUM OXIDE HYDROXIDE)

**#4730 13.7×11.5×5.7cm** **Refer to Chapter 12, Plate #12-32**

**Ambolotara, near Betafo, Madagascar**

This is an enormous cluster of four major octahedral betafite crystals. This specimen includes the largest and second largest crystals of the species. It came from the Sorbonne Collection, Paris, through dealer Al McGuinness, to Dr. Don Doell to Barlow. (See Chapter 12, page 315 and Chapter 5, page 138.)

***Plate #3-13, Specimen #1189*** **BORNITE – Carn Brea, Cornwall, England. A fine example of an uncommon crystal group from the classic locality. (VP)**

## BORNITE (COPPER IRON SULFIDE)

**#1189 12.0×7.0×5.0cm Plate #3-13**

**Carn Brea, Cornwall, England**

A superb specimen from a classic British locality, this piece consists of an intergrown mass of distinct and sharp bornite crystals with elongated stacks of sharply terminated bornite crystals in step growth to 2.5cm. Most bornite specimens show dull, rounded crystals, but these are lustrous and sharp.

The specimen shows a typical bluish metallic oxidation coating. This specimen came through dealer Bill Larson in an exchange with the British Museum of Natural History.

## BOURNONITE (LEAD COPPER ANTIMONY SULFIDE)

**#4915 18.0×16.0×10.0cm Plate#3-14**

**Yaogangxian, South Hunan Province, China**

Without question, this is one of the finest and largest bournonites recently recovered from China. This cabinet size specimen consists of an exceptionally large cluster of intergrown cylindrical crystals to 8.0cm on a flat matrix of quartz crystals. Resting next to the bournonite crystal group is a purplish gray fluorite crystal with minor fluorite overgrowths, measuring approximately 4.0cm on an edge. The luster of the velvety faces of the bournonites looks slightly brassy in reflected light. The clear to slightly milky quartz crystals, some doubly terminated, are typical prisms, the largest reaching 7.0cm. Pictured on the cover of *Rocks and Minerals*, Volume 67, Number 3, 1992. Acquired from Doug Parsons.

***Plate #3-14, Specimen #4915*** **BOURNONITE – Yaogangxian, South Hunan Province, China. A choice specimen. (VP)**

***Plate #3-15, Specimen #5610*** **CHALCOCITE – Levant Mine, Cornwall, England. One of the top Cornwall specimens. (VP)**

## CHALCOCITE (COPPER SULFIDE)

**#5610 5.0×3.7×3.0cm Plate #3-15**

**Levant Mine, Cornwall, England**

This spray of stout, lead-gray tabular diverging chalcocite crystals is said to be one of the top Cornwall specimens. This outstanding miniature contains five major crystals to about 2.0cm long by 1.0cm wide, aesthetically arranged on a base of smaller chalcocites. The crystals have a brilliant bluish gray luster. Minor pyrite and quartz are present on the base of the specimen. Pictured in *Mineralogical Record*, "Colorado Issue," Volume 10, November-December 1979, and labeled jalpaite.

### The Jalpaite That Wasn't

As a hunter tracks a grizzly bear, Barlow was on his drive to hunt down the best of the silver-bearing minerals.

He couldn't get out of his mind a picture of a jalpaite from Colorado pictured on page 331 in *Mineralogical Record*, the "Colorado Issue."

On the way to visit the Sweet Home Mine in August 1993, Barlow stopped to visit Rich Kosnar to talk minerals and trade stories, many not to be repeated. Barlow hoped to get a world-class specimen and perhaps a silver-bearing mineral — jalpaite, if it was available.

After several hours of conversation, Barlow asked about the availability of silver-bearing minerals.

Kosnar's wife Tresa asked Rich if he would ever part with his jalpaite, and he said, "No, not really." Finally, in the wee hours of the morning, with the jalpaite in hand, Barlow was ready to complete the purchase. But not only did Barlow question whether the specimen was from Colorado, he wasn't sure of it being jalpaite either.

A deal was made, though, and Barlow's confidence in Rich was sufficient for him to feel that if the specimen was not jalpaite, his money would be returned.

After several months of concern, Barlow felt it necessary to have the specimen X-rayed. Fortunately, Paul Pohwat of the Smithsonian agreed to take on the project. Many weeks were consumed and much testing was done on the jalpaite. The conclusions and the hard copy X-rays appeared to prove the specimen to be chalcocite. This was earth-shaking for many reasons.

A chalcocite of this size and quality was unheard of, except perhaps from the Levant Mine in Cornwall.

After examination by Bill Smith, Charlie Key, and several others, this chalcocite didn't seem to be from Mexico, Tsumeb, or Bristol, let alone Colorado.

After numerous conversations with Tresa and Rich, Barlow's money was returned and after a period of time negotiations began to confirm the specimen as a chalcocite. This was accomplished in November 1994.

The specimen was thought to be from the Caribou Mine, Boulder County, Colorado because Kosnar had purchased it from Pat Gross, who indicated that it came from Dr. R. Allen's collection and was formerly owned by William Turnby, a partner in the Caribou Mine (the New Jersey Mining Company). The book *Silver Saga* by DuWayne A. Smith, published by Pruett Publishing Company, indicates how it would be possible for a Cornwall chalcocite to get to Colorado and come through William Turnby. Turnby had children born in New Jersey, Colorado, and Cornwall, England. The Caribou Mine had many Cornish miners during its mining days.

"Is this the trail of the jalpaite that wasn't?"

*Plate # 3-16, Specimen #1836* **CHALCOCITE – Bristol, Hartford County, Connecticut. Superb, from a classic locality. (MH)**

## CHALCOCITE

**#1836 8.0×6.0×4.0cm Plate #3-16**

**Bristol, Hartford County, Connecticut**

An exceptional American classic, this impressive specimen is composed of a group of sharp, intergrown, twinned crystals of chalcocite. The crystals vary in size up to 1.5cm, but all show the "V" twinning characteristic of Bristol. Once owned by Dave Wilber, this specimen was acquired in 1984 from Larry Conklin. The deposit at Bristol was an important source of copper in the 19th century but is now obliterated by construction and oil storage sites.

*#1836 was pictured in* Rocks & Minerals *November-December 1995 as Robert B. Cook's "Connoisseur's Choice."*

Chalcocite is a very common copper sulfide ore, yet it is uncommon in fine crystals. The finest known have come from Cornwall, England and Bristol, Connecticut and are considered the classics. Fewer crystallized specimens have come from Butte, Montana; Messina, South Africa; and other copper deposits.

In the United States, the last major producer of superb chalcocites was the copper mine at Bristol, Connecticut. Its halcyon days were in the 1850s when it was operated by a group of Yale professors, including Benjamin Silliman. Important specimen production there ceased in 1859 as Silliman failed to convince the other owners to save specimens for study.

Since that time, only scattered and largely unimportant finds of chalcocite occurred in America until 1994. In that year, a new mine, the Flambeau, near Ladysmith, Wisconsin, began yielding remarkably fine crystallized chalcocites. The first important find was by Casey Jones, mining manager of Burminco Minerals, in California. (See Chapter 16 for complete account.) The best of his discovery rests now in the Barlow Collection and is described herein.

## CHALCOCITE

**#5390 3.2×2.2×1.4cm Plate #3-17**

**Flambeau Mine, Ladysmith, Rusk County, Wisconsin**

The specimen is a small miniature consisting of two sharp and brilliant crystals arranged in "V" form. The longer crystal is 3.2cm. Both are beautifully and sharply twinned along the vertical *c*-axis. The twins show a classic "X" pattern when viewed from the front. Both have a brilliant bronze patina of chalcopyrite that gives them a high metallic luster unlike that of most chalcocites.

The major crystal has a tight cluster of small sharp chalcocites jutting from the middle of the prism. The base of the crystal pair has another small chalcocite cluster protruding.

Other chalcocites found in the pocket show a range of crystal forms and colors due to the chalcopyrite patina. (See plate below.) This find is more than significant. It ranks already as having produced the best American chalcocites in nearly 150 years! Pictured on the cover of *Rocks and Minerals*, September-October, 1995.

*This find is more than significant — it ranks already as having produced the best American chalcocites in nearly 150 years!*

***Plate #3-17, Specimen #5390*** **CHALCOCITE – Flambeau Mine, Ladysmith, Rusk County, Wisconsin. Choice twins, from a new find. (MH)**

*Plate #3-18, Specimen #5720* **CHALCOCITE – 1010 Level, Flambeau Mine, Ladysmith, Rusk County, Wisconsin. One of the best specimens from the mine. (MH)**

## CHALCOCITE (COPPER SULFIDE)

**#5720 6.0×4.5×4.2cm Plate #3-18**

**1010 Level, Flambeau Mine, Ladysmith, Rusk County, Wisconsin**

An outstanding miniature specimen, this piece consists of a large, 5.0cm twinned, "spear-point" chalcocite crystal rising from a cluster of 2.0cm pseudohexagonal chalcocites. The crystals have a purplish bronze patina, producing a very aesthetic, colorful specimen. This specimen, named the "Flambeau Ace," is one of the best from the mine.

*An outstanding miniature specimen.*

## CHALCOCITE

**#5450 7.5×5.5×4.5cm Plate #3-19**

**Sunrise Pocket, 1010 Level, Flambeau Mine, Ladysmith, Rusk County, Wisconsin**

This is the finest crystal group recovered from the mine to date and may be the best chalcocite mined in America in this century. It has been named the "Flambeau Chief." The specimen is composed largely of five intergrown sixling chalcocite twins with typical pseudohexagonal form. They range from 1.5 to 3.5cm in diameter. The largest of the group sits in dominant position on the face of the specimen. All of the sixlings are made up of tabular chalcocite wafers each about 1mm thick, producing a laminated surface along the edges of the twinning. The faces of the sixlings are studded with small, to 5mm, tabular, blocky chalcocites.

Along the base of the specimen is a 2.5cm wide "chisel-point" termination of yet another chalcocite. The color of the entire mass of crystals is a dull, brassy yellow with purple iridescent highlights.

*Plate #3-19, Specimen #5450* **CHALCOCITE – Sunrise Pocket, 1010 Level, Flambeau Mine, Ladysmith, Rusk County, Wisconsin. Huge crystals, the "Flambeau Chief." (MH)**

> *. . . the "world's largest chalcostibite crystals."*

***Plate #3-20, Specimen #5843*** **CHALCOSTIBITE – St. Pons, Provence, France. (MH)**

## CHALCOSTIBITE (COPPER ANTIMONY SULFIDE)

**#5843 16.2×8.0×5.0 cm Plate #3-20**

**St. Pons, Provence, France**

A spectacular specimen with two subparallel, striated, tabular, 7.0cm crystals, and several smaller chalcostibites in an open 4.5cm wide vein lined with yellow-brown siderite crystals. The crystals are lead-gray in color reminiscent of stibnite. The specimen was pictured on page 138 of the *Mineralogical Record*, Volume 27, Number 2, 1996, when it was reported to show the "world's largest chalcostibite crystals."

*Plate #3-21, Specimen #2347* **CINNABAR – Erzberg, Steirmark, Austria. One of the finest from this location. (JS)**

## CINNABAR (MERCURY SULFIDE)

**#2347 9×2mm gem crystal on 3.5×2.5cm siderite matrix Plate #3-21**

**Erzberg, Steirmark, Austria**

This piece is a classic specimen thought to be one of the finest from this location. From the W. J. Knoblock Collection of Vienna, Austria. (Barlow "cut Bill Larson off at the pass." Refer to Chapter 17, page 390 for the rest of the story.) The specimen is a vivid ruby-red gem (9.0×2.0mm) of a cinnabar crystal with a prominent hexagonal form. The prism faces are 2.0mm with 9.0mm basal faces. The cinnabar crystal is nestled in 8.0–12.0mm greenish brown siderite rhombohedral crystals with curved faces. Several other small cinnabars are visible in the siderite.

## COPPER (NATIVE ELEMENT)

**#5208 25.0×12.5×12.5cm Plate #3-22**

**Ahmeek No. 2 Mine, Kearsarge Amygdaloid Lode, near Ahmeek, Keweenaw County, Michigan**

Found in 1910 by the mine captain, this specimen is the best of the lot from the Ahmeek No. 2 Shaft. It was subsequently given to a clerk in the accounting department of the Calumet and Hecla Mining Company.

This specimen is a long, reticulated intergrowth of greenish bronze, 2.0–4.5cm copper crystals with complex tetrahexahedral form. Some crystals are elongated nearly four times their width.

The rather open network of copper crystals is "framed" by bright green malachite enclosed in quartz, producing a very attractive color arrangement to this aesthetic large cabinet specimen.

*Plate #3-22, Specimen #5208* **COPPER – Ahmeek No. 2 Mine, Kearsarge Amygdaloid Lode, near Ahmeek, Keweenaw County, Michigan. A complex tetrahexahedral form. (MH)**

## COPPER

**#5050 9.0×8.0×5.5cm Plate #3-23**

**Point Prospect, Keweenaw County, Michigan**

This is arguably the best specimen from the recent find of large dodecahedral (or steep tetrahexahedral) copper crystals with vein prehnite in an outcropping of basalt east of Copper Harbor.

The specimen consists of a large grouping of well-defined crystals perched on a matrix of basalt and prehnite with a total size of 8.6×7.5×6.0cm. The crystals exhibit rough and pitted surfaces and curved crystal faces, typical for copper crystals from this locality. This complicates the exact determination of crystal habit. The largest single crystal is 6.0×5.3×5.0cm. Superior to other specimens from the locality by nature of its exceptional single crystal development, unusual aesthetic qualities, and the fact that the crystals occur on matrix, this is truly a world-class copper crystal specimen acquired after a long, heavy trading session with John Almquist. (See Chapter 11, page 283.)

***Plate #3-23, Specimen #5050*** **COPPER – Point Prospect, Keweenaw County, Michigan. (JS)**

## COPPER

**#4951 8.5×3.0×2.7cm Plate #3-24**

**Bogoslovsk District, Northern Ural Mountains, Sverdlovsk Oblast, Russia**

This impressive specimen, very lustrous, is from a classic Russian locality. It features an unusual group of very sharp, elongate, cubic crystals to 1.0cm in arborescent growth modified by the dodecahedron, octahedron, and one or more tetrahexahedra along one edge, combined with elongated, sharp octahedral crystals to 2.7cm. Such sharp, well-formed crystals, especially with a variety of crystal forms represented, make this an exceptional piece. The slightly tarnished color, large size and exquisite crystal development add to its fine character. Acquired from Mike Bergmann.

***Plate #3-24, Specimen #4951*** **COPPER – Bogoslovsk District, Northern Ural Mountains, Sverdlovsk Oblast, Russia. Exceptionally fine copper crystals. (JS)**

The finest rubies in the world, gems of luscious purplish red color termed *pigeon blood*, come from the Mogok, Burma deposits. Known today as the Mogok Stone Tract, located in Upper Burma, or Myanmar, this region was first worked over 500 years ago.

The crystals come from a complex group of metamorphic gneisses, schists, marble, and more, with the white marble being ruby's parent. Weathering not only released the rubies, along with other gem crystals, but also created caverns in the karst topography. Most gem crystals are recovered by the arduous task of hand-digging down to the gem-bearing gravels, then hand-washing and sorting. *R.W.J.*

**"THE BURMA BEAUTY"**

***Plate #3-25, Specimen #5872*** **CORUNDUM, var. ruby – Mogok, Myanmar. (MH)**

## CORUNDUM, VAR. RUBY (ALUMINUM OXIDE)

**#5872 8.0×4.5×4.1cm Plate #3-25**

**Mogok, Myanmar (formerly Burma)**

This specimen is a large, partially formed, deep raspberry-red hexagonal crystal with basal faces. The two lustrous prism faces are 5.0×4.0cm with numerous irregular indentations. Calcite and minor pyrrhotite also occur in the indentations. White carbonate remains in several depressions, producing interesting color contrasts. The basal faces are exceptionally lustrous, with well-developed trigonal growth patterns. Most of the crystal is somewhat translucent, with wavy "veils" passing through the crystal, but some gemmy areas are present. The color is quite uniform throughout this unusually large Burmese ruby.

Obtaining this number one world-class Burma ruby became a challenge to Bill Larson of Pala International, who is constantly pressed by Barlow to bring him only the finest specimens. Refer to Chapter 17, page 399 for this great story.

## CORUNDUM, VAR. RUBY

**#2286 4.0×2.3×1.8cm Plate #3-26**

**Khanabad, Afghanistan**

This intriguing specimen is essentially a complex group of four perfectly terminated ruby crystals standing parallel to one another. The color is a rich red. The prism faces are horizontally striated and very lustrous. A dab of white matrix contrasts nicely with the ruby color. This specimen came from author Peter Bancroft, who brought it back on one of his early visits to Afghanistan.

***Plate #3-26, Specimen #2286*** **CORUNDUM, var. ruby – Khanabad, Afganistan. A complex of four crystals. (MH)**

***Plate #3-27, Specimen #2990*** **CORUNDUM, var. sapphire – Ratnapura, Sri Lanka. A choice, doubly terminated gem. (VP)**

## CORUNDUM, VAR. SAPPHIRE

**#2990 11.0×5.5×4.3cm Plate #3-27**

**Ratnapura, Sri Lanka**

This is an exceptionally large, doubly terminated sapphire crystal of a form virtually unknown in Western collections. Measuring 11.0cm long and 5.5cm across the middle, the crystal shows the textbook tapering to a point at both ends.

The central body of the crystal is bluish yellow. The tapering ends are pleasant yellow. The faces are well striated perpendicular to the long axis yet are lustrous. The interior has large clear areas.

This specimen was acquired by dealer Bill Larson in Sri Lanka. When the piece arrived in this country, at least two prominent curators and one major collector lined up for a chance to acquire the piece for their museums. To their regret, however, they found themselves lined up behind Barlow. (Refer to Chapter 17 for the rest of the story.)

Sri Lanka, formerly Ceylon, is a ragged bit of metamorphic rock that juts out of the Indian Ocean. It is part of the crustal shield that includes India, another noted gem source. India, however, does not compare with the "Island of Gems."

Exactly where the gems of Sri Lanka formed is still under investigation. Suffice to say that once the gems formed, an amazing potpourri of species and value ended up in the alluvial and elluvial overburden.

Such gems as sapphire, in an array of colors; spinel, largely red but tending, at times, to the blue; zircon; peridot of enormous size; beryl; moonstone; garnet; plus an assortment of rare or collector stones, all join rich red ruby in completing a remarkable suite. Lovely cat's eye gems of chrysoberyl, superb star sapphires, and rubies are perhaps the crème de la crème of the gravel pits.

For the collector, the prizes of the gem gravels are the well-formed, quite complete crystals, particularly sapphire in blue or yellow hues, which managed, thanks to a high hardness, to hold their shape. *R.W.J.*

*Plate #3-28, Specimen #2740* **CREEDITE – Santa Eulalia, Chihuahua, Mexico. One of Mexico's finest examples. (VP)**

## CREEDITE
### (CALCIUM ALUMINUM SULFATE FLUORIDE HYDROXIDE HYDRATE)

**#2740 12.0×8.0×8.5cm**
**Plate #3-28**

**Santa Eulalia, Chihuahua, Mexico**

When a spectacular batch of creedite was brought up from Mexico in 1984, Barlow acquired almost the entire lot. These were the first creedite specimens exhibiting perfect, lustrous, gemmy violet crystals in large sizes. The entire surface of this specimen is covered with glassy, sharply terminated crystals of excellent uniform color to 3.5cm in length. The crystals spray randomly in all directions and are completely free of damage. This piece is one of the finest small cabinet specimens from that first lot. (See Chapter 13, page 343.)

***Plate #3-29***
***Specimen #5419*** **- left**
***Specimen #T365*** **- center**
***Specimen #5820*** **- right**
**CUMENGITE – Amelia Mine, Santa Rosalia, Baja California Sur, Mexico. (MH)**

## CUMENGITE (COPPER LEAD CHLORIDE HYDROXIDE)

**#5419 1.6×1.6×1.5cm**
**Plate #3-29 (left)**

**Amelia Mine, Santa Rosalia, Baja California Sur, Mexico**

This old specimen is a perfect crystal; clean and sharp with an indigo blue found with the old top-quality finds. It was held by an old Mexican family who turned it over to a southern European dealer who brought it to Barlow.

**#T365 2.5×2.0×1.8cm Plate #3-29 (center)**

**Amelia Mine, Santa Rosalia, Baja California Sur, Mexico**

A perfect sharp 1.5×1.5×1.2cm indigo blue crystal sits in its hard white 2.5×2.0×1.8cm matrix.

**#5820 1.2×1.1×0.9cm Plate #3-29 (right)**

**Amelia Mine, Santa Rosalia, Baja California Sur, Mexico**

This specimen is a rare perfect sharp single crystal. Very few single crystals were ever saved. This one followed the same path as #5419 from the old Mexican family. Very few museums in the world have one to equal this great specimen.

***Plate #3-30, Specimen #5080*** **CUPRITE – Mashamba West Mine, Shaba Province, Zaire. A most unusual single crystal. (MH)**

***Plate #3-31, Specimen #895*** **CUPRITE (coated by malachite on calcite) – Onganja, Namibia. One of the finest from this locality. (MH)**

## CUPRITE (COPPER OXIDE)

**#5080 4.0×3.5×2.5cm Plate #3-30**

**Mashamba West Mine, Shaba Province, Zaire**

This is a remarkably large and complex crystal from a locality that has produced numerous world-class specimens in recent years. The crystal faces are sharp and show an excellent metallic luster. The color is a deep, rich red with bright red internal reflections showing as the piece is rotated. The cubo-octahedral crystal is hoppered with a sharp cluster of smaller fine cuprites nested in the opening. Crystals from here and from Onganja, Namibia rank among the world's best, and this one is exceptional. This fine specimen was brought to this country by a French collector, Eric Asselborn, and acquired from Wayne Thompson, an Arizona dealer. (See Chapter 12, page 317.)

## CUPRITE (COATED BY MALACHITE ON CALCITE)

**#895 9.5×9.0×7.0cm Plate #3-31**

**Emke Mine, Onganja, near Seeis, Namibia**

This is one of the finest matrix specimens from a locality that produced hundreds of specimens during the early to mid 1970s. It consists of a single large octahedral cuprite crystal, 4.7cm across, attached in the center of a matrix of small, honey-colored calcite crystals. A second, smaller, 3.5×1.0cm cuprite shows the elongate distorted habit common in this species. As is typical of specimens from this locality, the cuprite crystals are covered with a coating of microcrystalline malachite. Acquired from the Frank Collins Collection.

## CUPROADAMITE (COPPER ZINC ARSENATE HYDROXIDE)

**#5132 6.1×6.9×4.2cm Plate #3-32**

**Tsumeb, Namibia**

Rich green fans and groups of euhedral crystals to 1.1cm nearly completely compose the specimen. Thin septa of quartz and rare tsumcorite separate the clusters into two areas. The crystals are very lustrous, sharp, and large for the species. Choice cuproadamites from Tsumeb are uncommon at best. This is an exceptional specimen. (See Chapter 12, page 301.) Acquired from Mike Ridding.

***Plate #3-32, Specimen #5132*** **CUPROADAMITE – Tsumeb, Namibia. A very choice, uncommon specimen. (MH)**

*Plate #3-33, Specimen #782* **EUCLASE – Saramenha, near Ouro Preto, Minas Gerais, Brazil. (MH)**

## EUCLASE (BERYLLIUM ALUMINOSILICATE HYDROXIDE)

**#782 5.5×2.5×1.3cm Plate #3-33**

**Saramenha, near Ouro Preto, Minas Gerais, Brazil**

The euclase crystals recovered from this locality usually tend to be smallish, incomplete fragments measuring less than 3.0cm in length. One magnificent exception is this large, virtually complete, single crystal recovered during the mid 1970s. This gemmy, glassy crystal has a pleasing pale bluish body color with a tinge of yellow in the central zone. It was acquired from collector Dave Wilber via dealers Milton and Hilda Sklar in 1975. Pictured in *Mineralogical Record*, Volume 7, Number 3, 1976, page 107.

*Plate #3-34, Specimen #4900* **FERRO-AXINITE – New Melones Dam, Calaveras County, California. (JS)**

## FERRO-AXINITE (CALCIUM IRON ALUMINUM BOROSILICATE HYDROXIDE)

**#4900 21.0×13.0×8.0cm Plate #3-34**

**New Melones Dam, Calaveras County, California**

This specimen is crowned with lustrous "root beer" colored ferro-axinite crystals to 6.0cm clustered across a large block of matrix. The crystals are gemmy, highly lustrous, and razor sharp. Rich brown-violet lights flash from the crystals when moved. The face of the matrix is covered with paler small axinites. The ferro-axinites found at this locality are North America's best, and specimens of this size and quality are quite rare. Collected by Jim Walker in knee-deep water.

*Plate #3-35, Specimen #4849* **FERRO-AXINITE – Rocher des Armentiers, Le Bourg d'Oisans, Isère, France. (MH)**

## FERRO-AXINITE

**#4849 12.0×11.0×4.5cm Plate #3-35**

**Rocher des Armentiers,**
**Le Bourg d'Oisans, Isère, France**

This specimen is a superb group of crystals to 2.5cm in length from a classic European locality.

The crystals, which completely cover the matrix, are bright, sharp, and of a rich plum color. Internal reflections abound in the crystals. Specimens such as these are highly coveted by connoisseurs, and the Barlow ferro-axinite ranks among the best in public or private hands. It was found in 1987 and obtained through the help of collector Eric Asselborn.

*Plate #3-36, Specimen #786* **FLUORAPATITE – Panasqueira, Portugal. Perhaps the best known specimen from here. (VP)**

## FLUORAPATITE
**(CALCIUM PHOSPHATE FLUORIDE)**
**#786 4.8×3.7×1.5cm Plate #3-36**
**Panasqueira, Portugal**

Composed of two elongated, lavender crystals joined at the base, this choice, "V"-shaped specimen is among the finest fluorapatite specimens ever recovered from this prolific locality. The major crystal is 4.8cm in length, while the smaller one measures 3.5cm. Both crystals are striated, with perfect flat terminations. The lavender color is intense within the body of each crystal and shades to a delicate gray at the terminations. This superb piece is well known, having passed through the hands of dealers Rick Smith and Dave Wilber before becoming a part of this collection. Pictured in *Mineralogical Record*, Volume 7, Number 3, 1976, page 109.

## FLUORAPATITE
**#3099 2.0×1.0×1.0cm Plate #3-37**
**Mount Apatite, Auburn, Androscoggin County, Maine**

This specimen is a spectacular gemmy purple 2.0cm fluorapatite on matrix from this classic locality. The fluorapatite crystal is exceptionally sharp, with multiple prism faces, the larger of which are prominently striated perpendicular to the length of the crystal. Complex pyramid terminations join the basal faces to the prism. The color grades from deep purple in the center to a paler shade at the ends of the crystal. The matrix consists of several intergrown terminated milky quartz crystals, a smaller purple apatite crystal, and minor brown mica. A truly classic specimen. Acquired from the Ramsey Collection.

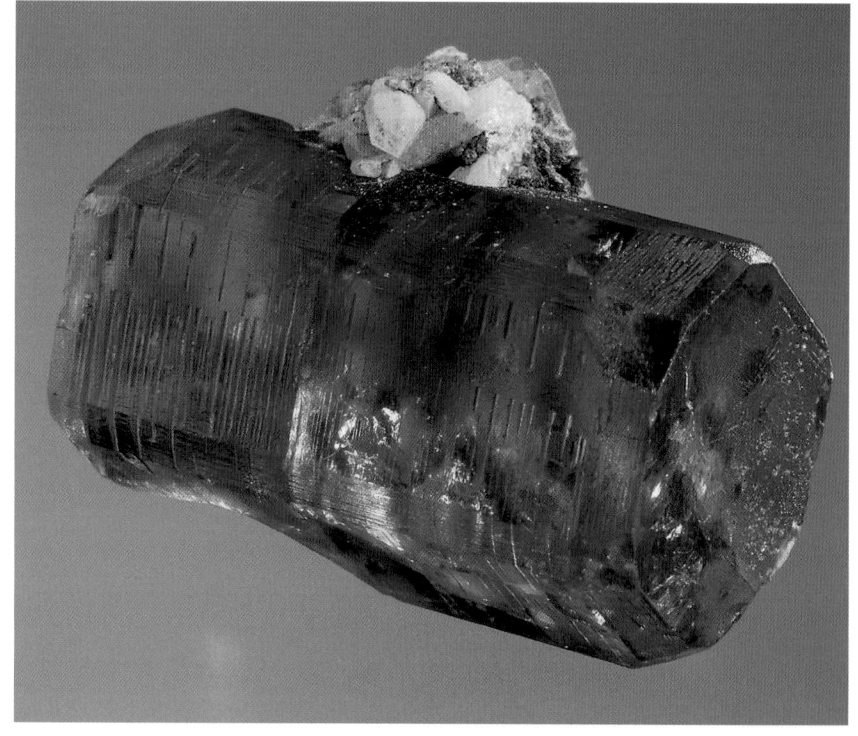

*Plate #3-37, Specimen #3099* **FLUORAPATITE – Mount Apatite, Auburn, Androscoggin County, Maine. (MH)**

***Plate #3-38, Specimen #4940*** **FLUORITE on quartz – North Face, Les Droites, bassin d'Argentiere, Massif du Mont Blanc, Haute-Savoie, France. The premier French pink fluorite. (VP)**

## FLUORITE (CALCIUM FLUORIDE) ON QUARTZ

**#4940 18.0×12.0×12.0cm Plate #3-38**

**North Face, Les Droites, bassin d'Argentiere, Massif du Mont Blanc, Haute-Savoie, France**

One of the premier specimens of the collection, this truly magnificent piece, reportedly collected from a fractured ice pocket at the 3600-m level, consists of a large single crystal of smoky quartz with many clusters of pink, octahedral crystals of fluorite (a total of 22) to 2.5cm, distributed across the surface. The fluorites are a strong reddish pink, and some of the octahedra penetrate each other. All are sharp and perfect, and they run nearly the full length and breadth of the quartz, with one 2.0cm crystal crowning the tip of the quartz crystal. This piece, even with minor repairs, is without peer, being absolutely one of the world's finest fluorite specimens. It was found August 3, 1989 by J. F. Charlet and R. Ghilini, two famous Chamonix, France strahlers. Obtained for the Barlow Collection with the help of French collector Eric Asselborn, this specimen was pictured on the cover of *Mineralogical Record*, Volume 23, Number 3, 1992.

*Plate #3-39, Specimen #5018* **FLUORITE – Chumar Barkmoor Nala, Pakistan. (VP)**

## FLUORITE

**#5018 18.0×11.0×7.0cm Plate#3-39**

**Chumar Barkmoor Nala, Pakistan**

This is an exceptional specimen of pink fluorite on muscovite, on a matrix of fine-grained granite. The specimen is terminated by a superb 8.0×7.0cm fluorite crystal that has a green core and pink exterior. Four smaller fluorites, also color zoned, form a line beneath the major crystal. The lustrous muscovite occurs in randomly oriented blades covering the matrix granite — a fine specimen from the famous Pakistan pegmatite district.

*Plate #3-40, Specimen #5780* **FORSTERITE, var. peridot – Indus River, Kohistan Area, Pakistan. (MH)**

## FORSTERITE, VAR. PERIDOT (MAGNESIUM SILICATE)

**Kohistan, Pakistan**

While forsterite is the most common member of the olivine group and is well known as the gem peridot, crystals of forsterite come from very few localities. The recent finds in the Kohistan area and in the Mansehra District are superb, well-formed gem crystals with a few on matrix. The following represent specimens in the Barlow Collection.

**#5780 5.0×4.5×2.5cm Plate #3-40**

**Indus River, Kohistan Area, Pakistan**

This superb crystal with well-defined faces has exceptional green color and transparency. Acquired from Wayne Thompson.

*Plate #3-41, Specimen #5821* - left; *Specimen #5822* - center; *Specimen #5823* - right **FORSTERITE, var. peridot – Kohistan, Pakistan. (MH)**

## FORSTERITE, VAR. PERIDOT

**#5821 6.0×4.0×3.4cm**
**Plate #3-41 (left)**

**Kohistan, Pakistan**

This specimen is a group of bright translucent green crystals to 6.0×3.0×3.0cm.

**#5822 8.0×6.5×3.0cm**
**Plate #3-41 (center)**

**Kohistan, Pakistan**

These four bright green clear crystals to 3.0×1.6cm are on a hard 8.0×6.5cm matrix with a clay bond.

**#5823 5.6×2.0×1.6cm**
**Plate #3-41 (right)**

**Kohistan, Pakistan**

This is a double crystal, bright green gem, and completely terminated.

**ANDRADITE,** VAR. DEMANTOID
(CALCIUM IRON SILICATE)
**#3281 3.4×2.1×2.6cm Plate #3-42**

**Campo Franscia, Val Malenco, Sondrio, Italy**

One of the finest specimens of demantoid garnet in existence, this superb miniature consists of two major intergrown crystals perched on a matrix of matted asbestos. The crystals are modified dodecahedra to 1.6cm with smoothly brilliant faces and sharp edges on a white matrix. It is estimated that the smaller of the two crystals would yield a 15-carat clean stone and the larger crystal would yield 12 carats, cut and polished (see sidebar). The specimen was originally collected by Dr. Francesco Bedogné and was part of the Morelli Collection in Italy. On a tip from Paul Desautels, Bill Larson purchased this famous specimen on April 14, 1972 from Morelli. Larson then sold the specimen to Robert Ramsey. Barlow acquired the specimen from Robert's son, John Ramsey, in 1987. Pictured on the cover of *Lapidary Journal*, December 1981. Also pictured in *Dizionario Mineralogico* by Marcella E. Antonio Antonucci, pages 160 and 161, and featured in *Rocks and Minerals*, "Connoisseur's Choice," Volume 68, Number 3, 1993, page 176.

***Plate #3-42, Specimen #3281*** **ANDRADITE, var. demantoid – Campo Franscia, Val Malenco, Sondrio, Italy. (VP)**

From a letter written to John Barlow by John Ramsey on August 4, 1992:

*As regards the large demantoid garnet* [#3281, Plate #3-42], *I did in fact examine it quite closely while it was in possession of my parents. By far and away it is the largest gem quality demantoid ever found. The cleaner of two crystals should cut a stone of approximately 15 carats. The 15 carat stone would be the largest stone to be cut out of this piece. It would be, to my knowledge, much larger than any demantoid ever cut. The other crystal while not as clean is larger and should be capable of cutting approximately 12 carats of stone in a total of 3 or 4 stones. In all, the specimen has the potential of close to 30 carats of cut stone. This, of course, is an approximation. Other cutters might have different ideas. Needless to say, this is a stone that should never be cut since it is such a fine mineral specimen.*

*As far as gem quality garnet specimens are concerned, this is probably the most important piece in the world. It is a rare variety of garnet. The piece is a matrix piece. The formation of the garnet is textbook perfect. The crystals are shiny and beautifully polished by nature. What more could a person want?*

*Plate #3-43, Specimen #787* **GROSSULAR, var. hessonite – Jeffrey Mine Asbestos, Quebec, Canada. A superb matrix piece from a noted locality. (VP)**

## GROSSULAR, VAR. HESSONITE (CALCIUM ALUMINUM SILICATE)

**#787 8.0×6.5×4.0cm Plate #3-43**

**Jeffrey Mine Asbestos, Quebec, Canada**

The Jeffrey Mine, one of North America's classic localities, has produced thousands of specimens of "hessonite" garnet plates over the last several decades. Only a very few specimens, however, feature truly gemmy crystals scattered across a matrix of pure white albite crystals.

This piece is perhaps the best of those few, and the contrast between crystals and matrix highlights the dodecahedral form of the honey-golden crystals. The major crystal, 2.0cm across, is above center on the matrix and shows light growth patterns on brilliant faces.

Above and below it are two paler gems of smaller size. At the lower left is another gem garnet, a perfect smaller image of the main crystal. This piece was pictured on page 149 of Peter Bancroft's *Gem and Crystal Treasures*, and was also featured on the Tucson Gem and Mineral Show program cover in 1993, and in *Mineralogical Record*, Volume 7, Number 3, 1976, page 11. Barlow acquired it from dealers Rick Smith and Dave Wilber.

## GOLD (NATIVE ELEMENT)

**#5060 4.5×2.5×1.6cm**
**Plate #3-44**

**94.44g**

**Coloma, El Dorado County, California**

This large, distorted, hoppered octahedral crystal was found in California's famous Mother Lode District around 1860, and it is one of the larger crystals still existing from that site. Prior to the turn of the century, it was acquired by a German museum and repatriated in 1992 with the help of French collector Eric Asselborn. Similar specimens are described in *Mineralogical Record*, 1987, Volume 18, page 36 after the *American Journal of Science and Arts*. (See Chapter 7, page 188.)

*Plate #3-44, Specimen #5060* **GOLD – Coloma, El Dorado County, California. A truly fine large crystal. (MH)**

***Plate #3-45, Specimen #G119*** **GOLD – Spanish Dry Diggings, El Dorado County, California. A beautiful gold and quartz combination. (MH)**

## GOLD

**#G119 9.0×8.0×3.5cm Plate #3-45**

**Spanish Dry Diggings, El Dorado County, California**

This specimen, which consists of a brilliant mass of sharp, small gold crystals to 0.5cm and weighing 338.7g, was originally donated by Mrs. R. L. Stuart to the American Museum of Natural History (#260) around the turn of the century. The diverse crystal forms include modified cubes, octahedra, and elongated twins combined with more complex rhombic dodecahedra. The crystals heavily encrust quartz crystals, among which they are seemingly woven. It is interesting to note that this specimen is strikingly similar to the 6.5g Fricot Nugget in the California State Mineral Collection obtained from Larry Conklin.

## GOLD

**#G206 10.3×6.0×1.5cm Plate #3-46**

**Ace of Diamonds Mine, near Liberty, Kittitas County, Washington**

This is an incredible group of microcrystals, wires, and dendritic plates to 1.5cm showing arborescent, or herringbone, structure due to multiple twinning and elongation of octahedra. The crystals are intergrown to form a solid network of gold. Bright and lustrous, it is an exceptional size and color.

This piece, acquired from the Ernest Butler Collection, has a bright, rich, gold color. Leached out of the enclosing calcite, it weighs 99.64g. (See Chapter 7, page 199.)

***Plate #3-46, Specimen #G206*** **GOLD – Ace of Diamonds Mine, near Liberty, Kittitas County, Washington. A remarkable mass of crystallized gold. (MH)**

*This is an incredible group of microcrystals, showing arborescent, or herringbone, structure.*

*Plate #3-47, Specimen #3731* **GOLD – Kilometer 88, Tiouiou, Zapata Field near Santa Elena, La Gran Sabana, Venezuela. An extraordinary and sharp crystal. (MH)**

## GOLD

**#3731 2.1×2.5cm Plate #3-47**

**Kilometer 88, Tiouiou, Zapata Field near Santa Elena, La Gran Sabana, Venezuela**

An extraordinarily fine 1.8cm crystal, this is truly one of the world's choice "thumbnail sized" gold specimens. This specimen (pictured in *Mineralogical Record*, Volume 18, Number 1, 1987, page 90) passed through the hands of the noted gem and mineral dealer Allan Caplan on its way to the Barlow Collection.

*An extraordinarily fine crystal, this is truly one of the world's choice "thumbnail sized" gold specimens.*

## GOLD

**#3730 2.5×1.9×1.5cm**
**Plate 3-48**

**Kilometer 88, Tiouiou, Zapata Field near Santa Elena, La Gran Sabana, Venezuela**

This exceptionally large, hoppered octahedral crystal of gold, along with #3731, is one of the finest gold crystals from this locality that the authors have ever seen. This magnificent crystal was acquired from Lawrence Conklin at the Detroit mineral show in 1983. Pictured in *Mineralogical Record*, Volume 18, Number 1, page 90.

*Plate #3-48, Specimen #3730* **GOLD – Kilometer 88, Tiouiou, Zapata Field near Santa Elena, La Gran Sabana, Venezuela. One of Venezuela's finest hoppered crystals. (MH)**

***Plate #3-49, Specimen #3756*** **GOLD – Lena River District, Yakutsk, Siberia, Russia. Superb cubes just slightly worn. (WW)**

## GOLD

**#3756 3.7×3.0×1.5cm Plate #3-49**

**Lena River District, Yakutsk, Siberia, Russia**

This is an exceptional group of slightly worn cubic crystals of impressive size from a classic Russian locality. It is pictured in *Mineralogical Record*, Volume 18, Number 1, 1987, page 91. From Theo Manos.

*This is an exceptional group of slightly worn cubic crystals of impressive size.*

***Plate #3-50, Specimen #5515*** **HAMBERGITE – Turakuloma Range, East Pamirs, Tadzhikistan. (MH)**

## HAMBERGITE

**(BERYLLIUM BORATE HYDROXIDE)**

**#5515 5.7×6.4×1.9cm Plate #3-50**

**Turakuloma Range, East Pamirs, Tadzhikistan**

This specimen is an aggregate of complex, twinned hambergite crystals. The largest crystal is 5.7cm tall. The other crystals are much smaller and are twins of the main crystal. The twinning makes the specimen look like a twinned cerussite. The piece has excellent luster, and both the front and rear faces are heavily striated. The terminations are unstriated and lustrous. The color is primarily opaque white with some off-white translucent areas. The piece is partially covered with a tan-colored druse of quartz or secondary hambergite. On one lower corner, a small broken tourmaline is attached, suggesting the specimen derived from a pegmatite, a common occurrence for hambergite.

This specimen is the better of only two fine pieces the authors have seen from Tadzhikistan during the past five years. The crystals are pristine and undamaged. Given its quality, location, and rarity, this piece is important both scientifically and mineralogically. Pictured on page 13 of the Russian magazine *World of Stones*, Issue 4, 1994. Acquired from Dave Wilber and Bryan Lees.

***Plate #3-51, Specimen #1543*** **HEMATITE with rutile on quartz – St. Gotthard, Switzerland. A classic hematite rose. (MH)**

## HEMATITE (IRON OXIDE)

**#1543 7.0×4.0×3.0cm Plate #3-51**

**St. Gotthard, Switzerland**

The specimen is composed of a subparallel stack of brilliant black hematite crystals to 4.5cm across, growing on and in the termination of glassy, colorless quartz crystals.

The hematite is thickly bladed, sharp edged, and perfect, with delicate facial striations. Typically, some of the hematite crystals have red, oriented rutile crystals on their surfaces. This is an exceptionally showy piece acquired from the collection of David Shauver of Stanton, California.

***Plate #3-52, Specimen #1090*** **HESSITE – Botés Mine, near Zlatna, Transylvania, Romania. Sharp crystals of an uncommon mineral. (VP)**

## HESSITE (SILVER TELLURIDE)

**#1090 8.5×4.0×5.0cm Plate #3-52**

**Botés Mine, near Zlatna, Transylvania, Romania**

The specimen is composed of hessite crystals to 2.0cm lying across the tops of numerous quartz crystals that make up the matrix.The most prominent crystal is a sharp, elongated prism measuring almost 2.2cm in length. It was acquired from the British Museum by Charlie Key, who sold it to Barlow. Pictured in Peter Bancroft's *Gem and Crystal Treasures*, page 450. (See Chapter 10, page 254.)

As Charles Key was saying, "I have just returned from the British Museum of Natural History and you are the first to see this world-class hessite." It wasn't what Charlie was saying, but what the specimen was saying, and it kindled Barlow's love for black, silver-bearing minerals. Here was a rare hessite with three crystal forms on white quartz. It became Barlow's hessite silver-bearing mineral specimen #1090 and has traveled to many shows with its pal, Barlow.

***Plate #3-53, Specimen #1411*** **HYDROXYLHERDERITE – Xanda Mine, Virgem de Lapa, Minas Gerais, Brazil. The world's best for the species. (VP)**

## HYDROXYLHERDERITE (CALCIUM BERYLLIUM PHOSPHATE HYDROXIDE)

**#1411  14.0×10.0×4.5cm  Plate #3-53**

**Xanda Mine, Virgem de Lapa, Minas Gerais, Brazil**

This huge (14.0cm), lustrous, gray-violet crystal of twinned hydroxylherderite is attached at the base to a corroded feldspar crystal studded with muscovite and schorl. It is one of the premier specimens in the collection. In 1976, a large pocket containing a handful of world-class matrix specimens was discovered at Virgem de Lapa, in Minas Gerais, Brazil. This piece, the finest specimen produced by that pocket, is the world's greatest specimen of this uncommon mineral. It was acquired by Barlow from dealers Ed Swoboda and Bill Larson in 1976. It is pictured on the cover of *Mineralogical Record*, Volume 10, January-February, 1979.

*Plate #3-54, Specimen #4816* **ILVAITE – South Mountain, Owyhee, Lemhi County, Idaho. An exceptionally large crystal group. (VP)**

## ILVAITE (CALCIUM IRON SILICATE HYDROXIDE)

**#4816 11.0×8.0×10.0cm**
**Plate #3-54**

**South Mountain, Owyhee, Lemhi County, Idaho**

Ilvaites from South Mountain were once quite common. They tended to be single crystals seldom exceeding 5.0cm in size. This lustrous black cluster, however, is very large for the species, 5.0×5.0×4.0cm.

It is a parallel group of sharply terminated crystals that form a thick bundle. The faces are brilliant, striated and are contrasted by minor white quartz druses sprinkled on the crystals. The matrix is composed of diverging and interlocking sprays of hedenbergite, the top surface of which holds the ilvaite skirted with bright minor quartz crystals.

This is an exceptional piece for the locality. Acquired from the Bill Garbon Collection.

## INESITE (CALCIUM MANGANESE SILICATE HYDROXIDE HYDRATE)

**#5035 30.0×16.5×6.0cm**
**Plate #3-55**

**Wessels Mine, Kalahari Manganese District, North Cape Province, South Africa**

This is a spectacular inesite of enormous size, one of the best from a relatively small find in May, 1992 at the Wessels Mine, Kalahari Manganese District, North Cape Province. It consists of red, gemmy crystals with minor natrolite-tobermorite covering the matrix solidly. Brought directly to Barlow by Charles Key. (See Chapter 12, page 311.)

*Plate #3-55, Specimen #5035* **INESITE – Wessels Mine, Kalahari Manganese District, North Cape Province, South Africa. A huge specimen of a colorful mineral. (VP)**

## JEREMEJEVITE
### (ALUMINUM BORATE OXIDE FLUORIDE HYDROXIDE)
**#2470 5.5×3.5×1.0cm Plate #3-56**

**Mile 72, near Cape Cross, Namibia**

While jeremejevite usually occurs as loose single crystals, this cluster of three crystals is a rare exception, the largest of its kind. This fine specimen consists of blue, slender crystals measuring 1.2cm, 2.5cm, and 5.5cm. The two smaller crystals jut randomly from the larger freestanding prism. Barlow acquired this specimen, one of three known clusters from this locality dug in 1972, while visiting dealer Sid Pieters in Windhoek in 1978.

*This cluster of three crystals is the largest of its kind.*

***Plate #3-56, Specimen #2470* JEREMEJEVITE – Mile 72, near Cape Cross, Namibia. An exceptionally large group of a rare species. (VP)**

***Plate #3-57, Specimen #5815* KERMESITE – Globe and Phoenix Mine, close to Que Que, Zimbabwe. (NB)**

## KERMESITE (LEAD OXYSULFIDE)
**#5815 13.3×8.0×6.5cm Plate #3-57**

**Globe and Phoenix Mine, close to Que Que, Zimbabwe**

Generally, this rare mineral occurs as an alteration of stibnite, frequently associated with other secondary antimony minerals or with native antimony as in British Columbia, Canada. Crystals are very flexible and fragile with a cherry-red color. They are translucent to opaque with adamantine luster. Generally, they look like red powder covering stibnite crystals, sometimes as radiating aggregates or hairlike; they form some nice specimens for micromounting.

Exceptional crystals appear in magnificent specimens in the Globe and Phoenix Mine, close to Que Que, now known as Zimbabwe. This deposit is closed, and it is now impossible to obtain kermesite specimens.

Before this fabulous discovery, the best specimens came from the stibnite deposit of Pernik in western Bulgaria. The quality and size of this specimen is the best of any known to the authors. In April, 1948 it was given to a European museum by N.A. Dumbleton.

*Plate #3-58, Specimen #4923* **LAZURITE – Koksha Valley, Badakshan Province, Afghanistan. (MH)**

## LAZURITE

**(SODIUM ALUMINUM PHOSPHATE)**

**#4923 6.5×5.0×5.0cm Plate #3-58**

**Koksha Valley, Badakshan Province, Afghanistan**

A superb pseudomorph after nepheline. The specimen consists of a sharp 2.5×2.2×1.2cm grouping of six vivid blue lazurite crystals on a white marble matrix. The lazurite crystals all have the hexagonal outline typical of nepheline. The major crystal is 1.5×1.2cm and is surrounded by a number of smaller crystals.

*Plate #3-59, Specimen #2496* **LEADHILLITE – Tsumeb, Namibia. A choice, large crystal of a rare species. (MH)**

## LEADHILLITE

**(LEAD SULFATE CARBONATE HYDROXIDE)**

**#2496 4.5×4.0×0.8cm Plate #3-59**

**Tsumeb, Namibia**

This superb, glassy crystal of leadhillite with sharp prism boundaries is rare indeed, one of the finest ever to be preserved. It is a clear crystal from a very limited source. (See Chapter 12, page 303.)

In the hands of a lapidary genius, this only known flawless crystal of leadhillite could become a "one of a kind in the world" gem — a fate that the authors hope will never happen.

## LEGRANDITE

**(ZINC ARSENATE HYDROXIDE HYDRATE)**

**#1797 8.0×7.0×2.6cm Plate #3-60**

**Mina Ojuela, Mapimi, Durango, Mexico**

Larger legrandite crystal groups certainly exist, but this yellow diverging spray of needlelike crystals to 3.0cm, attached horizontally in a limonite vug, is among the most pleasing cabinet specimens seen by the authors.

The crystals start at a common point and fan out to 5.0cm wide. The crystals are sharp, lustrous, and undamaged. This piece passed through the collections of dealers Keith Proctor and Dave Wilber before being acquired by Barlow. (See Chapter 13, page 338.)

*Plate #3-60, Specimen #1797* **LEGRANDITE – Mina Ojuela, Mapimi, Durango, Mexico. A fine spray of a colorful mineral. (MH)**

Plate #3-61, Specimen #5133 LEITEITE – Tsumeb, Namibia. A superb specimen of a rare mineral. (VP)

## LEITEITE (ZINC ARSENITE) WITH LUDLOCKITE, CHALCOCITE, AND QUARTZ

**#5133 10.5×6.0cm Plate #3-61**

**Tsumeb, Namibia**

This extremely large composite crystal mass of gemmy leiteite has sprays of rare ludlockite to 1.0cm. The leiteite has a silky luster. With a major crystal of 10.5cm, this specimen has set new standards for the crystal size of this species. It is considered one of the finest found to date because of its size, clarity of the crystals, and other associated minerals. A very important Tsumeb specimen. (See Chapter 12, page 303.) Acquired from Jergen Henn.

*Plate #3-62, Specimen #3282* **LIBETHENITE – Rokana Mine, Broken Hill, Zambia. A very select, large single crystal. (JS)**

## LIBETHENITE (COPPER PHOSPHATE HYDROXIDE)

**#3282 3.6×3.6×2.0cm Plate #3-62**

**Rokana Mine, Broken Hill, Zambia**

An extraordinarily large, lustrous, single crystal, this specimen was brought to dealer Sid Pieters by the miner who collected it. Barlow acquired the specimen shortly thereafter. (See Chapter 12, page 320.)

*Plate #3-63, Specimen #5381* **LUDLAMITE – 160m Level, Huanuni Mine, Oruro Dept., Bolivia. Ludlamite perfectly set on pyrite. (MH)**

## LUDLAMITE (IRON MAGNESIUM MANGANESE PHOSPHATE HYDRATE)

**#5381 8.0×7.5×5.5cm Plate #3-63**

**160m Level, Huanuni Mine, Oruro Dept., Bolivia**

This aesthetically arranged cluster of fine green ludlamites was recently obtained from Bolivia. Much larger crystals are available, but this specimen has marvelous eye appeal. The matrix is crystallized massive pyrite showing bright, sharp cubic crystal faces both fore and aft. The ludlamites are three classic tapered euhedal crystals to 2.0cm, formed in diverging clusters to give the well-known "bow-tie" effect. All crystals are perfectly terminated and undamaged, with a fine, mottled green color and bright luster. The crystals sit perfectly at the apex of the pyramid-like pyrite matrix. Acquired from Tony Jones.

## MAGNETITE (IRON OXIDE)

**#5265 13.0×5.0×8.0cm Plate #3-64**

**No. 4 Mine, Zinc Corp. of America, Balmat, St. Lawrence County, New York**

This piece, by far the finest cabinet specimen the authors have ever seen from this recent find, has nine major, lustrous, black magnetite crystals, to 2.5cm on edge, freestanding on matrix. The unusual, nearly equally sized pseudocubic crystals rest on a whitish matrix of hardened clay and granular halite. This specimen was secured by Barlow from collector Bill Pinch in 1992, then cleaned by dealer Cal Graeber to reveal its hidden beauty.

*Plate #3-64, Specimen #5265* **MAGNETITE – No. 4 Mine, Zinc Corp. of America, Balmat, St. Lawrence County, New York. Sharp, fine crystals from a recent find. (MH)**

## MALACHITE
### (COPPER CARBONATE HYDROXIDE)
**#5594 9.0×6.0×4.0cm Plate #3-65**

**Shangulowe, Shaba Province, Zaire**

This outstanding primary malachite displays lustrous dark green crystals to 9.0×3.0cm. The sample is roughly diamond shaped with the malachite crystals fanning out from the long axis of the specimen, a very unusual form for malachite. The bottom of the specimen has 0.5-1.5cm hemispherical depressions, suggesting that the malachite crystals formed on earlier botryoidal malachite. The specimen was obtained by Barlow from dealer Gilbert Gauthier.

***Plate #3-65, Specimen #5594*** **MALACHITE – Shangulowe, Shaba Province, Zaire. An outstanding primary malachite. (MH)**

***Plate #3-66, Specimen #937*** **MICROCLINE (AMAZONITE, var. microcline) – Reeser Claim, Park County, Colorado. (MH)**

## MICROCLINE (AMAZONITE, VAR. MICROCLINE)
### (POTASSIUM ALUMINUM SILICATE)
**#937 8.7×7.3×5.3cm**
**Plate #3-66**

**Reeser Claim, Park County, Colorado**

A classic microcline crystal with vivid greenish blue color perched on a matrix of fine-grained bladed white albite. The microcline crystal measures 5.0×5.0×3.5cm and has a 1.3cm wide band of white feldspar coating one face on the amazonite crystal. A very aesthetic specimen. Acquired from Richard Kosnar.

## MIMETITE
### (LEAD ARSENATE CHLORIDE)
### #1277 5.0×4.0×4.5cm Plate #3-67
**Tsumeb, Namibia**

Without question, the world's finest crystals of mimetite have been found at the Tsumeb Mine. One of the finest examples is this specimen featuring several freestanding, thumb-sized, translucent, honey-colored crystals to 2.5cm, attached to a fist-sized brownish rock matrix. The color of the crystals grades from pale yellow to a richly colored honey-orange at the sharp terminations. The faces show multiple growth patterns and are brilliant, lightly striated, and undamaged. Acquired from Prosper J. Williams at Tucson in 1976.

***Plate #3-67, Specimen #1277*** **MIMETITE – Tsumeb, Namibia. One of Tsumeb's superior mimetites. (VP)**

## MONTEBRASITE (LITHIUM ALUMINUM PHOSPHATE FLUORIDE HYDROXIDE)
### (Probably amblygonite)
### #943 4.3×4.0×2.0cm Plate #3-68
**Linopolis, Minas Gerais, Brazil**

Large euhedral crystals of this mineral are rare. Most crystals recovered tend to be irregular and highly etched. This choice crystal is by far the sharpest, cleanest, most complete crystal the authors have ever seen. The specimen consists of two distinct crystals intergrowing without matrix. Not only is the major crystal quite gemmy, but most of the crystal faces are highly lustrous and distinct.

The color of the crystals ranges from pale to sulfur-yellow. The terminations are sharp, undamaged, and perfect. Some etching is noted along the boundary of the two prisms. Barlow was able to acquire this piece by outmaneuvering prominent dealers, collectors, and curators during a mineralogical "feeding frenzy" in the booth of dealers Milton and Hilda Sklar on set-up day at the 1974 Lincoln, Nebraska show.

***Plate #3-68, Specimen #943*** **MONTEBRASITE – Linopolis, Minas Gerais, Brazil. One of the world's finest examples of this species. (VP)**

*Plate #3-69, Specimen #3138* **NAMBULITE – Kombat Mine, Namibia. A superb specimen of a very rare mineral. (JS)**

## NAMBULITE

**(LITHIUM SODIUM MANGANESE SILICATE HYDROXIDE)**

**#3138 5.0×4.0×3.0cm Plate #3-69**

**Kombat Mine, Namibia**

Another well-crystallized rarity acquired from dealer Sid Pieters, this specimen is a cluster of crystals and crystal fragments cemented together by selenite. The major crystal on this specimen measures 2.0×1.1×0.4cm. It is lustrous, lightly striated, and sharp. The smaller crystals are as perfect and lustrous. The color is a rich, dark red. Barlow was able to acquire the specimen after the pocket was stored away and finally procurred by Sid Pieters. Years of persistence produced four fine matrix specimens. (See Chapter 12, page 321.)

## NATROLITE

**(SODIUM ALUMINUM SILICATE HYDRATE)**

**#2200 8.5×6.5×3.0cm Plate #3-70**

**Surface near Aurora Mine, Charcas, San Luis Potosi, Mexico**

Sitting nearly centered on a crystallized calcite matrix is a perfect snow-white sparkling spherical crystal aggregate of natrolite to 3.0cm across. A smaller near-perfect sphere sits just above it. Few natrolites from any source possess the aesthetic quality of this piece. Although Mexico is not noted for zeolites, this one ranks as one of Mexico's best and compares well worldwide. (See Chapter 13, page 346.) From Bill Larson, 1976.

*Plate #3-70, Specimen #2200* **NATROLITE – Surface near Aurora Mine, Charcas, San Luis Potosi, Mexico. (VP)**

## NICKELINE (NICKEL ARSENIDE)

**#5275 11.0×9.0×4.5cm Plate #3-71**

**Pohla Mine, Erzgebirge, Saxony, Germany**

Nickeline is a mineral that rarely occurs in well-formed crystals. This specimen, which consists of two 5.0cm spheres of intergrown pseudohexagonal crystals to 1.5cm, is an exception.

A third aggregate sphere is 2.5cm across. One unusually sharp crystal on the upper left corner of the matrix measures almost 1.0cm across, perhaps one of the world's largest single nickeline crystals. Recently, a small batch of about eight specimens appeared on the market, all of which were vastly superior to anything previously known. This piece was acquired from the German dealer Manfred Schwartz and is among the top three specimens from the find.

***Plate #3-71, Specimen #5275*** **NICKELINE – Pohla Mine, Erzgebirge, Saxony, Germany. (MH)**

***Plate #3-72, Specimen #5224*** **PARAVAUXITE – Contacto Vein, Siglo XX Mine, Llallagua, Bolivia. Large paravauxites are rare. (MH)**

## PARAVAUXITE (IRON ALUMINUM PHOSPHATE HYDROXIDE HYDRATE)

**#5224 12.5×7.0×2.0cm Plate #3-72**

**Contacto Vein, Siglo XX Mine, Llallagua, Bolivia**

An extremely fine specimen consisting of razor-sharp, glassy, pale green, bladed crystals of paravauxite to 2.9cm long. They are well placed over a plate of tan rock matrix. Interspersed are lustrous pale brown clusters of the mineral childrenite. The paravauxite crystals are freestanding to flat lying; some are doubly terminated. Collected in 1950, this specimen is easily one of the finest paravauxite specimens brought up from Bolivia by dealer Richard Kosnar in the late 1970s out of the Ahlfeld Collection. The Siglo XX Mine is noted for exceptional specimens of this mineral.

## PROUSTITE
### (SILVER ARSENIC SULFIDE)
**#1649 6.9×4.5×3.0cm Plate #3-73**

**Dolores Tercera Mine, Chañarcillo, Chile**

Because of its rich color, crystal form, and rarity, many mineralogical connoisseurs consider well-crystallized specimens of proustite to be the most desirable of all of the metallic minerals. This piece features a cluster of three prominent, lustrous, ruby-red crystals to 3.5 cm with numerous intergrown crystals surrounding them. As proustite tends to darken on exposure to light, a fine ruby-red example like this is a joy to behold. The specimen was once a part of the Harvard Mineralogical Museum Collection (#97184) and was obtained by Barlow from Gene Schlepp, an Arizona dealer. (See Chapter 10, page 260.)

***Plate #3-73, Specimen #1649***
**PROUSTITE – Dolores Tercera Mine, Chañarcillo, Chile. A very fine example of a popular silver mineral. (VP)**

***Plate #3-74, Specimen #1671*** **PYRARGYRITE – Samson Mine, St. Andreasberg, Harz Mountains, Germany. A superb group of a silver sulfide. (MH)**

## PYRARGYRITE
### (SILVER ANTIMONY SULFIDE)
**#1671 8.5×8.0cm Plate #3-74**

**Samson Mine, St. Andreasberg, Harz Mountains, Germany**

Originally a part of the collection of the noted mineralogist Lazard Cahn, this specimen consists of a fist-sized block of rich matrix, partially covered with flattened white calcite crystals. Perched near the center of the matrix is a contrasting group of extremely lustrous, deep red pyrargyrite crystals to 3.0cm. They are sharp and lightly striated. Pyrargyrite crystals from this classic locality are among the world's finest, and this is a particularly aesthetic group. (See Chapter 10, page 262.)

***Plate #3-75, Specimen #992*** **QUARTZ, vars. rose on smoky – Sapuçaia Pegmatite, Governador Valadares District, Minas Gerais, Brazil. A superb crown of rose quartz. (VP)**

## QUARTZ, VARS. ROSE ON SMOKY (SILICON DIOXIDE)

**#992 11.0×7.5×7.0cm Plate #3-75**

**Sapuçaia Pegmatite, Governador Valadares District, Minas Gerais, Brazil**

This remarkable specimen is composed of a halo of deep, rose-pink quartz crystals to 3.5 cm, arranged in parallel overgrowths, crowning and enclosing a single crystal of smoky quartz. Originally in the Peter Bancroft Collection, this piece was part of a major discovery of world-class rose quartz crystals and was recovered in the 1960s. Pictured on page 222 of *Gem and Crystal Treasures* by Peter Bancroft.

*This piece was part of a major discovery of world-class rose quartz crystals recovered in the 1960s.*

## RHODOCHROSITE

**(MANGANESE CARBONATE)**

**#5144 6.0×5.0×5.5cm Plate #3-76**

**Good Luck Pocket, Sweet Home Mine, Alma, Park County, Colorado**

Until quite recently, top-quality specimens of rhodochrosite from Colorado have been quite scarce. Recent mining here has brought to light some truly amazing crystal specimens that exceed any known. This piece features two rich red gem crystals, the larger 3.5cm across, perched atop a matrix of sphalerite. They are remarkably undamaged. The crystals have a simple rhombic form, slightly penetrate each other, and have very sharp clean mirror faces, completely damage free. The unusually gemmy crystals and their rich red color help rank this piece among the finest hand specimens recovered from the Good Luck Pocket in 1992 during the landmark reopening of one of America's premier rhodochrosite localities. From Bryan Lees, Collector's Edge.

The search for Colorado's finest rhodochrosite at the Sweet Home Mine was carried out by Brian Lees of the Collector's Edge. For the first time, a new technique, ground penetrating radar (GPR), was applied to the search for crystal pockets. GPR has the capability of imaging potential crystal pockets as deep as 150 feet within granitic-type rock. The cost of using GPR is offset by the savings in man-hours and exploration time. Coupling GPR with traditional mining methods has brought to light the world's finest rhodochrosites, crystals to 12.5cm and crystal groups exceeding one foot, at the Sweet Home Mine. *R.W.J.*

***Plate #3-76, Specimen #5144*** **RHODOCHROSITE – Good Luck Pocket, Sweet Home Mine, Alma, Park County, Colorado. (VP)**

## RHODOCHROSITE

**#5140 21.0×12.5×6.5cm**
**Plate #3-77**

**Sweet Home Mine, Alma, Park County, Colorado**

A world-class specimen of rhodochrosite, this strikingly beautiful cabinet-sized piece consists of more than a dozen intergrown translucent, red, glassy crystals to 4.5cm on an edge. They cover entirely a matrix of tetrahedrite, pyrite, and other sulfide minerals. The specimen is remarkable for its intense color, superb condition, and mirrorlike luster. One of the top cabinet specimens mined in 1992 from the Good Luck Pocket. The piece was obtained from Bryan Lees, Collector's Edge.

*One of the top cabinet specimens mined in 1992 from the Good Luck Pocket.*

***Plate #3-77, Specimen #5140***
**RHODOCHROSITE – Sweet Home Mine, Alma, Park County, Colorado. (VP)**

## RHODONITE (MANGANESE SILICATE)

**#1186 3.0×2.0×3.0cm Plate #3-78**

**Broken Hill, New South Wales, Australia**

Complete gem crystals of rhodonite, such as this individual, are exceedingly rare. This crystal, perhaps the finest outside Australia, was acquired by dealer Dave Wilber from renowned Australian collector Albert Chapman in the 1960s. Chapman later remarked that the piece was far too good to be in an American collection. Barlow acquired the crystal from Wilber during his legendary "Detroit Fire Sale" of 1977. The crystal is tabular, transparent gemmy, and of deep pink color. It has sharp-edged, lightly striated faces that are bright and lustrous.

***Plate #3-78, Specimen #1186***
**RHODONITE – Broken Hill, New South Wales, Australia. (VP)**

## SCHEELITE (CALCIUM TUNGSTATE)

**#1346 4.5×6.0×3.5cm Plate #3-79**

**Morro Velho Mine, Minas Gerais, Brazil**

A superb bright red, gemmy, aesthetic cluster of sharp scheelite crystals on a matrix of lustrous dolomite. The scheelite has a brilliant luster that creates a spectacular specimen. The matrix dolomite is in lustrous, small, rhombohedral crystals. The specimen was featured as the frontispiece in the book *Gem and Crystal Treasures* by Peter Bancroft, 1984.

***Plate #3-79, Specimen #1346*** **SCHEELITE – Morro Velho Mine, Minas Gerais, Brazil. (MH)**

*Plate #3-80, Specimen #5225* **SILVER – Bulldog Mine, Creede, Mineral County, Colorado. One of the more aesthetic silver specimens. (MH)**

## SILVER
**(NATIVE ELEMENT)**
**#5225 6.5×3.5×6.5cm**
**Plate #3-80**

**Bulldog Mine, Creede, Mineral County, Colorado**

This pleasingly aesthetic specimen is composed of a tangle of freestanding bright wire silvers intergrown through clusters of small, bright, quartz crystals typical of the district. The silver wires curl in all directions. The quartz crystals and silver wires are perched on a matrix composed of massive black sulfide minerals. (See Chapter 9, page 230.) This specimen was acquired from Rich Kosnar.

*Plate #3-81, Specimen #5061* **SILVER (coated with argentite) – Himmelsfürst Mine, Freiberg, Saxony, Germany. A truly superb old German silver. (MH)**

## SILVER
**(COATED WITH ARGENTITE)**
**#5061 6.0×5.0×3.0cm**
**Plate #3-81**

**Himmelsfürst Mine, Freiberg, Saxony, Germany**

Essentially a stout, acanthite-coated silver wire that loops around itself several times, this specimen was found in 1882. It is an exceptionally solid wire for a German locality.

The specimen was obtained from a French collector in the 1970s through an exchange with a German museum. Fine silvers from the centuries old mines at this locality are quite rare. (See Chapter 9, page 218.)

## SILVER

**#1154 8.0×7.5×5.5cm**
**Plate #3-82**

**Kearsarge Lode, Houghton County, Michigan**

This specimen consists of a 6.0×6.0×4.0cm array of curved silver crystals to 3.5cm surmounting a rock matrix and associated with minor amounts of crystallized copper. The sharp, curving crystals are outstanding. This piece is among the best of the thousands of specimens of crystallized silver produced from this classic locality. (See Chapter 17, page 393.) From Tony Otero.

***Plate 3-82, Specimen #1154***
**SILVER – Kearsarge Lode, Houghton County, Michigan. The "Silver Bridge," a choice Michigan specimen. (VP)**

## SILVER

**#1157 4.0×2.5×1.5cm Plate #3-83**

**Kongsberg, Norway**

This specimen is a superb miniature from the classic locality for crystallized silver. While wire silver is fairly common from Kongsberg, individual crystals, particularly the sharp, slightly modified cubic crystals that compose this specimen, are far more rare. The faces are smooth and quite bright, with the largest crystal reaching 1.3cm. This choice specimen was acquired from Bill Larson of Pala International at the Detroit Show in the late 1970s. (See Chapter 9, page 225.)

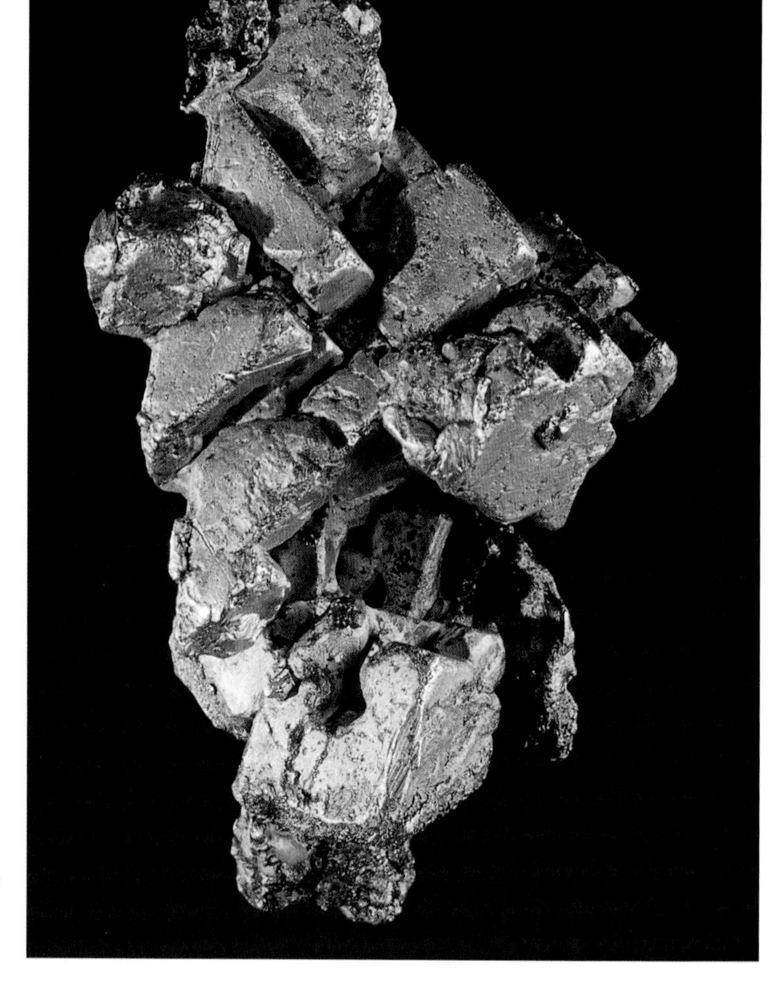

***Plate #3-83, Specimen #1157***
**SILVER – Kongsberg, Norway. Choice, rare cubes of silver. (MH)**

***Plate #3-84, Specimen #1930*** **SILVER – Quincy Mine, Kearsarge Lode, Houghton County, Michigan. Certainly the best silver in the Barlow Collection. (MH)**

## SILVER (WITH COPPER)

**#1930 9.0×8.6×4.8cm Plate #3-84**

**Quincy Mine, Kearsarge Lode, Houghton County, Michigan**

An extremely aesthetic group of bright arborescent silver crystals to 2.0cm associated with highly oxidized, slightly foliated crystalline copper that serves as a contrasting matrix for the silver crystals. Some of the larger silver crystals are gracefully curved. All are well formed, sharp, and show the classic arborescent form. This specimen originally resided in the collection of Katherine Jensen and was acquired from her by Richard Kelly of Rochester, New York. Gene Schlepp of Western Minerals brought the specimen to Barlow's attention, and Barlow acquired it from Kelly in April 1979. (See Chapter 9, page 232.)

Marc Wilson, author of Chapter 11 in this book, writes, "This world-class specimen of sharp, elongated, twinned tetrahexahedral silver crystals on a copper matrix is one of the finest examples from the Kearsarge Lode, Michigan, possibly second only to the 'Buffalo' in the A.E. Seaman Museum."

## SMITHSONITE (ZINC CARBONATE)

**#5438 9.5×6.0×8.5cm Plate #3-85**

**Berg Aukus Mine, Grootfontein, Namibia**

Smithsonite is a widespread mineral usually in low quantity in the world's zinc mines. Broken Hill, New South Wales, Australia, and Tsumeb, Namibia, however, are the reigning giants for having produced great crystals in quantity. This specimen, formerly in the David Wilber Collection and in the Richard Webster Collection, was pictured on page 51 in Peter Bancroft's book *The World's Finest Minerals and Crystals*, and was regarded then as the best smithsonite specimen known, being included among 100 best of species specimens pictured in that book.

The main crystal on the specimen is gigantic, reaching 5.5cm. The mineral is a beautiful translucent light amber color, and the surface of the crystal is composed of abundant vicinal faces. The major rhombohedral face is aesthetically positioned (facing the front of the specimen) and shows a subtle four-petaled rosette pattern. The intersection of the rhombohedral faces reveals a subparallel offset of crystals, producing the satiny reflectivity of the vicinal faces. The sides and back of the specimen have similar major smithsonite crystals that would be welcome in any major museum's exhibit as separate specimens. (See Chapter 12, page 307.)

***Plate #3-85, Specimen #5438*** **SMITHSONITE – Berg Aukus Mine, Grootfontein, Namibia. One of the world's great crystallized smithsonites. (MH)**

*Plate #3-86, Specimen #5774* **SPODUMENE – Pala, San Diego County, California. (MH)**

**SPODUMENE, VAR. KUNZITE**
**(LITHIUM ALUMINUM SILICATE)**
**#5774 11.5×5.5×2.0cm Plate #3-86**
**Pala, San Diego County, California**

This is an historic kunzite crystal from a classic locality. The specimen is a gemmy, rather etched, kunzite with a fine lavender color along the *c*-axis but colorless and transparent across the crystal. The basal part of the crystal contains partial coating and several 3-5 mm cavities containing iron-stained clay. The top of the crystal is a strongly etched, serrated surface with some herringbone patterns and some small cavernous areas. The main part of the crystal is deeply striated due to the etching. This specimen was from the original find by Kunz and was part of the Tiffany Collection. It was sold to J.P. Morgan, who donated it to the Natural History Museum of Paris in 1911. Barlow acquired it from Brad Van Scriver.

*This specimen was from the original find by Kunz and part of the Tiffany Collection. It was sold to J.P. Morgan, who donated it to the Natural History Museum of Paris in 1911.*

*Plate #3-87, Specimen #4848* **STOLZITE – +22nd Level, Sainte-Lucie Mine, Saint-Léger-de Peyres, Lozère, France. A remarkably large stolzite crystal. (JS)**

## STOLZITE (LEAD TUNGSTATE)

**#4848 5.0×4.5×1.5cm Plate #3-87**

**+22nd Level, Sainte-Lucie Mine, Saint-Léger-de Peyres, Lozère, France**

This specimen is essentially one exceptionally large crystal of stolzite, perhaps the largest in existence. It consists of a stack of parallel crystalline plates, one behind the other. The lower left third of the crystal aggregate is intermixed with matrix, but the huge exposed crystal face is clean, lustrous, and virtually complete. This is by far the largest crystal of stolzite the authors have ever seen. It was found in 1990 and obtained from collector Eric Asselborn.

## SZENICSITE (COPPER MOLYBDATE)

**#5630 10.0×7.8×8.0cm Plate #3-88**

**Tierra Amarilla, Atacama, Chile**

This szenicsite is reputed to be the finest example of this rare copper molybdate ever found. The dark green bladed crystals to nearly 3.0×1.7×0.6cm are unusually large for this species. The intergrown crystals form an attractive rosette in a cavity in the porous, granular matrix of powellite, molybdenite, and brochantite. The szenicsite cluster forms a crown across the entire specimen. The material was first collected in January, 1993 by Terry and Marrissa Szenics and was tentatively identified as the mineral lindgrenite. It was sent to the Harvard Mineralogical Museum for identification, where it was determined that it was a new mineral, szenicsite. Both Carl Francis, of Harvard, and Terry Szenics have indicated this specimen is the finest found.

*Plate #3-88, Specimen #5630* **SZENICSITE – Tierra Amarilla, Atacama, Chile. The finest example of this rare copper molybdate ever found. (MH)**

# Discovery of Szenicsite

A letter to Barlow from Terry and Marissa Szenics:

*John, you asked me to recite the story of finding the new mineral named szenicsite.*

*The discovery was a bit of luck, some intuition, and yes, experience and knowledge. Both Marrissa and I (co-discoverers) were fortunate enough to be at the mine when the new material was encountered. When I saw the material, my background in handling materials told me this was something different and worth further investigation. In spite of having to get to the 1993* [Tucson] *show, I felt it was very important that we return to the mine and do some collecting before we left. Actually, John, we were on a buying trip in January of 1993 in the region around Copiapo, Chile.*

*While we were in the town of Tierra Amarillo one afternoon, we decided to visit a small copper/molybdenite mine nearby, expecting to find molybdenite crystals to take to Tucson.*

*When we checked over the ore pile for molybdenite, I spotted a forest-green mineral with a hint of yellow-green. It was covered with a sticky, brick-red clay. The color to me was unusual, and the bladed crystals I had not seen before. The crystals were 2.0cm sheaf-like micaceous aggregates. They had a pearly rather micaceous appearance. Associated with the unknown green mineral were drusy lime-green crystals I thought to be cuprian powellite. I thought the green mineral looked like lindgrenite.*

*Lindgrenite is very rare, so Marrissa and I set about collecting what we could of the material before having to leave to go home. In spite of the severe heat, we decided to stay another day. We collected all that day, finding a number of interesting specimens, including some powellites pseudomorphed after molybdenite and, of course, fine specimens of the "green mineral."*

*When we got back to Santiago we washed all of the specimens indicating concentrations of lindgrenite. With the exciting thought of finding great specimens, I decided to go back to that little mine near Tierra Amarillo which held us up again in getting to Tucson. It was a tough trip, 900 kilometers to get to the mine, but Marrissa and I made it. We went underground and saw that the miners had already passed the zone where molybdenite was, a zone in which the new material had occurred in small vugs. However, we went out on the ore piles again and brought out the finest of the new mineral specimens. None of the miners seemed interested in the new colored material and hadn't I gotten into the piles, it would have all gone to the crusher.*

*I asked Marrissa to revisit the mine periodically while I headed for Tucson. I rendezvoused with Harvey Seigel of Aurora Minerals in New York enroute to Tucson. We decided that Dr. Carl Francis of Harvard Museum should see these specimens and run an analysis for us. Initial testing by Dr. Francis proved the mineral to be something new, as it was an orthorhombic copper hydroxyl molybdate, not monoclinic lindgrenite. On further testing, Dr. Francis informed me by telephone in Tucson that it was a new mineral and that he was going to recommend it be named szenicsite.*

*John, this is one of the proudest moments of Marrissa's and my life; to be co-discoverers of another new mineral, and Marrissa's comment to me tells the whole story: "Querido, tu lo merecias." "Dear, you deserved it!"*

*John, I am very happy that your collection has the number one specimen and Harvard the large museum piece.*

*Signed, Terry and Marissa Szenics*

*Plate #3-89, Specimen #2088* **TITANITE – Sapucaia, Minas Gerais, Brazil. (MH)**

## TITANITE
### (CALCIUM TITANIUM SILICATE)
**#2088 10.0×8.5×4.0cm Plate #3-89**

**Sapucaia, Minas Gerais, Brazil**

A superb matrix specimen of titanite with a twinned gemmy yellow-green 4.0×1.0 crystal along with two gemmy 2.0×1.0 titanites on a matrix of prominently striated greenish black epidote and pinkish gray feldspar crystals. The epidotes range from 0.5 to 2.5cm long in a jackstraw pattern and are studded with the more equant feldspar crystals. The specimen evidently represents part of the wall of a cavity because all of the crystals show crystal faces. Acquired by Barlow while visiting collector Jacinto Gauen Neto at his home in Teofilo Otoni, Brazil in 1979.

*Plate #3-90, Specimen #1420* **TOPAZ – Ghundao Hill, Katlang, Mardan District, Northwest Frontier Province, Pakistan. One of the world's great topaz specimens. (VP)**

## TOPAZ
### (ALUMINUM SILICATE FLUORIDE HYDROXIDE)
**#1420 8.0×6.0×5.5cm Plate #3-90**

**Ghundao Hill, Katlang, Mardan District, Northwest Frontier Province, Pakistan**

Reddish pink topaz crystals are exceedingly rare, especially large matrix specimens. In recent years, however, several dozen superb crystals were recovered from this locality in Pakistan. Commonly, smaller (less than 2.0cm) crystals tend to have the most intense color, while larger crystals are usually quite pale. However, this 4.0cm crystal is exceptional in that it is of the deepest intense raspberry-pink color. Much of the crystal is gem clear. The prism faces are lightly striated, and the termination is perfect. The vast majority of the crystals from this locality are recovered as loose single crystals. Even loose, single crystals comparable in quality with this one are extremely rare. Crystals of this quality that are still attached to their original matrix, as is this one, are virtually nonexistent. The crystal lies fully exposed on the matrix above a pedestal of quartz crystals and another 1.5cm crystal of the same intense red topaz color.

This specimen, by far the best of its type, was obtained from the Lunia family in Jaipur, India. It appeared on the cover of *Mineralogical Record*, January-February, 1995. (Refer to Chapter 17, page 395 for the rest of the story.)

*Plate #3-91, Specimen #4680* **TORBERNITE – Musonoi Mine, Shaba Province, Zaire. A beautiful, undamaged specimen of a lovely mineral. (MH)**

## TORBERNITE (COPPER URANYL PHOSPHATE HYDRATE)

**#4680 15.5×7.0×6.0cm Plate #3-91**

**Musonoi Mine, Shaba Province, Zaire**

Acquired from French dealer extraordinaire Gilbert Gauthier, this fine matrix specimen was once a part of the Sorbonne Collection in Paris. The matrix is completely covered with deep green, lustrous crystals up to 2.0cm. Groups of single crystals interleave to form rosettes measuring almost 3.0cm across. Torbernite is a very soft mineral, easily damaged, yet this specimen is damage free. The crystals are bright and show a strong pearly luster. Only a few truly superb pieces have come from this mine, which has produced the finest torbernites in the world. This is one such specimen. (See Chapter 12, page 324.)

## TOURMALINE, VAR. ELBAITE (SODIUM LITHIUM ALUMINUM BOROSILICATE HYDROXIDE FLUORIDE)

**#5432 11.0×7.0×13.0cm Plate #3-92**

**Chia Mine, near Governador Valadares, Minas Gerais, Brazil**

A recent discovery, the specimen has a black schorl core covered with an abraded, etched quartz. The core tourmaline grades into a gemmy rose red and is more than 6.0cm long. The crystal is capped with a 4.5cm quartz crystal restricting terminal growth. The red crystal shades to green as it only partially successfully develops around and beyond the restricting quartz. This later growth is as parallel crystals. A thin layer of green elbaite, with quartz prisms, overgrows the rubellite. (See Chapter 4, page 117.) Acquired from Jerry Cole.

*Plate #3-92, Specimen #5432* **TOURMALINE, var. elbaite – Minas Gerais, Brazil. (MH)**

***Plate #3-93, Specimen #2440*** **TOURMALINE, var. elbaite – Tourmaline Queen Mine, Pala, San Diego County, California. The famous "Postage Stamp" tourmaline. (VP)**

## TOURMALINE, VAR. ELBAITE

**#2440 13.0×9.0×8.0cm Plate #3-93**

**Tourmaline Queen Mine, Pala, San Diego County, California**

This is the famous "Postage Stamp" tourmaline. For a complete discussion of its remarkable story, see Chapter 4, pages 105–106. Front and rear cover pages of this book show the U.S. stamp sheets. The front cover page pictures the only known major error sheet. Note the color wash left to right and, most important, the mineral name (tourmaline) is missing. The rear cover pages picture "first-day covers" and a full normal sheet. Both are signed by collector Lillian Turner and die carver Leonard Buckley. Pictured on the cover of *Rock and Gems*, March, 1976. (Refer to Chapter 4, page 102.)

## TOURMALINE, VAR. ELBAITE

**#790 10.5×4.0cm Plate #3-94**

**Stewart Mine, San Diego County, California**

This specimen is a doubly terminated parallel growth of at least six crystals having a lively rose-red color. The multiple crystal growth gives it a corrugated appearance. It is damage free, showing no point of attachment, with a prominent basal termination and a minor, complex, slightly etched pyramidal termination. The prism faces are slightly dull and typically striated, with a cluster of violet lepidolite and elbaites near the base. Acquired from the Dave Wilber Collection. Pictured in *Mineralogical Record*, Volume 7, Number 3, 1976, page 108. (See Chapter 4, page 108.)

***Plate #3-94, Specimen #790*** **TOURMALINE, var. elbaite – Stewart Mine, San Diego County, California. A beautiful single crystal on matrix. (MH)**

***Plate #3-95, Specimen #5232*** **TOURMALINE, var. elbaite – Otjua Mine, Karibib, Namibia. Five distinct colors can be seen in this lovely prism. (MH)**

## TOURMALINE, VAR. ELBAITE

**#5232 10.0×1.7cm Plate #3-95**

**Otjua Mine, Karibib, Namibia**

While bicolor and even tricolor tourmaline crystals are fairly common, crystals that exhibit five distinct color zones are exceedingly rare. Attached at the base to a matrix of lilac-colored lepidolite crystals, this tourmaline crystal starts out as a dark smoky brown, then a golden-cognac, followed by deep aquamarine-blue, then a rich burgundy, terminating in a light green color. (See Chapter 4, page 120.)

*Crystals that exhibit five distinct color zones are exceedingly rare.*

*Plate #3-96, Specimen #1176* **WARDITE – Blow River, Yukon Territory, Canada. A choice specimen with excellent crystals. (VP)**

## WARDITE
### (SODIUM ALUMINUM PHOSPHATE HYDROXIDE HYDRATE)

**#1176 4.5×3.0×1.5cm**
**Plate #3-96**

**Blow River, Yukon Territory, Canada**

This fine specimen consists of four wardite crystals up to 2.2cm on an edge, spaced across a shard of brownish rock matrix. These sharp, glassy, greenish yellow crystals are among the largest and finest crystals known of this specimen and come from a remote locality, now closed. A large plate was acquired by Barlow from dealer G.E. Penekis in 1975. Barlow trimmed the plate to this superb specimen.

## WELOGANITE
### (STRONTIUM ZIRCONIUM CARBONATE HYDRATE)

**#1531 5.0×3.0×2.0cm Plate #3-97**

**Francon Quarry, St. Michael, Quebec, Canada**

Weloganite crystals tend to look like stacks of slightly offset dinner plates that alternately widen and taper toward the termination. This fine, large, typically complex crystal is uncommon as it is still attached to the rock matrix on which it formed. It is a pale yellow color, perfect in form and condition, and 4.0cm in length. This specimen was secured from Dr. Don Doell in 1977 as a small hand-size geode. By means of a long, tedious, careful trimming project, Barlow brought to light this superb specimen.

*Plate #3-97, Specimen #1531* **WELOGANITE – Francon Quarry, St. Michael, Quebec, Canada. A superb euhedral crystal of a rare species. (VP)**

**Plate #3-98, Specimen #978** WOLFRAMITE – Siglo XX Mine, Llallagua, Bolivia. Excellent, large crystals. (MH)

## WOLFRAMITE
### (IRON MANGANESE TUNGSTATE)
**#978 17.0×15.0×7.0cm Plate #3-98**

**Siglo XX Mine, Llallagua, Bolivia**

Acquired from dealer Bill Larson at the 1978 Detroit Show, this outstanding specimen consists of a plate of lustrous quartz crystals surrounding a 6.0cm tall wolframite crystal of exceptional quality, one of three brilliant, sharp, lustrous wolframites. Perched on and among the quartz crystals are numerous disc-shaped, contrasting tan siderite crystals that are typical of this famous tin-producing mine.

## WURTZITE (ZINC IRON SULFIDE)
**#1359 4.0×4.0×2.5cm**

**Potosi Mine, San José Vein, Llallagua, Bolivia**

Aesthetic crystal groups of wurtzite are exceedingly rare, particularly of this size. This superb miniature specimen is composed of shiny, bladed crystals intergrowing to form a solid cluster on a quartz crystal matrix. Barlow obtained it from the collection of the late Dick Jones, Arizona dealer.

## ZOISITE, VAR. TANZANITE
### (CALCIUM ALUMINUM SILICATE HYDROXIDE)
**#5810 6.0×5.0×3.5cm Plate #3-99**

**Merelani Hills, near Arusha, Umba Valley, Tanzania**

The usual litany of superlatives fails to convey the jolt that even jaded collectors of superb minerals receive when they first lay eyes on this specimen. Clearly, this matrix specimen grew unimpeded in a vug, which is most unusual, as the vast majority of tanzanites are imbedded completely in matrix. Here we have two brilliant, perfectly formed, gem crystals of tanzanite measuring 5.0×3.7×1.0cm and 4.5×3.5×1.5cm aesthetically perched on a mass of bright, metallic black crystals of graphite! Associated with the crystals of graphite are minor, albeit choice, free-growing crystals of light green diopside, and a large twin carbonate crystal, probably dolomite or possibly ankerite (ergo the evidence of a vug as opposed to an etched piece). Either of these tanzanite crystals would rank as a major specimen by itself, but together in this grouping with these wonderful associated species, in such size and perfection, not to mention the very serious gem values therein, well . . . any adjective you choose would seriously diminish the reality of what truly qualifies as one of the world's great mineral specimens. Purchased from Tuckman Mines and Minerals Ltd., which secured it from Husseinali Manti. *C.L.K.*

**Plate #3-99, Specimen #5810** ZOISITE, var. tanzanite – Merelani Hills, near Arusha, Umba Valley, Tanzania. (MH)

***Plate #3-100, Specimen #4937*** **ZOISITE, var. tanzanite – Merelani Hills, near Arusha, Umba Valley, Tanzania. Trichroism at its best. (VP)**

## ZOISITE, VAR. TANZANITE

**#4937 5.5×4.4×3.0cm Plate #3-100**

**Merelani Hills, near Arusha, Umba Valley, Tanzania**

This truly magnificent tanzanite, the gem variety of zoisite, was recently recovered from the Merelani claims in Tanzania and is known as the "Sleeping Beauty of Tanzania." Exceptionally trichroic, this crystal appears deep red when viewed down the *c*-axis, sapphire-blue when viewed perpendicular to the *c*-axis, and amethystine when rotated 90°. The crystal's complex termination suggests that the crystal may be a twin. Mama Kuku brought it to Mumbasa, Tanzania where it was sold to Apichart of Thailand and where Krupp bought it and returned it to Idar-Oberstein for cutting. Jergen Henn bought and sold it to Bill Larson, who brought it to Barlow at the 1991 Denver Show. Pictured in three views on the cover of *Mineralogical Record*, Volume 24, Number 3, 1993. Also pictured in *Gems & Gemology*, Summer 1992, page 94, and on page 68 of *Gemstones of East Africa* by Peter Keller. (See Chapter 12, page 325.)

## REFERENCES

Bancroft, P. (1973) *The World's Finest Minerals and Crystals*, 1973, Viking Press, New York, 176 pp.

Bancroft, P. (1984) *Gem and Crystal Treasures*, Western Enterprises, Fallbrook, California, and Mineralogical Record, Tucson, Arizona, 488 pp.

Cheilletz, A., et al. (1994).

Smith, D. *Silver Saga*, Pruett.

# PART III

## *THE GEM CRYSTAL SPECIES*

### *Introduction*

Gem crystals have long been of great fascination to John Barlow. Ever since he experienced the thrill of mining gem tourmalines in San Diego County, California, John has worked to assemble a superb collection of that species. A culmination of this interest may be seen in his collection as he helped mine out, then install, a whole gem tourmaline pocket in one wall of his mineral display room.

Since tourmaline is most often found in gem pegmatite deposits, it was only natural that John's interest quickly expanded to include beryl and the other species found in such an environment. With a high interest in gem pegmatite minerals, it was obvious that all gem crystals should also become part of John's interest and collecting effort.

The crystallography of gem crystals tend to center, although not exclusively, on one system — the hexagonal. Crystallographic descriptions provided by the authors in the following chapters are more easily understood with the aid of a few drawings. The editor recommends that the reader refer to:

Klein, C., and C. Hurlbut (1993) *Manual of Mineralogy* (according to J.D. Dana), 21st ed., John Wiley and Sons, New York, 681 pp.

Berry, L.G., B. Mason, and R.V. Dietrich (1983) *Mineralogy*, 2nd ed., W.H. Freeman and Co., San Francisco, California, 561 pp.

*Let the "forgotten man" become a good collector,*
*a master of his subject,*
*and the members of at least one group will make him feel*
*that he has, after all, a place of importance in his world.*

*- Douglas and Elizabeth Rigby*

CHAPTER 4

# Tourmaline

*by Vandall T. King*

The precious gems — diamond, emerald, ruby, and sapphire — all are household words. The hundreds of other gem materials in use today are called semiprecious. If the semiprecious gem happens to be one of the monthly birthstones or an accepted substitute for a particular traditional birthstone, it is unquestioned as a "worthy" gem. Tourmaline is not a birthstone, and it has only recently become a well-known semiprecious gem.

Tourmaline is the name applied to a variety of substances sometimes quite dissimilar in appearance. Tourmaline has been facetiously referred to as the "garbage can mineral," a reflection of its diverse chemistry.

The principal names given to species of the tourmaline group include schorl (sodium-iron-rich), dravite (sodium-magnesium-rich), uvite (calcium-magnesium-rich), liddicoatite (calcium-lithium-rich), and the reigning gem queen of the group, elbaite (sodium-lithium-rich). Additional, rare, members of the group exist but are not treated here.

The first four mentioned members of the group are not extensively represented in the Barlow Collection. The first three occur in black to very darkly colored crystals. The fourth, liddicoatite, is a mineral of very limited occurrence. Elbaite, however, is a featured species of the collection.

*Tourmaline has been facetiously referred to as the "garbage can mineral", a reflection of its diverse chemistry.*

Elbaite shines pre-eminently in the gem world because it occurs abundantly as beautiful transparent gem crystals, often bright and multicolored. The focus of the Barlow subcollection of tourmalines is elbaite, with 356 specimens devoted to representing it; fewer than 15 specimens represent the remainder of the group. Because of the importance of elbaites from California, they will be described first, followed by tourmalines worldwide and, finally, other tourmaline varieties.

## The Experiences That Direct a Collection's Future

When a collector realizes that he is destined to collect a particular series of objects, in this case tourmalines, life has a way of stepping in and directing the choices that are to be made. The immediately following sections reveal the experiences that by "chance" led up to Barlow's acquisition of the "Postage Stamp Tourmaline" and his choosing to assemble a subcollection of tourmalines.

Within any collectible category, there are rare or unique items that achieve a notoriety that spills into the public's awareness. In mineralogy, there are relatively few specimens that have obtained such general or public recognition or awareness. The "Postage Stamp Tourmaline" is one such specimen.

On June 13, 1974, in conjunction with the National Federation of Gem and Mineral Clubs' show at Lincoln, Nebraska, the United States Post Office issued a series of four setenant (all four grouped on one sheet) stamps depicting minerals of the United States. The four chosen subjects were petrified wood, rhodochrosite, amethyst, and tourmaline. The tourmaline stamp has been cataloged as Scott #1539.

The selected four specimens were from the collection of the Smithsonian in Washington, D.C. The shapes of the specimens were, however, artistically rendered, not photographically reproduced. The renderings' colors were significantly different from those of the "real McCoys," but, by their being chosen, uniqueness was conferred on these pieces, despite the unimportance of each specimen for pure science, or even their minimal significance as representatives of their genre.

Each mineral specimen is unique. The products of nature differ from the products of man wherein "catalog collectibles" such as coins, stamps, paper weights, guns, etc. contain relatively few unique items. Minerals offer an infinite selection of major, minor, and reference pieces, with only a few true duplicates. Within mineral collecting the taste of the collector and, secondarily, his serendipity, comes through.

Barlow, a stamp dealer beginning at age twelve, attended the first day of issue ceremonies of the mineral stamps. At that time, possession of the specimen depicted on the stamp was the farthest thing from Barlow's mind, though he did display the eagerness and enthusiasm that we all have when first seeing a fine example of our specialty. The true seeds of desire were soon to be sewn, and these bore fruit, as you will later read.

## A Visit to the Tourmaline Queen Mine and the Gem Pocket Reconstruction

A scant three weeks after the Lincoln show, Barlow enjoyed an extended visit with three of his daughters in California. Knowing that active pegmatite mining was occurring at the Tourmaline Queen Mine near Pala, California, he took a short trip that was to change forever his place in mineralogy and that culminated with his ownership of the "Postage Stamp Tourmaline."

The following anecdote was recounted by Barlow in 1993:

*My wife Dorothy and I went to Hawaii for a vacation and planned to visit our daughters in California. Dorothy and I flew into San Francisco on the first of July and, truthfully, I had hoped, if possible, to get down to Fallbrook in the Pala pegmatite district to see Bill Larson.*

*On arrival in San Francisco, I called Bill and he explained that he was so happy to hear from me because it seemed that every time I got to Fallbrook or even just called him, something good was happening. Larson revealed that it looked as though his mining company was about to enter a very, very prolific area in the Tourmaline Queen pegmatite dike and that while they didn't have the blast results of the day before, there were indications of some large gem pockets.*

*Larson's question was, of course, "John, do you want to go diggin'?" and I said, "By all means, when do we start?" Larson replied, "How soon can you get here?" And I said, "I'll be there tomorrow morning."*

*It turned out to be one of the greatest experiences in my life, because I spent three days underground in the Tourmaline Queen Mine with Bill Larson, Peter Bancroft (noted author), Ed Swoboda (famous specimen miner), and Mike Gray (expert gem cutter).*

*By the end of the first day, July 2, the group had revealed numerous "windows" (evidence of pockets) and potential pockets. Fortunately, there were enough indications to give an opportunity for each of the "miners" to work his own pocket. I quickly learned how sharp freshly mined quartz is, and I learned how to protect my hands after the first day my hands were cut with the "Red Badge of Courage" of my enthusiasm.*

*During this time, Bancroft said to me, "John, you really don't know how lucky you are. There are great geologists, mineralogists, professors, and collectors who would give their left arm to be doing what you are today." At that time, I didn't give much weight to the statement Pete made, but, of course, realized later what a tremendous experience it was and what a great affect it was to have on my knowing more about pegmatite dikes, tourmalines, and associated minerals.*

*I always liked red, and rubellite tourmaline was my color. Red was an important color in the logo of one of my first companies when I was fresh out of college.*

*One of the appealing things about tourmaline was the challenge to seek out and collect high-quality pieces from all over the world. There are so many variations of color, crystal form, and associations that it was as great a challenge as trying to hunt down record book African trophies or North American rams.*

*The 1974 mining experience at the Tourmaline Queen Mine turned out to be one of my most prolific diggings, second only, perhaps, to the previous big find of the "Blue Caps" in 1972, but it was significant and very fortunate for me to be present on the scene, not only participating in it, but also having an opportunity to make a selection of the find for my own collection.*

*I joked, tongue in cheek, to Larson, "Of course, I can keep all the pieces that I find and like," knowing full well I would have to pay for them. Larson responded that he could not answer that right away, as he would have to consult with Ed Swoboda as they were 50-50 partners. Swoboda might want to make his selection first with Larson and him dividing the discovery before making any sales decisions. "But I'll tell you this, John, if Ed says you can have your pick I will do the same."*

*Fortunately, on the third morning rendezvous at the Tourmaline Queen Mine, I was told that they agreed that I would have the opportunity to select whatever specimens I wanted from the find, and that they would establish a reasonable price. This was fortunate, not only from the standpoint of acquisition, but also for the resulting impression it made on me, greatly affecting my future activities in mineral collecting.*

*At this point, I should tell you of the gem pocket bought and reconstructed in my home. The concept came out on the third day while returning from the mine with Pete Bancroft in the rear of the pickup truck after we had opened some pockets. I asked Bancroft, "Has anybody ever built (reassembled) a true tourmaline gem pocket?" The response was, "John, so many people have talked about it, they all talked about it, but nobody has really ever done it." "You're kidding!," I responded. He admitted, "That is true, John." The deal was set. "If that's the case, I am going to do it." Pete immediately said, "John, if you do it, I will come to Appleton and help you reconstruct it and put it together."*

*Some photographs were necessary to record the gem pocket once it was exposed, but when we started, only one picture was made before my camera battery went dead. Mike Gray saved the day when he said, "I brought my camera and I'll take pictures." And he did.*

*Before we left Fallbrook, I selected a lot of pocket pieces to be shipped to Appleton including a 55 gallon (250 liters) drum filled with the necessary "country rock," lepidolite, quartz, and pocket clay.*

*Pete and Virginia Bancroft came out to Appleton and spent about a week. With the help of my house man, my wife Dorothy, and several of my daughters, we started the gem pocket reconstruction on two carpenter's horses.*

*It was great fun and experience until it came time to move it into place in the wall of the mineral room. Remember the guy who built his boat in his basement and couldn't get the boat out when it was finished? Same problem! A construction crew had to be hired to move and install the gem pocket, a company I incidentally acquired only several years later.*

Barlow's experience was profound. The reconstructed gem pocket has had seminal value. It was not merely a personal trophy of the "one that didn't get away." It has become an inspiration for the many geology students who have been brought to his personal museum on field trips each year. More importantly, it has become the inspiration for similar gem pocket reconstructions now seen in major public museums.

Bancroft was right. Few geologists, mineralogists, or gemologists, professional or amateur, get to see a gem pocket in situ. The gems seen in the jewelry store give no hint as to their mode of occurrence in nature. A gem tourmaline mine might seem to exist only on another planet were it not for photographs. Barlow had been to such a locality and had collected there. The construction of a representative gem pocket does not simply involve the purchase of the blasted debris and the application of some glue, however. Neither is it similar to the dismantling of an ancient castle with serially numbered blocks awaiting reassembly. The product involved a combination of both approaches with the application of experience with many such pockets and the imagination to select the salient features of the pocket environment to be represented.

The actual gem pocket reconstruction occupies a frame measuring 1.25×1.0m (Plate #4-1). It has three-dimensional relief and carries the information that a factual reconstruction must have. The right hand side of the gem pocket model consists of the ordinary rock that surrounds the gem pocket and abruptly grades into an assemblage of minerals that is interpreted to have been the replacement of the pre-existing pegmatite by corrosive fluids which were very much enriched in rare elements. The most obvious replacement minerals of the Tourmaline Queen Mine are cleavelandite, a white feldspar that is partially stained by rust and tan clays, and a purple-to-lilac mica called lepidolite. These minerals formed a mosaic providing the matrix on which the gem crystals grew. The majority of the crystals were quartz, but in the case of the Tourmaline Queen Mine, red tourmalines, as well as pale pink beryl crystals, were also abundant in the gem pocket. The pocket was not solid; rather, it was filled by debris resulting from the scouring action in the pocket by turbulently and effervescently generated water vapor. The debris consisted of sandy grains of quartz, feldspar, lepidolite, tourmaline, etc. surrounding larger, less pulverized grains

***Plate #4-1, Specimen #0500*** **RECONSTRUCTED GEM POCKET, 1.25×1.0m. (MH)**

ranging up to relatively undamaged gem crystals. The late-stage regrowth of the gem pocket's crystals included the growth of other minerals such as clays and rusty iron oxides. All are represented in the reconstructed pocket.

## "The Postage Stamp Tourmaline"

**#2440** **Refer to Chapter 3, Plate #3-93**

To return to the "Postage Stamp Tourmaline" story, the opportunity to acquire it is condensed from personal recollections of Barlow and Dave Wilber as well as from Shedenhelm.

In the same 1972-74 time period when activity at the Tourmaline Queen Mine was greatest, another important collector, Dave Wilber, was assembling a museum suite of tourmaline crystals. At one time, Wilber had over 100 specimens from the Tourmaline Queen Mine, but these have subsequently been distributed to various museums, public and private.

Near the end of Wilber's building of his Tourmaline Queen collection, the "Postage Stamp Tourmaline" stamp was unveiled. Wilber, too, was at Lincoln for the first day of issue ceremonies. Wilber inquired of Paul Desautels, then curator of the Smithsonian, on the provenance and history of the specimen. The "Postage Stamp Tourmaline" was from the Tourmaline Queen Mine and was discovered in an early mining episode in 1913. It was apparent that the notoriety of this specimen had become an unsuspected reality. The Smithsonian had provided the hitherto undistinguished and scientifically unrevealing specimen to the museum of the University of Texas at Austin through normal exchange channels long before the stamp was announced, much less issued. (Actually, the specimen was initially seen by Wilber, in December of 1967, in the New Jersey shop of Charlie Key and Rick Smith, "two young guys who were making big waves." Wilber saw it in their display, but he was told that it had been spoken for by Paul Desautels of the Smithsonian.) It was only at the February 1974 Tucson, Arizona mineral show that the U.S. Post Office announced the designs of the mineral stamps. Wilber was able to provide a favorable enough exchange with the University of Texas to secure the prize in September 1974.

The first-day ceremonies, the successful collecting at the mine, and the subsequent gem pocket reconstruction provided a basis for Barlow's wanting to add the "Postage Stamp Tourmaline" to his collection. Barlow eventually communicated with Wilber (in 1983) concerning the availability of the specimen and discovered that his inquiry came at the right time. "Yes, perhaps a deal could be made."

Wilber traveled to Appleton with the specimen along with several additional major specimens including an essentially flawless aquamarine crystal (32.0×6.0×6.0cm, now Barlow #2443). Over several hours, negotiations progressed. The beryl was acquired, several major and minor specimens were selected, and the tourmaline deal was made. The "Postage Stamp Tourmaline," mined in 1913, became Barlow #2440.

### *Description of the Postage Stamp Tourmaline*

**TOURMALINE, VAR. ELBAITE**
**(SODIUM LITHIUM ALUMINUM BOROSILICATE HYDROXIDE FLUORIDE)**
**#2440 13.0×9.0×8.0cm**

The "Postage Stamp Tourmaline" has many qualities to commend it as a "specimen of rank." Although neither the largest nor the "finest" tourmaline — either worldwide or even from its parent Tourmaline Queen Mine of Pala, San Diego County, California — the reason that this specimen was chosen for illustration on a U. S. postage stamp was not merely the "luck of the draw." This specimen offers an artistic quality that would raise any specimen above the norm.

The physical dimensions of the specimen are in an attractive proportion. The tourmaline is at the center of the specimen and is flanked by a well-terminated, translucent, light smoky quartz crystal (9.0×5.0×5.0cm) with an isolated smaller quartz crystal (6.0×4.0×3.0cm) of matching quality to balance the cluster.

The three crystals are positioned on a high mound of cleavelandite, which in turn is accented by a small apron of rose-purple lepidolite. The quartz crystals are very lightly frosted and etched, which diminishes the usual high luster of the quartz in such a way that the tourmaline becomes the focus of the viewer.

The tourmaline crystal itself is triangular in aspect, having a combination of three-sided and six-sided prisms. These result in prominent closely spaced vertical striations that add character to the crystal. The top of the crystal is capped by a simple basal face. The color zonation is entirely typical of specimens from the Tourmaline Queen Mine: grayed rose-pink on the central prism with a suggestion of gray-lilac with minimal pink near its base. A contrasting, nearly equally apportioned, complexly color-zoned "cap," 1.0cm, grades up from the grayed rose-pink central prism quickly (through several millimeters) into a paler version of itself and then into a pale green layer sharply separated from a prominent terminating light blue cap. Close inspection reveals that the blue cap, 4mm thick, is a color doublet itself, composed of blue and light purple layers. The "Postage Stamp Tourmaline" is the crowning piece of an extensive collection of tourmalines described herein.

## CALIFORNIA

## TOURMALINE, VAR. ELBAITE

The gem mines of San Diego County have been studied extensively and reported in the scientific literature. For tourmalines, the better known Californian mines include the Himalaya, Tourmaline Queen, Tourmaline King, Pala Chief, and Stewart.

**Himalaya Mine, Mesa Grande, San Diego County, California**

**#1001 11.0×6.5cm**

Himalaya Mine elbaite tourmalines have a classic color combination that specialists quickly learn to recognize, but each locality in the world, including this one, seems to have a few aberrant specimens. This crystal, 10.0×2.5cm, shows a familiar habit — a corrugated prism zone with a crenulated rounded triangular green cap shading to emerald, olive, and bluish green on an uncertain smoky-pink zone, 3.0–4.0cm, shading to a variegated light smoky-gray central zone, 3.5cm. This zone has faint tints of pink and green and rapidly changes into a dark "impenetrable" smoky to smoky-green to the end. The elbaite is partly encased in a fragment of a rusty quartz crystal.

**#1501 8.0×3.0cm**

**Himalaya Mine, Mesa Grande, San Diego County, California**

This specimen is a beautiful and unique doubly terminated gemmy elbaite crystal. The crystal has basal faces in combination with pyramidal faces on both ends, unusual for tourmaline. On one end, the termination has a thin green zone, 2–3mm, which abruptly changes into a cinnamon-pink then progresses into a rose-pink. Finally, a diagonal, vibrant gemmier green zone continues to the end of the crystal with minor development of pink color on the thin surface of the termination. While the color zoning results in a beautifully colored crystal, the green end was naturally broken or split as well as slightly divergently offset during its formation. It subsequently healed during regrowth. The result is an unusual forked crystal. Found in 1958.

**#1833 8.0×4.5cm**

**Himalaya Mine, Mesa Grande, San Diego County, California**

This is a gemmy gray-pink to gray-red elbaite crystal (2.0×2.0×1.5cm) with a somewhat "dished" to scalloped surface indicating regrowth of the low-angled pyramidal termination. The crystal is partially embedded in marvelously tightly intergrown parallel rose-lilac lepidolite crystals to several millimeters, producing a "beaded" texture overgrowing a large, coarse, single pseudohexagonal muscovite crystal. The side of the beaded muscovite/lepidolite crystal shows a terraced effect due to the external lepidolite growth.

**#1843 10.3×3.0cm and 9.5×3.0cm**

**Himalaya Mine, Mesa Grande, San Diego County, California**

This piece consists of an interesting "V" cluster of two nearly identical rose elbaite crystals with some silky pink tube formation along the prism. Both crystals have complex pyramidal terminations at their junctions with opposing dominant basal terminations facing outwards at the extremities, and are overgrown with layered lilac lepidolite. This is considered a major specimen from this locality.

**#2432 11.0×5.0cm**

**Himalaya Mine, Mesa Grande, San Diego County, California**

This is an extraordinary gemmy doubly terminated deep rose-red elbaite crystal with complex pyramidal termination on one end and a basal pedion on the other. The lower basal termination caps a 2mm green zone, while a 1.8cm irregular color zone of pinkish gray cuts through the upper prism zone below the slightly etched terminating pyramids. This piece was mined in 1983 and is large for the locality.

## TOURMALINE, VAR. ELBAITE

### #2435 8.7×3.4cm

**Himalaya Mine, Mesa Grande, San Diego County, California**

Partly coated with white drusy stilbite (uncommon in pegmatites), this gemmy rose elbaite crystal is capped by a 2mm green zone just below a basal termination. The lower third of the crystal rapidly grades into a muted light green. One end contains numerous healed fractures and "threads" in a pink zone next to an oblique skeletal regrown termination. This specimen has minor lepidolite attached.

***Plate #4-2, Specimen #4731*** **TOURMALINE, var. elbaite – Himalaya Mine, Mesa Grande, San Diego County, California. (MH)**

### #4731 12.5×8.5cm Plate #4-2

**Himalaya Mine, Mesa Grande, San Diego County, California**

This specimen is a doubly terminated, deep rose-to-gray gemmy elbaite crystal with an interrupted, etched, complex pyramidal and basal termination on a large creamy white-to-ash-gray, deeply etched, Baveno twinned, microcline crystal. The microcline is selectively overgrown by cleavelandite and lilac-to-purple crystals of lepidolite. Three smaller elbaites complete the piece.

***Plate #4-3, Specimen #4801*** **TOURMALINE, var. elbaite – Himalaya Mine, Mesa Grande, San Diego County, California. (MH)**

### #4801 5.5×9.0cm Plate #4-3

**Himalaya Mine, Mesa Grande, San Diego County, California**

This is a subparallel-to-intersecting, terminated-to-etched, multicolored elbaite, rose-pink to slightly gray-green. Crystals are up to 5.0cm on, and very slightly included in, a clear-to-cloudy quartz crystal measuring 6.0cm, with minor cleavelandite. An aesthetic matrix specimen.

### #2612 7.5×4.0cm

**Himalaya Mine, Mesa Grande, San Diego County, California**

This is an exceptional specimen. It is a trigonally terminated, deeply striated elbaite crystal with gemmy rose-colored etched termination zone and gemmy gray to gray-green prism. What makes it exceptional is the overgrowth of milky-white stilbite crystals. The major stilbite is a 4.5cm "bow-tie" form. This is an important specimen from the 1983 find.

### #3222 2.5×2.1×1.5cm

**Himalaya Mine, Mesa Grande, San Diego County, California**

This specimen is a unique recrystallized shard of tourmaline, probably an original pocket fragment now overgrown. The specimen consists of what looks like a somewhat crenulated wafer of milky fibrous elbaite, overgrown at the edges and corners by parallel growth gemmy, rose-colored elbaite crystals.

*Plate #4-4, Specimen #4802* **TOURMALINE, var. elbaite – Himalaya Mine, Mesa Grande, San Diego County, California. (MH)**

## TOURMALINE, VAR. ELBAITE

**#4802 14.3×3.3cm Plate #4-4**

**Himalaya Mine, Mesa Grande, San Diego County, California**

This piece is an unusual and instructive doubly terminated elbaite crystal (14.0×3.3cm). The opposite terminations are typical: basal on the rose-red end and complexly pyramidal on a green and grayed-rose, coarsely multiple-terminated end. Some coarsely lacy-to-crenulate white cleavelandite decorates the mid-section of one side of the specimen. The main crystal exhibits two growth periods that can be explained as consisting of a basally terminated gemmy rose-red crystal with a thin, 2mm, green zone and a gemmy, medium smoky-gray zone. Fracturing of the main crystal has been naturally healed by white elbaite, which consequently curved and offset the primary crystal. The second growth period is evidenced by a slightly silky appearing, very offset, color-zoned green and mauve coarse parallel growth of elbaite. The crystal is clear evidence of the upheavals it has experienced. (See Foord, 1977.)

*Plate #4-5, Specimen #1415* **TOURMALINE, var. elbaite – Pala Chief Mine, San Diego County, California. (VP)**

**#1415 11.0×8.4cm Plate #4-5**

**Pala Chief Mine, San Diego County, California**

This is a deep, intense rose-red, gemmy, deeply striated single elbaite crystal with a minor gray 5mm color zone beneath the basal pedion termination. This magnificent single crystal, with characteristics indistinguishable from those of elbaite from the nearby Tourmaline Queen Mine, weighs over 1150g and is a fine example from this mine.

**#790 10.5×4.0cm**

**Refer to Chapter 3, Plate #3-94**

**Stewart Mine, San Diego County, California**

This doubly terminated, gemmy, intense rose-red elbaite is made up of six or more deeply corrugated parallel growth crystals precisely sutured together. It has a prominent basal termination at one end and a minor deep red complex and slightly pyramidal termination at the other. Gray-lilac pseudohexagonal lepidolite crystals are clustered on the lower prism area with small elbaite crystals and minor cleavelandite. This specimen is one of the best from this locality.

## TOURMALINE, VAR. ELBAITE

**#1476   1.4 to 1.2 cm**

**Stewart Mine, San Diego County, California**

Suite of five pink flawless nodular elbaites and a single pastel green nodule (1.0×0.8×0.5cm). This specimen is very unusual for the locality and is representative of the most desirable structure sought by gem cutters to produce the finest gems. Nodules form when a zone in the crystal remains unaffected after late corrosive fluids attack and disaggregate a flawed tourmaline. Nodules must be flawless to survive, otherwise the flaws would serve as pathways for the corrosive fluids.

**#1484   2.7×1.8×1.2cm**

**Stewart Mine, San Diego County, California**

This is a rare achroite-zoned elbaite crystal with an asymmetrically grown termination. The crystal has color zoning ranging from pale pink just under the termination through what looks like a full 1.0cm, transparent, but flawed, achroite zone. Under magnification, the achroite reveals itself to be about one half pale pinkish gray imperceptibly grading into essentially colorless tourmaline, which quickly grades into pink, then pinkish cinnamon.

**#446   12.0×5.5cm   Plate #4-6**

**Tourmaline Queen Mine, Pala, San Diego County, California**

This piece is a blue-capped basal termination on rose-red, deeply striated, prismatic elbaite crystal (8.7×5.5cm) with similarly appearing, but minor subparallel growth elbaite crystals. The termination of the main crystal is partly enclosed by granular lepidolite and small quartz crystals. This specimen is from the 1972 "best ever" Tourmaline Queen Mine find, and was the first tourmaline purchased by Barlow in 1973 from Bill and Karla Larson.

***Plate #4-6, Specimen #446***
**TOURMALINE, var. elbaite – Tourmaline Queen Mine, Pala, San Diego County, California. (MH)**

**#922   16.0×12.0cm   Plate #4-7**

**Tourmaline Queen Mine, Pala, San Diego County, California**

This is a large, basally terminated, gemmy rose-red elbaite crystal, 8.7×6.5cm, capped by a gray-to-gray-green 1.0cm zone intersecting an etched-to-frosted light-smoky-to-smoky-citrine quartz crystal. The elbaite consists of two major and several minor parallel-to-subparallel crystals. Dark lilac lepidolite crystals to 3mm are clustered among corrugations in the main elbaite crystal. This crystal is the largest of those self-collected July 4, 1974, so it is affectionately called the "Barlow Buster."

***Plate #4-7, Specimen #922*** **TOURMALINE, var. elbaite – Tourmaline Queen Mine, Pala, San Diego County, California. (VP)**

*Plate #4-8, Specimen #1513* **TOURMALINE, var. elbaite – Tourmaline Queen Mine, Pala, San Diego County, California. (MH)**

## TOURMALINE, VAR. ELBAITE

**#1513 15.5×8.0×7.0cm Plate #4-8**

**Tourmaline Queen Mine, Pala, San Diego County, California**

This specimen is a basally terminated 4.0×3.0cm elbaite crystal with a 1.0cm blue cap and a 1mm red zone above a gray and green blended color zone on a deep rose-colored zone, 2.0cm, in a gray-to-pink crystal. The core of the lower part of the crystal is concentrically color zoned: a dark blue core with pink rind. Clear-to-pale quartz crystals, to 2.0×4.0cm, and minor cleavelandite and lepidolite cover the lower crystal area.

**#1515 15.0×8.0cm, 2kg Plate #4-9**

**Tourmaline Queen Mine, Pala, San Diego County, California**

This is an enormous blue-capped, rose-red, single elbaite crystal with basal termination and ancillary subparallel growth of smaller similar crystals. One of the largest crystals from a 1972 discovery, the specimen was formerly part of the David Wilber Collection.

**#1288, #1289, #1290 3.5 to 2.5×1.0cm**

**Tourmaline Queen Mine, Pala, San Diego County, California**

This suite of three similar color-zoned mauve-to-blue-gray elbaite crystals shows nodular structure on the pink end of the crystals, very unusual for the locality.

**#906 10.0×5.5cm**

**Tourmaline Queen Mine, Pala, San Diego County, California**

This piece consists of two elbaite crystals on a matrix of quartz crystal fragment. The two elbaite crystals show typical bicolor zonation, blue-gray on rose-red, but the unusual feature of this specimen is that the larger elbaite is a cross-sectional fragment of a basally terminated crystal with the cross-section regrown enough to show new crystal faces. The indentations in the matrix suggest a breaking away of part of the specimen by scouring and breaking action in the pocket during formation. The other unusual feature of the specimen is a small undamaged basally terminated elbaite crystal (5.0×0.9cm) half jutting out of the quartz in an unsupported fashion. The larger elbaite may have been directly hit by some churning debris while the nearby smaller crystal escaped any calamity.

*Plate #4-9, Specimen #1515* **TOURMALINE, var. elbaite – Tourmaline Queen Mine, Pala, San Diego County, California. (MH)**

***Plate #4-10, Specimen #1658*** **TOURMALINE, var. elbaite – Tourmaline Queen Mine, Blue Pocket #2, Pala, San Diego County, California. (MH)**

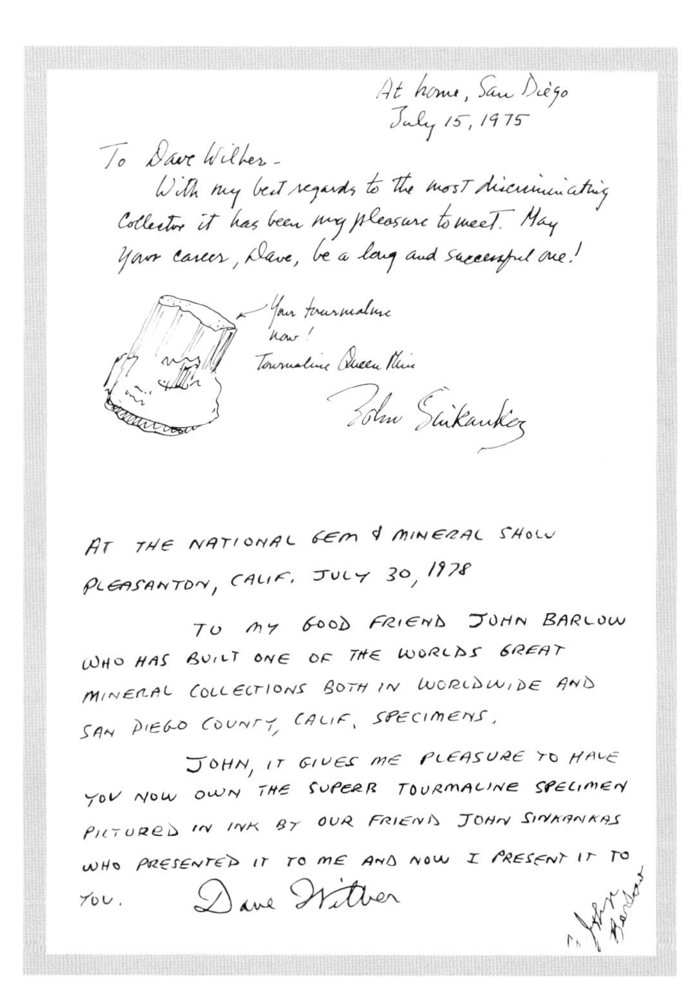

At home, San Diego
July 15, 1975

To Dave Wilber -
With my best regards to the most discriminating Collector it has been my pleasure to meet. May your career, Dave, be a long and successful one!

Your tourmaline now!
Tourmaline Queen Mine

John Sinkankas

AT THE NATIONAL GEM & MINERAL SHOW
PLEASANTON, CALIF. JULY 30, 1978

TO MY GOOD FRIEND JOHN BARLOW WHO HAS BUILT ONE OF THE WORLDS GREAT MINERAL COLLECTIONS BOTH IN WORLDWIDE AND SAN DIEGO COUNTY, CALIF. SPECIMENS.

JOHN, IT GIVES ME PLEASURE TO HAVE YOU NOW OWN THE SUPERB TOURMALINE SPECIMEN PICTURED IN INK BY OUR FRIEND JOHN SINKANKAS WHO PRESENTED IT TO ME AND NOW I PRESENT IT TO YOU.

Dave Wilber

John Barlow

**Front end sheet of "Gemstones of North America" by Dr. John Sinkankas, 1959. (MH)**

## TOURMALINE, VAR. ELBAITE

**#1658 5.5×6.0cm Plate #4-10**

**Tourmaline Queen Mine, Blue Pocket #2 Pala, San Diego County, California**

This is an unusual, asymmetrical, interrupted growth of blue-to-gray-to-dark-blue elbaite, 4.0×3.8cm, with etched basal termination on a wide base of elbaite coated and occluded by cleavelandite. A rare brown manganotantalite crystal is associated. The specimen is a twin sister to the one shown in Figure 56 of *Gemstones of North America* by John Sinkankas.

---

## AFGHANISTAN

**#1100 9.0×8.0×7.0cm**

**Kureghal Mine, Nuristan Valley, Pegmatite District, Afghanistan**

This is a large, single, gemmy-to-translucent color-zoned, frosted, basally terminated elbaite crystal from the Kureghal Mine. The green basal zone grades into cloudy pastel green into gray-pink.

**#1802 3.5×1.5cm**

**Nuristan Valley, Pegmatite District, Afghanistan**

Magnificent and rare var. achroite (pastel-zoned elbaite) with multiple growth terminations. The specimen is a transparent, flawless, and untinted achroite crystal with a central pastel green zone grading into slightly cloudy pink. This is not a large specimen, but it is as fine as an achroite gets from a classic achroite locality.

## TOURMALINE, VAR. ELBAITE

**#2064 8.0×2.2cm**

**Mawi, Laghman, Afghanistan**

This single, doubly terminated crystal from Mawi has a complex pyramidal termination and gemmy multicolored prism and lower basal termination. The color zoning in the crystal is exceptional. A pink termination graduates into a pale-gray-to-cloudy zone, changing quickly into a yellow-green "phantom" zone, highlighting a slightly cinnamon-pink crystal outline in the base of the specimen. A view down the basal termination shows a rose-red core delineated from a pink rind by a thin dark yellow-green color stripe.

**#2193 4.0×1.5×1.5cm**

**Mawi, Laghman, Afghanistan**

This distinctively color-zoned, transparent but slightly flawed elbaite crystal also originates from Mawi. The entire crystal seems to consist of a pastel-pink central zone (one half of the crystal) that quickly grades to essentially colorless achroite. A deep but vibrant blue tinge is seen along fractures in a few random parts of the crystal where some minor silky tubes occur, which concentrate on the complexly surmounted, hopper-growth termination. Blue color fringes occur along fractures and along surface pits of the crystal, creating an astounding, but perfectly natural, appearance.

**#2406 9.0×3.0×3.0cm Plate #4-11**

**Nuristan Valley, Afghanistan**

This specimen is a remarkable cluster of nearly-colorless-to-pastel-pink, nine-sided prism combination elbaite crystals to 2.5cm, with frosted basal terminations, clustered randomly — two vertical, several inclined, and several prone columns. These are on one end of an etched-to-somewhat-fibrous slab of colorless-to-pale-lilac kunzite. This specimen is not only highly aesthetic, but also a rare association of elbaite-kunzite.

***Plate #4-11, Specimen #2406*** **TOURMALINE, var. elbaite – Nuristan Valley, Afghanistan. (MH)**

**#2441 10.0×4.5cm Plate #4-12**

**Kureghal Mine, Nuristan Valley, Afghanistan**

The Kureghal Mine produced this uniformly bright pink, basally terminated, deeply striated elbaite crystal of exceptional color. It has numerous side crystals to several centimeters, some "drusy" similar pink tourmalines, and cleavelandite overgrowths. One exceptional terminated side crystal, 5.0×2.3cm, grew transverse to the main crystal and shows numerous naturally healed dislocations and segmentations producing a nearly 25° curvature!

***Plate #4-12, Specimen #2441***
**TOURMALINE, var. elbaite – Kureghal Mine, Nuristan Valley, Afghanistan. (VP)**

## TOURMALINE, VAR. ELBAITE

**#4732 4.8×2.0cm**

**Nuristan, Afghanistan**

This is a uniquely sharp and striation-free prismatic, transparent, and flawless deep yellow-green elbaite crystal. The prism faces are remarkably smooth and lustrous. They are surmounted by an equally unusual set of three parallel growth, etched, pyramidal terminations with a corrugated six-faced hopper depression between the isolated apices. This is a very attractive specimen.

Specimen #4733, another Nuristan specimen in the collection, is an exceptionally transparent and nearly flawless greenish blue elbaite with frosted and indistinct double terminations.

**#4985 4.7×4.5×3.7cm**

**Paprook, Nuristan, Afghanistan**

This is an outstanding "skeletal" single elbaite crystal with a rose-colored, trigonal pyramid termination grading into a 1.0cm green zone and then into a 7.0cm olive-green zone. A minor unusual milky-lilac zone, to 5mm, ends on a partial basal termination. The central skeletal tube, 3.0+cm deep×2.0×1.5cm, is caused by selective deposition of granular lepidolite on the center of termination. The specimen has been repaired.

***Plate #4-13, Specimen #4986*** **TOURMALINE, var. elbaite – Paprook, Nuristan, Afghanistan. (VP)**

**#4986 6.7×5.5cm Plate #4-13**

**Paprook, Nuristan, Afghanistan**

This is an excellent association of a basally terminated (with minor pyramidal modifications) elbaite crystal, 5.0×2.7cm, on a large open white cleavelandite rosette that has a tan selective overgrowth of drusy carbonatian apatite. The elbaite from the Paprook is particularly lustrous and shows the typical gemmy color-zoning characteristic of the region, beginning with a 1mm clear-to-palest-pink terminal zone sharply separated from a slightly yellow-green zone by a paper thin smoky zone. The base of the crystal is not all smoky, but rather consists of a smoky-green core with a thick transparent green rind.

**#2601 8.3×1.0cm, 20.7g**
**Plate #4-14**

**Kunar Province, Afghanistan**

This single, nearly flawless, greenish-blue elbaite (indicolite) crystal has a partially etched pyramidal termination. It is an exceptional gem crystal from Kunar Province.

***Plate #4-14, Specimen #2601*** **TOURMALINE, var. elbaite – Kunar Province, Afghanistan. (MH)**

**#4913 18.0×8.5×6.5cm**

**Kunar Province, Afghanistan**

This large, slightly divergent cluster of gray-pink elbaite crystals from Kunar Province is capped by light green zones grading into light pink at the simple pyramidal terminations. Although the composite crystal spreads outward from the base towards the termination, the essentially parallel character of the component crystals is revealed by the coalescing and nearly coalescing nature of the successively offset terminations across the crystals.

## BAHIA, BRAZIL

### UVITE (CALCIUM MAGNESIUM ALUMINUM BOROSILICATE HYDROXIDE)

**#5266 9.0×7.5×6.0cm**

**Brumado, Bahia, Brazil**

Brumado produces two dramatically different colors of uvite. This specimen consists of uniformly dark grass-green tabular uvite crystals to 1.0–2.0cm terminated by simple three-faced pyramids sutured by chevronlike six-sided prisms. The specimen consists of more than a hundred similar uvites intergrown with 1.0–2.0cm clear-to-white rhombohedral magnesite crystals and twins with numerous cream-to-light-orange-brown crystals (to 3mm).

***Plate #4-15, Specimen #5777*** **TOURMALINE, var. dravite – Pedra Pit, Brumado, Bahia, Brazil. (MH)**

### DRAVITE (SODIUM MAGNESIUM ALUMINUM SILICATE HYDROXIDE)

**#5777 6.5×4.5×3.5cm Plate #4-15**

**Pedra Pit, Brumado, Bahia, Brazil**

This specimen consists of an aesthetic grouping of about eight grayish-green dravite crystals highlighted by small lustrous white dolomite crystals. The dravites range from less than 0.5cm to 2.0cm in diameter and from 1.0cm to 3.0cm in length. Most crystals have a sharp rhombohedral termination with a prism that grades into a fibrous, parallel growth lower prism. Some dravite terminations have 5mm dolomite crystals embedded in the faces. A superb grouping of dravite from this famous locality, the piece comes from Carlos Barbosa's collection through collector Eric Asselborn.

## MINAS GERAIS, BRAZIL

Brazil is justly famous for being a prodigious supplier of tourmaline specimens as it has a great number of pegmatite mines that produce the gem. While the variety of Brazilian specimens is equally famous, beautiful rubellite and distinctly color-zoned crystals from Brazil are the specialty of the Barlow Collection. Green monochromatic crystals, which are more common than the multicolored specimens, are also represented. The names of individual mines (if known) are included in the specimen descriptions.

***Plate #4-16, Specimen #1539*** **TOURMALINE, var. elbaite – Virgem de Lapa, Minas Gerais, Brazil. (MH)**

### TOURMALINE, VAR. ELBAITE

**#1539 8.5×3.5cm Plate #4-16**

**Virgem de Lapa, Minas Gerais, Brazil**

This specimen is an outstanding cluster of two virtually identical gemmy pale-mauve-to-pink elbaite crystals, to 7.5×1.2cm, with translucent gray tips, nested and accented by zoned (cloudy-clear-cloudy) pink-to-gray-pink "bull's-eye" lepidolite crystals to 5mm. While the two elbaite crystals are not in contact with each other, they are nearly parallel, and in a fortuitous quirk of nature, one crystal is situated slightly above the other, making this one of the most aesthetically composed specimens of its species. The piece was purchased from Jean Pierre Cand in a dark parking lot in 1977.

### The Pegmatites of Brazil

The rich gem pegmatites of Brazil rank as the most prolific producers known. The most important gem crystals found here include various beryls, with aquamarine the most abundant; tourmaline, with elbaite extremely common; topaz, sometimes in huge sizes; spodumene, with lilac-colored kunzite the prize of that litter; unusual or rare gem minerals like brazilianite; and collector minerals like hydroxylherderite, eosphorite, euclase, and the like. The Barlow Collection contains a broad selection of Brazilian pegmatite minerals, some of which are the world's best. *R.W.J.*

***Plate #4-17, Specimen #1670*** **TOURMALINE, var. elbaite – Sao Josê da Safira Mine, Minas Gerais, Brazil. (MH)**

## TOURMALINE, VAR. ELBAITE

**#1670 5.0×2.4cm Plate #4-17**

**São José da Safira Mine, Minas Gerais, Brazil**

This piece is a gemmy and classic watermelon-zoned, basally terminated, subparallel crystal cluster, 4.5×2.0cm. It shows a green-to-bluish-green cap with a pinkish-purple core showing through a transparent green rind.

### Jonas Mine, Itatiaia, south of Conselheiro Peña, Minas Gerais

Everyone familiar with Brazilian pegmatite gems has seen or heard of the great find of raspberry-red tourmalines here in 1978. The richness of their color, the gem clarity of the crystals, and their huge size is unmatched. To top it off, they are aesthetically beautiful specimens. For this reason, many superb pieces survived the gem cutter's wheel.

Discovered on Good Friday, the pocket was so immense that one could walk into it. Loose crystals lay in the pocket clay. Others remained attached to the walls.

The largest crystal is reputed to be "The Rocket" — over a meter long and weighing about 660 kg. Other breathtaking specimens were recovered along with several millions of dollars in gem crystals.

***Plate #4-18, Specimen #1932*** **TOURMALINE, var. elbaite – Jonas Mine, Itatiaia, Minas Gerais, Brazil. (MH)**

## TOURMALINE, VAR. ELBAITE

**#1932 5.0×3.8cm Plate #4-18**

**Jonas Mine, Itatiaia, Minas Gerais, Brazil**

A uniform burgundy-red gemmy pair of parallel, basally terminated elbaite crystals, this specimen has gray-purple lepidolite crystals to 7mm overgrowing the prism.

**#2218 5.0×2.0cm**

**Santa Rosa Mine, Itambacuri, Minas Gerais, Brazil**

A single color-zoned elbaite crystal with etched and furrowed termination, this piece grades from a dense smoky-green to transparent dark-yellow-green then into a medium deep blue central zone. The deep blue color continues as a thin rind overgrowing a dark but translucent indigo core. Fine striations on the prism produce a surface diffraction effect.

## TOURMALINE, VAR. ELBAITE

**#1903 6.0×1.8cm**

**Jonas Mine, Itatiaia, Minas Gerais, Brazil**

This is a doubly terminated, strikingly dark, rose elbaite crystal, 6.0×1.8cm, with light-to-bright-rose highlights on the terminations. There are sutures where the crystal was fractured during the formation process and then healed by regrowth. The segmented crystal shows an amazing 25° curvature. A doubly terminated elbaite crystal, 2.7×1.0cm, grew transversely across one end of the major crystal and shows partial skeletal development where drusy pink elbaite later partly covered the prisms of both the main and secondary crystals. The rich color is indicative of the locality.

**#2530 3.4×4.5cm**

**Lavra do Verdinha, Marilac, Minas Gerais, Brazil**

This specimen is a flattened greenish-blue-to-black elbaite with hexagonal prism and simple pyramid, overgrown completely on one side by a coating of uniform rose quartz crystals, 3–4mm each. This is an exceptional association because rose quartz crystals ordinarily overgrow only albite or quartz.

**#2392 4.0×1.2cm**

**Minas Gerais, Brazil**

This is an outstanding, lustrous, bicolored elbaite crystal terminated by lightly etched complex pyramidal faces. The gemmy crystal is a deep, nearly emerald green on its upper third and is separated from a rose-red zone on its lower two thirds by a thin pastel-green to murky-pink transition zone.

Golconda Mine, west of Governador Valadares, near Coroaci, is world-renowned for its huge deep green tourmalines. Heavy equipment is employed there, as opposed to the primitive pick and shovel methods of the *garimpero* (individual or small group miners).

There are three Golcondas, the earliest mining starting about 1908. Mica was the first product, but gems soon held sway. Choice indicolites (blue tourmaline) came from Golconda II, while Golconda III produced a single pocket of gem-quality green tourmaline, the best ever found in Brazil. It was a huge pocket, yielding nearly a ton of crystals.

Perhaps the most recognizable tourmalines from this locality have a pale pink prism tipped with a rich grass-green zone. They have figured frequently in gem dealer ads and are classic to the deposit.

***Plate #4-19, Specimen #2604*** **TOURMALINE, var. elbaite – Golconda Mine, Governador Valadares, Brazil. (MH)**

**#2604 14.0×7.5cm Plate #4-19**

**Golconda Mine, Governador Valadares, Brazil**

This coarse jackstraw cluster of dark olive-green elbaite crystals shows an internal silky "flash" of lighter olive as light reflects from extensively developed, minutely etched, tubes. The major crystal is 12.5cm long. Crystal cross-sections reveal large hollow tubular cores, to 1.5cm, infilled by tightly grown gray-to-lilac lepidolite flakes. The surface of the jackstraw clusters is abundantly sprinkled with small feathery-to-serrated silvery-gray stacks of muscovite crystals to 4mm. The piece also has minor associated cleavelandite.

**#2310 3.7×0.7cm**

**Lavra do Verdinha, Marilac, Minas Gerais, Brazil**

This elbaite tourmaline is a scarce color combination of a green rind on a milky-white core sometimes called a "cucumber crystal." The crystal's lower quarter is cucumber-zoned and slowly grades toward the basally terminated end with increasing freedom from flaws and cloudiness. The termination maintains a few unusual green highlights. The elbaite is in almost parallel contact with a transparent healed quartz shard that has regained its crystalline shape after having broken during a violent gem pocket event. Despite the quartz crystal having regained its faces due to regrowth, the faces remain quite asymmetrical.

## TOURMALINE, VAR. ELBAITE

**#5432 11.0×7.0×13.0cm**

**Refer to Chapter 3, Plate #3-92**

**Chia Mine, near Governador Valadares, Minas Gerais, Brazil**

Brazil has a justified reputation as the world's premier source of elbaite crystals. Among the population of great specimens, this recently discovered rarity still commands respect. The elbaite resides on matrix, having survived the throes of turbulence frequently associated with gem pocket formation. The tourmaline portion consists of a core of black schorl that is a barely translucent smoky greenish gray on its exterior. This core is covered with a cloudy translucent quartz crystal that was subsequently much abraded, nicked, and etched by corrosive hot water in the gem pocket. After these events occurred, the beauty of nature was expressed through on-going crystallization.

The ordinary smoky tourmaline began to grow as a nearly flawless cherry-to-rose-red rubellite for a distance of more than 6.0cm with a thickness identical to the schorl nucleus (4.0×3.5cm). A sudden shift in the nutrients available for crystallization followed. Smooth faced, unusually bright and transparent individual quartz crystals overgrew the matrix quartz crystal, sprinkling the elbaite's prism. One large transparent quartz crystal (4.5×2.4cm) grew on the termination of the rubellite, effectively preventing most of the elbaite termination from further growth.

Coincident with the quartz growth, the intense cherry-to-rose-red rubellite gave way to gemmy medium-green that graded into a gray-olive-green termination. The final termination is a three-faced crystal spike (3.6×2.6×1.8cm) extending from one edge of the rubellite zone. This large spike is convoluted in cross-section and is nearly a hollow tube. A look down the "tube" shows a base of quartz crystals that prevented the elbaite from growing.

On another part of the elbaite termination, numerous parallel, small (to 1.0×0.2×0.2cm), terminated spikes of elbaite successfully out-competed the large quartz crystal. Thin discontinuous white-to-tan patches of clay mask the interface of the earlier formed elbaite and quartz with the later bright quartz. Additionally, a thin, discontinuous, skin of green elbaite overgrown on the rubellite is punctuated by the clear quartz crystals showing only rubellite contacts at the interface. Some of the quartz crystals on the rubellite prism are stubby and in parallel growth, and many of the quartz crystals have small, rarely seen faces — the rare faces on the major crystal indicate that it has "left-handed" symmetry.

Such a specimen as this requires a summary: the gemmy red rubellite crystal with a green spiked termination juts out of a quartz crystal and is in turn coated by quartz crystals.

**#2413 4.7×1.5cm**

**Minas Gerais, Brazil**

This extraordinary prismatic elbaite crystal is terminated by a steep, complex, and modified termination. The upper half of the crystal is a gray-pink that carefully grades through a thin light blue and medium green zone to a dark smoky-green in the lower half.

**#2637 3.7×0.7cm**

**Minas Gerais, Brazil**

This outstanding small tricolored elbaite crystal has a deep raspberry-red zone beneath the pyramid termination, followed by a nearly flawless pale yellow central zone with a smoky-pink zone on the remainder of the crystal. Faint phantom striping in the smoky-pink zone runs parallel to the termination.

**#4686 15.5×15.0cm**

**Barro do Salinas, Minas Gerais, Brazil**

This is a large, very aesthetic, dark green hexagonal elbaite crystal. The complex pyramidal termination is capped by pink highlights. The base of the crystal is profusely fringed by light purple pseudohexagonal lepidolite crystals to 2.7cm. The upper two thirds of the elbaite is dark gemmy green, while the lower third has a dark red core. An external zone, just above the attached lepidolite, consists of four transverse, closely spaced, curved incipient fractures healed by white fibrous elbaite. The elbaite interpenetrates a coarse layered cluster of light smoky quartz crystals.

**#5096 17.0×4.0cm**

**Coronel Murta Mine, Araçuai, Minas Gerais, Brazil**

This specimen is a very dark red, single elbaite crystal with numerous small patches of etched tubular inclusions showing rose-red highlights. The entire specimen is bordered by dense cauliflower-like aggregates of muscovite crystals with several large gray-purple pseudohexagonal lepidolite crystals to 4.0×3.0×1.5cm.

**TOURMALINE,** VAR. **ELBAITE**

**#4693 16.0×18.0×14.0cm**

**Barro do Salinas, Brazil**

These large and magnificent hexagonal dark green elbaite crystals, 13.5×3.3cm and 4.6×1.6cm, have a thin raspberry-red overgrowth of color on their terminations. These watermelon-zoned green elbaites have deep burgundy-red cores with only a thin green veneer. The matrix consists of a medium (18.0×14.0cm) smoky quartz crystal with a terminal bifurcated summit showing parallel overgrowth of secondary quartz crystals. Numerous watermelon-zoned elbaite crystal sections shoot through the base and side of the quartz.

## CHINA, People's Republic of

**UVITE**

**#4811 8.2×7.3×6.2cm**

**Altai Mountains, China**

This piece is a large dark smoky-brown crystal with a thin "gemmy" exterior showing nearly equal development of the pyramid faces. It has a small basal face and deeply striated prism zones.

*Plate #4-20, Specimen #1651*
**TOURMALINE, var. elbaite – San Piero di Campo, Elba, Italy. (MH)**

## ITALY

**TOURMALINE,** VAR. **ELBAITE**

**#1651 5.0×3.5cm Plate #4-20**

**San Piero di Campo, Elba, Italy**

This is a single, pyramidally terminated, elbaite crystal, 1.5×0.8cm, on small quartz crystals and albite matrix. Coloration is typical of this site, the original elbaite locality. The slightly etched terminal zone is nearly colorless achroite and grades into to greenish-yellow with the lower portion a dark smoky-green.

### The Island of Elba — A Famous Tourmaline Locality

The Italian island of Elba is known not only as the place of Napoleon's first exile, but also as the place where the best gem tourmalines of central Europe were discovered. Throughout the 19th century, collectors and miners uncovered outstanding multicolored specimens in the district centered near the village of San Piero di Campo. Eventually, four important locations were uncovered in pegmatitic segregations in granite, producing a variety of pegmatite minerals including pale pink beryls. Two interesting minerals were discovered together with the tourmaline and were named castorite and pollucite in allusion to the mythological twins Castor and Pollux.

Although colored tourmalines had been well known previously from Sri Lanka, Russia, and even Brazil, Vladimir Vernadsky in 1913 followed the lead of American mineralogists Riggs (1888) and Clarke (1899) and others and chose the name elbaite, after the famous locality, to designate the lithium-bearing tourmalines that were so colorful and so eagerly sought. Elbaite was already well represented in the world's museums, but Vernadsky's name was slow in being accepted.

Beautiful crystals to more than 10cm long with bright pink, green, yellow, or colorless zones are much sought after. The black color zones of many specimens, including their terminations, are probably schorl. The Elba deposits produce virtually no specimens now; the examples seen today are recycled antiques, reminding us of bygone days.

## TOURMALINE, VAR. ELBAITE

### MOZAMBIQUE

**#2315 8.5×5.0cm**

**Alto Ligonha, Mozambique**

This is an excellent single, red-gray-red zoned, singly terminated, elbaite crystal. The bright red termination is etched and shows a few indentations where small tubular inclusions intersect the surface, and there are a few arcuate, naturally healed fractures cutting across the crystal apex. The termination grades irregularly into a central gemmy light ash-gray zone, which abruptly changes to a bright purplish red to the end of the crystal.

*Plate #4-21, Specimen #4844* **TOURMALINE, var. elbaite – Alto Ligonha, Mozambique. (MH)**

**#4844 6.0×2.8×2.7cm Plate #4-21**

**Alto Ligonha, Mozambique**

This single multicolored hexagonal elbaite crystal is terminated by complex pyramids. The termination grades from a pink mottled and crackled zone several millimeters thick to a gemmier irregular blue-green zone about 1.0–1.5cm thick. The remainder of the prism contains facetable areas grading from pastel sherry to yellow and yellow-green.

### NAMIBIA

**#4908 8.5×2.5×2.0cm**

**Otjua Mine, Karibib, Namibia**

This single, color-zoned elbaite crystal has a severely etched partial termination grading from a 1.0cm mauve zone to light gray to smoky-gray zones. These are followed by a cinnamon-red zone continuing to the lower basal termination. One side of the crystal is coated by a flat layer of pale pink, finely crystallized lepidolite that at some point forms hollow pseudomorphs. This specimen is one of the best from this deposit.

*Plate #4-22, Specimen #4982* **TOURMALINE, var. elbaite – Otjua Mine, Karibib, Namibia. (MH)**

**#4982 9.8×5.0×3.0cm Plate #4-22**

**Otjua Mine, Karibib, Namibia**

This piece is a highly aesthetic cluster of three dark red elbaite crystals with patchy bright carmine-red highlights. It is reputed to be the second-best specimen from this mine. The two main crystals are in parallel growth, with basal terminations. A third complexly pyramidal crystal is positioned on the side of the cluster, with its termination in full view. The cluster is overgrown on the lower prism by thin pinkish lilac lepidolite crystals to 6mm, with some lepidolite accenting the intersections of the main crystals.

## TOURMALINE, VAR. ELBAITE

**#5232 10.0×1.7cm**
**Refer to Chapter 3, Plate #3-95**

**Otjua Mine, Karibib**

Gem elbaite crystals from Karibib are scarce, and most of the large rubellite crystals known from there are very dark to cloudy. This crystal is more transparent than usual and is remarkable in many ways. Its most striking feature is the development of at least five major color zones along its length. The base of the crystal is a cloudy but fine, slightly gemmy purplish red with a few minor rose-purple "threads" indicating incipient fracturing and subsequent healing. The remaining 60% of the crystal is nearly equally apportioned with distinctive additional color zones. The rubellite base quickly becomes gemmier and less flawed as it grades into a pleasing and ever-more-transparent sherry-to-cinnamon-pinkish orange. A portion of this zone could be cut into a multicarat flawless gem. Just above this zone, the elbaite has a beautiful steely-blue zone whose base against the gemmy zone outlines a phantom crystal termination. The blue grades through a barely perceptible transition to a gemmier crimson zone.

The termination consists of an abrupt transition to gemmy grayish to slightly yellowish green with a slightly pinkish-to-olive-green cap. The terminal green zone is well bounded by smooth crystal faces.

One of the most unusual features of the crystal is the transition from crimson to green. The crystal seems to have been broken by nature at the upper end of the red zone; scalloped to flat indentations there indicate regrowth with the green "well-formed" zone, which forms a tall (1.0cm) ridge across the width of the crystal as a kind of cupola giving the aspect of a reverse scepter crystal. The rubellite base of the crystal is attached to an open cluster of uniform dark lilac pseudohexagonal lepidolite crystals sprinkled with occasional pink elbaite shards and crystals (to several millimeters each). This is a remarkable crystal from any location, and exceptional from this site.

***Plate #4-23, Specimen #2313*** **TOURMALINE, var. elbaite – Hyakule Mine, Chainpur, Nepal. (MH)**

## NEPAL

**#2313 9.0×3.2cm Plate #4-23**

**Hyakule Mine, Chainpur, Nepal**

This is an exceptional, gemmy, color-zoned elbaite crystal terminated by a simple pyramid, with minor tiny etched markings on the lower prism. The color zonation in the termination is mottled with a pale green color, surrounded by irregular pale sherry zones that grade into the main shaft of the crystal, which is a 4.0cm zone, pale sherry in color. This zone grades into a pinkish-lilac base. This specimen is an outstanding crystal from a rare locality.

**#3637 5.0×5.0×4.5cm**

**Kathmandu District, Nepal**

This etched elbaite crystal section shows a rare yellow-to-greenish yellow-translucent color with minor irregular zones of pink. The specimen was collected high in the Himalayas.

Plate #4-24, Specimen #3610 TOURMALINE, var. elbaite – Hyakule Mine, Chainpur, Nepal. (MH)

**TOURMALINE,** VAR. ELBAITE
**#3610 7.0×1.0cm Plate #4-24**
**Hyakule Mine, Chainpur, Nepal**

This feathery multicolored frosted elbaite shows a unique "Star-of-David" cross-section consisting of two superimposed triangular forms, the prism and the basal termination. The association is completed by an etched, nearly 2.0cm danburite crystal on the termination and a matrix for the tourmaline of a partial 4.0×3.0cm quartz crystal.

## NIGERIA

**#4918 9.0×3.0×3.0cm Plate #4-25**
**Jos, Plateau Province, Nigeria**

This parallel to subparallel, multiple growth, gemmy elbaite crystal group has etched multiple pyramid terminations. It has a deep-green-to-olive-green upper color zone, grading into a light green, with an abrupt transition to tannish pink. Tiny etch tubes make it appear cloudy.

Plate #4-25, Specimen #4918 TOURMALINE, var. elbaite – Jos, Plateau Province, Nigeria. (JS)

Plate #4-26, Specimen #2862 TOURMALINE, var. elbaite – Stak Nala District, Haramosh Range, Pakistan. (MH)

## PAKISTAN

**#2862 13.0×9.0cm Plate #4-26**
**Stak Nala District, Haramosh Range, Pakistan**

This specimen is a matrix assemblage of basally terminated green-to-olive elbaite crystals to 4.0cm, on fans and rosettes of white cleavelandite, associated with clear quartz crystals to 6.5cm. As is typical of the locality, the gemmy light green portions of the crystals generally occupy the upper one third of the crystal and the remaining portions are mostly dark smoky-green. The terminations of the tourmalines sometimes show clear and pink zonation. Associated 1mm clear blades appear to be bertrandite.

## TOURMALINE, VAR. ELBAITE

**#2863 7.5×7.5×7.3cm**

**Stak Nala District, Haramosh Range, Pakistan**

Three dark olive-green elbaite crystals to 3.0×2.0cm are placed among a completely crystallized assortment of gem pocket crystal species. A prominent pink milky-to-translucent tabular hexagonal fluorapatite crystal, 2.2×1.8cm, abuts the dominant elbaite crystal, which is 3.0×2.0cm. In turn, both species grow on, in, or among colorless stout quartz crystals to 5.0cm. The base of the specimen consists of two varieties of albite. The majority of the matrix is composed of open and scalloped clusters of white cleavelandite with two white, wedge-shaped-to-blocky valencianite crystals to 2.5cm. Tiny muscovite crystals to several millimeters and tan drusy carbonate apatite selectively coat the cleavelandite.

***Plate #4-27, Specimen #4843*** **TOURMALINE, var. elbaite – Stak Nala District, Haramosh Range, Pakistan. (MH)**

**#4843 7.0×0.8cm Plate #4-27**

**Stak Nala District, Haramosh Range, Pakistan**

This piece is a gemmy, pyramidally terminated, multicolored elbaite crystal with a doubly terminated quartz crystal. The unusual color zonation of the tourmaline grades from pale pink to colorless to light olive, to pale green, gradually changing into varying intensities of yellow-green. This superb crystal is an outstanding specimen.

## TOURMALINE, VAR. SCHORL (SODIUM IRON ALUMINUM BOROSILICATE HYDROXIDE)

**#2550 4.0×3.5cm**

**Gilgit District, Haramosh Valley, Pakistan**

Pocket schorl crystals are unusual in pegmatites as they generally are encased in a late growth of quartz or feldspar rather than remaining in open spaces. The Gilgit pegmatites have produced fine pocket-formed schorl in unusual abundance. This specimen shows a parallel growth of one large 2.7×1.0cm crystal with a smaller 2.0×1.0cm crystal on cleavelandite and muscovite crystals, 1.7×1.7cm. Barlow bought it at the 1984 *Mineralogical Record* auction in Tucson.

**#4782 11.5×10.7×6.0cm**

**Stak Nala District, Haramosh Range, Pakistan**

An enormous, simple pyramidal, rose-lilac elbaite termination with a small basal face rests on a short corrugated prism. The crystal is concentrically color-zoned on the base, has a dark smoky-green core rhythmically banded amber-to-olive, and a rose-to-lilac rind.

## TOURMALINE, VAR. ELBAITE
## RUSSIA

**#4781 7.0×3.0cm Plate #4-28**

**Lake Baikal District, Siberia, Russia**

This is a calcium-bearing tourmaline intermediate between elbaite and its calcium-dominant relative liddicoatite. It is a single, remarkably color-zoned, gemmy crystal, terminated by dominant basal termination and minor trigonal pyramids. The terminal end and lower prism show gradational color zoning from dark and light pinkish rose. The central and lowermost prism zones show a pinkish sherry to almost amber-yellow. Some surface indentations appear on the crystal where former cleavelandite, etc. interrupted crystal growth.

***Plate #4-28, Specimen #4781*** **TOURMALINE, var. elbaite – Lake Baikal District, Siberia, Russia. (JS)**

**#5388 3.8×1.5×1.0cm Plate #4-29**

**Srapulka, Central Ural Mountains, Russia**

This fine, gemmy, and bright faced rubellite crystal has a mauve-colored apex gently grading to cinnamon-pink through the main shaft of the crystal, with the lower third of the crystal resuming the mauve tones. A rounded rosette of muscovite (2.2×2.2cm) is attached to the lower half of the crystal. The termination consists of a lightly etched triangular basal face and a major and minor set of bright pyramidal faces that slope toward the prism. The specimen was self-collected by noted seismologist Terry Wallace 3km west of a seismic station called "Iris NVK," which he set up. Wallace traded the specimen to Barlow for another specimen.

***Plate #4-29, Specimen #5388*** **TOURMALINE, var. elbaite – Srapulka, Central Ural Mountains, Russia. (MH)**

## TOURMALINE, VAR. ELBAITE
## UNITED STATES

***Plate #4-30, Specimen #1921*** **TOURMALINE, var. elbaite – Gillette Quarry, Haddam Neck, Middesex County, Connecticut. (MH)**

**#1921 7.0×5.0cm Plate #4-30**

**Gillette Quarry, Haddam Neck, Middesex County, Connecticut**

The Gillette Quarry yielded fine elbaites decades ago. This specimen is dominated by a gemmy, dark grass-to-olive-green-to-greenish black, trigonal elbaite crystal with moderately steep trigonal pyramid. This is an unusual, highly curved crystal showing one major naturally healed dislocation and several small closely spaced dislocations healed with an opposite curvature, producing a slightly sigmoid shape. The main elbaite traverses a well-terminated, smoky-to-gray, gemmy quartz crystal, 5.2×2.4cm. There are also three minor dark green elbaite crystals, terminated, etched, or incomplete, investing quartz crystals and projecting from them. This is a small but important specimen from a classic locality now completely beyond the reach of the collector.

**#1849 5.5×7.0cm**

**Dunton Quarry (Plumbago Mine), Newry, Oxford County, Maine**

This unusual, multiple growth elbaite crystal shows steep pyramid terminations with clear lilac zones gradually blending lilac to blue-green color zones with rose-to-grayish rose deeply striated prisms. Healed fractures are present, some with silky tubular inclusions typical of the locality.

***Plate #4-31, Specimen #2969*** **TOURMALINE, var. elbaite – Dunton Quarry (Plumbago Mine), Newry, Oxford County, Maine. (MH)**

**#2969 7.8×4.5cm Plate #4-31**

**Dunton Quarry (Plumbago Mine), Newry, Oxford County, Maine**

This gemmy, steeply terminated, hexagonal, dark apple-green elbaite with pale green apex shows shallow-to-deep random lineations and striations, typical of specimens from Dunton Quarry.

## TOURMALINE, VAR. ELBAITE

**#3163 2.7×1.5cm Plate #4-32**

**Helen Beryl Mine, Custer, Pennington County, South Dakota**

This piece is a trigonally terminated, prismatic, bright black schorl crystal formed at the junction of light gray quartz and salmon-pink microcline. This Barlow favorite was purchased from Vern Stratton's father in 1983.

*Plate #4-32, Specimen #3163* **TOURMALINE, var. elbaite – Helen Beryl Mine, Custer, Pennington County, South Dakota. (MH)**

## MADAGASCAR

This huge island off the east coast of Africa is rich in pegmatites. The suite of minerals found here is, to some extent, typical of worldwide pegmatites. It includes gem beryl, vars. aquamarine and morganite, spodumene, var. kunzite, watermelon tourmalines; and a host of uncommon minerals including danburite, chrysoberyl, some garnet species, allanite, columbite, and more.

Liddicoatite is a calcium-lithium tourmaline of restricted occurrence first identified from Madagascar. In the Antsirabé district, it forms as sometimes murky-to-smoky-brown, smoky-green, or smoky-gray zones within more brightly colored elbaite, a sodium-lithium tourmaline.

Calcium is not a coloring agent; it merely coincides in abundance with the growth periods of the brown, gray, or, more commonly, pink or green liddicoatite. The zoned liddicoatite tourmalines are most intriguing. They look like elbaites and were long considered as such. However, tests showed them to contain zones of material richer in calcium than sodium. The calcium lithium tourmaline member is now called liddicoatite, as opposed to the sodium lithium member elbaite. Interestingly, a given crystal may have developed rhythmically with alternating zones of elbaite and liddicoatite, resulting in a crystal composed of two distinctly different tourmaline varieties.

## LIDDICOATITE (CALCIUM LITHIUM ALUMINUM BOROSILICATE FLOURIDE HYDROXIDE)

**#2318 7.2×2.0cm**

**Mount Bity Pegmatite District, Madagascar**

This single, multicolored, basally terminated, elbaite crystal is modified by a low-angle pyramid face. The gemmy variegated red-to-sherry-pink upper and lower zones are separated by a gemmy light steel-gray central color zone about 1.0–1.5cm thick. The lower end has concentric color zoning with a red core sharply separated from a lighter gray-pink rind.

**#2202 3.5×5.0cm**

**Antsirabé District, Madagascar**

This is a rare, mirror-faced, well-terminated, dark but gemmy liddicoatite with elbaite zones. The outer edges of this elbaite are green, olive, brown, pinkish brown, and pink. The prism faces have prominent striations, while the prominent basal termination is modified by mirror-faced pyramid faces. The lower portion of the crystal was sliced off into six polished wafers that show the progressive development of color phantoms unobservable by any other means. (See page 126.)

## LIDDICOATITE

### Antsirabé District, Madagascar

Six liddicoatite/elbaite tourmaline crystal slices from the termination of #2202 all have the typical bulging triangular shape of that piece. All show an extremely complex, finely zoned rind that consists of an almost paper-thin colorless skin with microscopic pink streaks, followed by a 3-5mm murky green but translucent zone with rhythmic lighter and darker fine lines exactly paralleling the external distorted triangular shape. Inside the murky green zone, the color changes to yellow-green and abruptly changes on a microscopic scale to pink and back to green two or three times ending with pink. All of these color changes occur through a thickness of only 5mm. The slices are dominated by a large internal gemmy red core. The slice from just below the terminal end has a relatively featureless core of dark pink grading to light port-wine-red, while the next slice shows a gemmy dark burgundy-red truncated triangular zone with three carmine-to-pink zones extending toward the prism edges. The next four slices reveal ever shrinking Burgundy-red triskelions on a magenta field, each with its own variation.

## REFERENCES


Bancroft, P. (1984) *Gem and Crystal Treasures*, Western Enterprises, Fallbrook, California, and Mineralogical Record, Tucson, Arizona, 488 pp.

Clarke, F.W. (1899) The constitution of tourmaline, *American Journal of Science*, 4th Ser., 8:111-121.

Dietrich, R.V. (1985) *The Tourmaline Group*, Van Nostrand Reinhold, New York, 300 pp.

Foord, E.E. (1977) Famous mineral localities: the Himalaya dike system, Mesa Grande, San Diego, California, *Mineralogical Record*, 8:461-474.

Foord, E.E., and L.A. Wright (1951) Gem and lithium-bearing pegmatites of the Pala district, San Diego County, California, *California Division of Mines Special Report 7-A*, 72 pp.

London, D. (1986) Formation of tourmaline-rich gem pockets in miarolytic pegmatites, *American Mineralogist*, 71:396-405.

Sinkankas, J. (1981) *Emerald and Other Beryls*, Chilton Books Co., Radnor, Pennsylvania, 664 pp.

Sinkankas, J. (1959) *Gemstones of North America*, Van Nostrand Reinhold, New York, 675 pp.

Stern, L.A., and G.E. Brown, Jr. (1986) Mineralogy and geochemical Evolution of the Little Three pegmatite-aplite layered intrusive, Ramona, California, *American Mineralogist*, 71:406-427.

Vernadsky, V. (1913) Über de Chemische Formel der Turmaline, *Zeitschrift für Krystallographie*, 53:273-288.

CHAPTER 5

# Beryl and Pegmatite Minerals

*by Vandall T. King*

The Barlow Collection encompasses a number of strong subcollections, each relating in some way to another. Tourmalines represent a significant subcollection within the main collection. Beryls are also well represented in the collection, with 180 cataloged specimens. Though most beryls occur in pegmatites, two varieties, emerald and red beryl, do not. The important suites of emerald and red beryl complement the beryls that come from pegmatites.

The crystal form of beryls is less complex than that of many other minerals. The six-sided cross-section is a dominant feature of beryls; they are frequently elongated and usually have flat terminations. Nonetheless, their color, transparency, common association with other minerals, and surface features enthrall even the novice.

The color varieties of beryl are so traditional as gems that the nomenclature is almost automatically understood — emerald (green), red beryl (red) and aquamarine (blue). Some varietal names are more recent, such as morganite (pink), heliodor (yellow), and goshenite (colorless).

## BERYL, VAR. AQUAMARINE (BERYLLIUM ALUMINUM SILICATE)

The color of aquamarine is literally expressed in the translation of the name, "water of the sea." The color of the sea is highly variable, ranging from the traditional sea-green to the blue hue of the reflected sky on the water's surface. In recent times, the term aquamarine has applied only to blue stones because of their abundance. Yet purely green crystals are also aquamarines.

### AFGHANISTAN

**#2184 5.0×2.2cm**

**Nuristan, Afghanistan**

This is a complex, gemmy, light aquamarine-blue crystal. The transparent portion of the crystal has an interesting straight milky "thread" composed of minute tubes extending from the basal termination to the opposite end of the crystal. Numerous 1×2mm chambers entrap immovable bubbles. The prism is somewhat "dished" and streaked due to etching.

### BRAZIL

**#843 8.4×6.2cm**

**Teofilo Otoni, Minas Gerais, Brazil**

This specimen is a large, gemmy, heavily frosted, nearly uniformly blue crystal. The prism faces show elliptical-to-rounded rectangular etch pits, while the basal termination has a straight, curved, and hexagonal etch pattern. An interesting movable bubble with a 7mm travel path can be seen through one prism face.

**BERYL, VAR. AQUAMARINE**
**#1183 13.2×3.8×3.2cm**
**Refer to Chapter 3, Plate #3-5**

**Minas Gerais, Brazil**

This essentially flawless 242g sea-green beryl crystal has a complex etched termination. The surfaces of the prism have delicate, patchy, shallow, pointed-to-elliptical etch pits. The termination consists of several bipyramid faces, some with chevronlike dihexagonal bipyramids and basal termination. Severe etching has caused an "elephant-hide" furrowing as a gross feature superimposed on the symmetrical etch patterns on the basal faces.

**#1542 4.5×4.5×3.9cm Plate #5-1**

**Minas Gerais, Brazil**

This light green, flawless crystal cluster is in parallel growth with lighter colored veils. Each crystal has mirror-faced prisms and is 18-sided due to a combination of prism forms. The basal terminations are glassy but flecked by minute etch pits, while the terminal edges are partly scalloped and rounded due to corrosion.

***Plate #5-1, Specimen #1542*** **BERYL, var. aquamarine – Minas Gerais, Brazil. (MH)**

**#2442 31.2×6.0×5.0cm, 2,100g**
**Refer to Chapter 3, Plate #3-4**

**Minas Gerais, Brazil**

This is a giant, essentially flawless greenish blue crystal. The termination is complex and extensively etched by elongated, irregular indentations. Just below the center of the crystal are brilliant-to-dull-black inclusions. Some are three-dimensional dendrites, while others are thick pseudohexagonal plates, suggestive of ilmenite. Yet another inclusion appears to be a rutile twin. Several orange-red irregular flakes of color also spot the prism. Despite its inclusions, this giant aquamarine is a phenomenal gem crystal.

**#2092 8.2×2.2×1.9cm**

**#2093 7.0×2.5×2.0cm**

**Refer to Chapter 6, Plate #6-2**

**Jaqueto Mine, Minas Gerais, Brazil**

This is a nearly matched pair of highly etched nearly flawless blue crystals. The etching is so severe that it obscures the natural hexagonal shape of the beryl and forms lightly frosted to transparent terraces, producing a doubly tapered rodlike shape.

**BERYL, VAR. AQUAMARINE**

**#5272 11.5×4.4×2.9cm Plate #5-2**

**Boca Rica, Minas Gerais, Brazil**

The specimen is a gemmy, blue-green, smoothly etched, and much-furrowed hexagonal prism steeply "terminated" by two conical summits. The off-center edge summit is overgrown by a rectangular corrugated column as well as a selvage of white beryl. The whiteness of the column is actually due to microscopically thin hollow tubes in the overgrowth. The beautiful blue-green aquamarine crystal is transparent and nearly flawless.

*Plate #5-2, Specimen #5272* **BERYL, var. aquamarine – Boca Rica, Minas Gerals, Brazil. (MH)**

## NIGERIA

**#5131 25.0×3.7×3.5cm** **Refer to Chapter 3, Plate #3-6**

**Near Jos Plateau, Nigeria**

This phenomenal, gemmy, doubly terminated, deep aquamarine-blue single crystal has complex pyramidal terminations and is artistically etched. The simple hexagonal prism is capped by first-order pyramid faces that are modified by a chevronlike bipyramid and a small, flat basal face. The prism faces show growth patterns indenting the resinous-to-oily-lustered faces. A smaller crystal, 3.7×1.4cm, projects from one end of the main crystal. This is one of the finest aquamarine crystals recovered from the locality to date. It was acquired from dealers Jergen Henn and Mike Ridding in 1992.

## NORTHERN IRELAND

**#4852 3.4×1.1×1.0cm**

**Mourne Mountains, County Down, Northern Ireland**

This specimen is a transparent and nearly flawless, etched, intense blue crystal. Extensive rounding from etching makes the faces indistinct. The apex of the crystal is jagged due to deep furrowing. The etch pits on the prism are elongated rectangles, while the etch marks on the pyramids are overlapping scalloped lobes. This classic specimen was found about 1790.

## PAKISTAN

**#5162 13.0×10.0cm Plate #5-3**

**Inna Nala, Fighar Village, Lower Nagar, Pakistan**

A partly doubly terminated, pale blue, 4.6×2.7cm aquamarine crystal that is nearly flawless lies in a bed of tabular silvery muscovite crystals to 2.8×2.6cm. The aquamarine is flanked on one side by a rose-pink 2.7×2.6cm fluorapatite crystal showing a hexagonal cross-section. Several irregular pink fluorapatites are embedded in muscovite on the other side of the beryl.

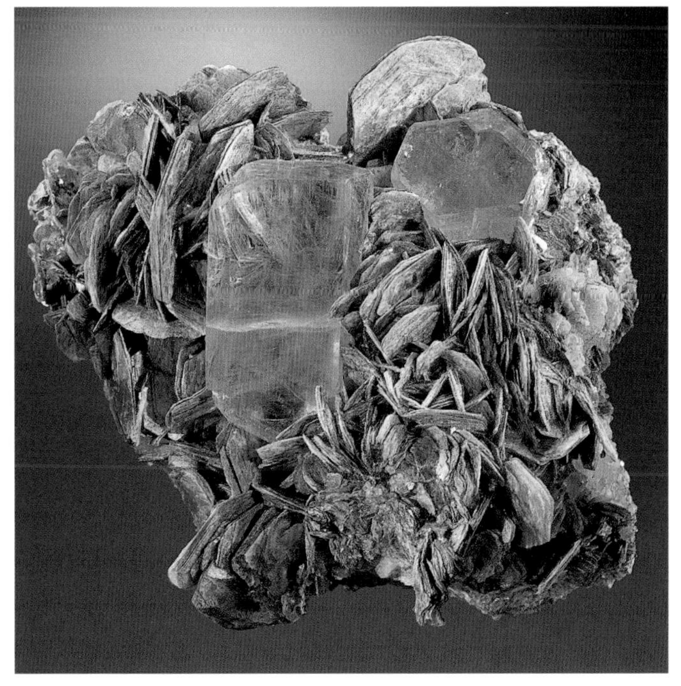

*Plate #5-3, Specimen #5162* **BERYL, var. aquamarine – Inna Nala, Fighar Village, Lower Nagar, Pakistan. (MH)**

## BERYL, VAR. AQUAMARINE

### RUSSIA

**#1545 7.9×3.1cm Plate #5-4**

**Mursinka, Ekaterinburg (Sverdlovsk), Russia**

This light blue, simple prismatic, basally terminated, vertically striated beryl crystal shows the classic thin color zones. Closely spaced blue zones alternate with nearly colorless zones parallel to the upper terminated end. The center of the crystal is divided by a central blue layer several millimeters thick, with 1.0+cm colorless zones above and below the midsection of the crystal. This is a companion specimen to topaz #1546 from the same location.

***Plate #5-4, Specimen #1545*** **BERYL, var. aquamarine – Mursinka, Ekaterinburg, (Sverdlovsk), Russia. (MH)**

### UNITED STATES

**#5814 12.5×3.5×3.7cm**
**Plate #5-5**

**Sheeprock Mountains, Tooele, Tooele County, Utah**

Rare localities exist for outstanding crystals of all kinds. These localities are "rare" as the production of specimens is generally small and usually generates no publicity. Frequently, the mining activity produces only one major specimen among a lot of "also rans." The Sheeprock Mountains are such a locality. They produced this outstanding aquamarine, which by any standard is significant. The intense color is more than striking. The crystal is color-zoned deep ice-green on the prism faces to deep steel-blue in the "core." The distribution of color in this aquamarine is disguised by the crystal's spectacular etching. The birth of a crystal does not ensure its survival. The fluids that generate a crystal can also be its demise as changing chemical composition, temperature, and pressure can result in an unfriendly corrosive medium "intent" on recapturing its former possessions. The original size of the beryl can only be imagined. The presence of two remnant prism faces with gemmy deep ice-green zones gives a frame of reference as to the crystal's original size. If the pattern of color zoning were symmetrical, the steel-blue "core" would suggest that over half of the crystal was etched away, leaving a lustrous, overlapping parquet texture of multilevel, varied rectangles, some of which have chamfered corners. Deep clefts in this blue-green gem crystal further accentuate the parallel with ice, especially where a skim of aquamarine obscures the interior from view. The color along the length of the crystal is a deep blue-green, more reminiscent of gem tourmaline than of beryl. This is one of North America's important aquamarines. Acquired from the Al Buranek Collection through Cal Graeber.

***Plate #5-5, Specimen #5814*** **BERYL, var. aquamarine – Sheeprock Mountains, Tooele, Tooele County, Utah. (MH)**

## Emeralds of North Carolina

Emerald, the "queen" of the beryls, occupies a prominent place in Barlow's collection, including fine specimens from most of the major source areas. Several outstanding pieces come from the emerald mines of North Carolina, the only emerald deposits in the United States.

The North Carolina deposits, first worked about 1875, have yielded many emeralds, most of which are not of gem quality. Those emeralds that have been cut show excellent color. The other North Carolina emeralds remain as crystal specimens and may be found in museum and private collections. (Note: Barlow specimen #4950 [see Chapter 3, Plate #3-9, page 43] is from the Hiddenite-Emerald Mine and #5376 [see Chapter 3, Plate #3-10, page 43] is from the Crabtree Mine, a typical North Carolina emerald locality.)

The North Carolina deposits are unique in that some are considered to be pegmatites while others appear to be vein-type deposits. The hiddenite deposits are hydrothermal veins where mineralized fluids were injected into the fractured metamorphic rock. The minerals found with emeralds here are rutile, pyrite, muscovite, and hiddenite, a rare type of spodumene. On the other hand, the Crabtree deposit at Spruce Pine appears to be an emerald-bearing pegmatite.

The Colombian deposits are hydrothermal veins in strongly folded and tilted sedimentary rocks. Carbonaceous black shales are important as the source of most emeralds. The green beryls can be found with calcite, dolomite, pyrite, fluorite, quartz, apatite, and the interesting rare-earth calcium carbonate, parisite. *R.W.J.*

## BERYL, VAR. EMERALD

### BRAZIL

**#2943 16.0×6.0×4.0cm**
**Plate #5-6**

**Bahia, Brazil**

A remarkable "log jam" of doubly and singly terminated light to dark green crystals to 6.0×1.2cm rests on a matrix of black, fine-grained biotite schist. A milky quartz vein cuts the specimen through its thinnest dimension with emeralds on both sides.

***Plate #5-6, Specimen #2943*** **BERYL, var. emerald – Bahia, Brazil. (MH)**

**#5375 10.0×7.5×7.0cm**

**Bahia, Brazil**

A nearly pure, exceptional cluster of Bahian emerald crystals to 5.0×3.4×3.0cm rests on very minor black biotite matrix. The upper quarter of many of the crystals is gemmy, with some zones having a slightly bluish emerald color. This piece is a very important regional specimen.

### COLOMBIA

**#2114 10.0×5.9cm**

**Muzo, Colombia**

This partly doubly terminated 2.7×1.2cm emerald tops a bed of drusy, clear to gray, 3mm calcite crystals. Several typical small pyrite crystals and a very rare pair of widely spaced caramel-brown 4×5mm and 1×2mm pseudohexagonal parisite-(Ce) crystals are associated with the emerald. The matrix consists of unusual columnar-to-fibrous milky-white calcite and a thin veneer of black carbonaceous shale vein wall.

*Plate #5-7, Specimen #2349* **BERYL, var. emerald – Muzo, Colombia. (MH)**

**BERYL, VAR. EMERALD**
**#2349 5.4×2.3×4.4cm Plate #5-7**
**Muzo, Colombia**

This fine miniature specimen shows a prominent gemmy and intense emerald-green 3.0×0.8cm crystal on a bed of interlocking, low-angle, pyramidal, gray, 2×3mm calcite crystals and a carbonaceous shale matrix. The emerald shows two major fractures where it is slightly dislocated. A secondary, 1.7×0.7cm, emerald crystal intersects the base of the major crystal.

**#2348 1.4×0.9cm**
**Muzo, Colombia**

This gemmy to facetable, uniformly deep green, 1.4×0.9cm crystal has a small, parallel growth, 5×4×1mm emerald emerging from the termination of the larger crystal. Minor emerald crystals intersect the lower portion of the main crystal, all lying on a druse of calcite, quartz, and pyrite crystals on carbonaceous shale.

**#5100 11.0×9.0×9.0cm Plate #5-8**
**Cosquez Mine, Otanché, Boyaca, Colombia**

From the Cosquez Mine, this outstanding bright gemmy and uniformly colored crystal is partly facetable. It is aesthetically canted toward two differently formed, complex, light smoky-gray calcite crystal groups. A wall of 1.8×3.8cm trigonally striated, low-angle rhombohedral crystals of calcite with additional complex calcite crystals rest on a carbonaceous shale matrix. The position and contrast of the intensely colored emerald crystal and gray calcite is very pleasing.

*Plate #5-8, Specimen #5100* **BERYL, var. emerald – Cosquez Mine, Otanché, Boyaca, Colombia. (MH)**

**#5160 7.0×4.5×4.0cm**
**Refer to Chapter 3, Plate #3-8**
**Cosquez Mine, Otanché, Boyaca, Colombia**

This unusual, richly colored dark emerald-green crystal, to 2.6×2.0×0.8cm, is clustered in a "V" arrangement on a tightly intergrown, cloudy, light gray cluster of crude calcite crystals.

**#5207 5.4×4.3cm**
**Cosquez Mine, Otanché, Boyaca, Colombia**

This choice, almost completely facetable, parallel growth of two 1.1×1.1cm deep emerald green crystals juts out of light smoky calcite cleavages and carbonaceous shale.

**NIGERIA**
**#5020 9.5×1.0×1.3cm**
**Refer to Chapter 3, Plate #3-7**
**Near Jos Plateau, Nigeria**

This specimen is a single, partially doubly terminated, elongated essentially flawless crystal with a dark uniform emerald-green color that has a slight bluish cast. A small cloudy zone exists at the base of the crystal. The upper first-order pyramidal termination has some deeply etched zones and comes to a sharp point, while the lower termination has only a slightly developed pyramid with a basal face. This outstanding specimen is one of the best single Nigerian emerald crystals found to date.

## BERYL, VAR. EMERALD

### UNITED STATES

**#300 6.1×1.6cm**

**Adams Property, Hiddenite-Emerald Mine, Hiddenite, Alexander County, North Carolina**

This fine, satiny etched, and doubly terminated, cloudy, light emerald-green crystal has numerous small dark red-orange rutile crystals, some twinned, embedded in and included in its prism faces: a classic example of hiddenite specimens. The etched edges are finely striated.

**#4950 9.0×5.5×2.7cm**
**Refer to Chapter 3, Plate #3-9**

**Adams Property, Hiddenite-Emerald Mine, Hiddenite, Alexander County, North Carolina**

This exceptional crystal group is described and pictured in Chapter 3, Plate #3-9. It is one of the finest North American "true" emeralds, perhaps second or third best from the locality.

**#5376 6.7×4.3×2.3cm**
**Refer to Chapter 3, Plate #3-10**

**Crabtree Mine, Spruce Pine, Mitchell County, North Carolina**

This specimen is described and pictured in Chapter 3, Plate #3-10.

***Plate #5-9, Specimen #5767*** **BERYL, var. emerald – Between Rist and Ellis Mines, Alexander County, North Carolina. (MH)**

**#5767 7.0×4.0×4.5cm Plate #5-9**

**Between Rist and Ellis Mines, Alexander County, North Carolina**

This specimen was mined April 11, 1995. The emerald crystal sits in a garden of quartz, amethyst, rutile, and muscovite. The crystal is small but gemmy and dark emerald-green unlike other Alexander County emeralds.

### RUSSIA

**#5433 9.0×5.5×2.5cm Plate #5-10**

**Tokovaya, Ekaterinburg (Sverdlovsk), Ural Mountains, Russia**

Fine-quality Russian emeralds have rarely been preserved as specimens, as emerald of almost any quality can be fashioned into jewelry. Russian emeralds of deep color and faceting grade are almost unknown in the West, but, as Plate #5-10 shows, at least one fine specimen escaped the gem cutter's wheel.

The main crystal group measures 5.0×2.5×2.5cm and consists of terminated hexagonal prisms in parallel contact with irregularly terminated emeralds interrupted by biotite schist. The emerald is a relatively uniform deep color with numerous multicarat facetable areas. Some large patches of emerald with biotite inclusions are also evident. The crystal group is offset by 3mm; biotite schist fills the separation. Minute brassy sulfides spot the walls of the impression of another emerald crystal.

***Plate #5-10, Specimen #5433*** **BERYL, var. emerald – Tokovaya, Ekaterinburg (Sverdlovsk), Ural Mountains, Russia. (MH)**

## BERYL, VARS. HELIODOR, MORGANITE, AND GOSHENITE

Heliodor, morganite, and goshenite are modern names of varieties of beryl. Heliodor (sun-gilded) originated about 1914 as a commercial name to market some newly found yellow to orange-yellow beryls from Namibia. This name has become synonymous with the earlier term golden beryl. Goshenite was first used for a pale pink beryl found at Goshen, Massachusetts by Charles Shepard in 1844. The locality yielded more colorless beryls than pink ones, however, and the name was eventually applied to signify colorless specimens. Morganite, a pink beryl, was named for J. Pierpont Morgan, wealthy financier, philanthropist, and mineral collector, when some deep pink gem beryls were discovered in Pala, California about 1911.

### BERYL, VAR. GOSHENITE
### CALIFORNIA

**#987 8.0×5.0cm Plate #5-11**

**Tourmaline Queen Mine, Pala, San Diego County, California**

This specimen is a faintly pink, transparent, nearly complete, 3.5×2.3 complex crystal. The crystal is attached to a rubellite and to smoky-colored elbaite with a thin gray rind and studded with small purple lepidolite crystals. Barlow collected this piece in 1974.

***Plate #5-11, Specimen #987***
**BERYL, var. goshenite – Tourmaline Queen Mine, Pala, San Diego County, California. (VP)**

### BRAZIL

**#5273 7.0×4.5×2.0cm Plate #5-12**

**Resplendor, Minas Gerais, Brazil**

This specimen consists of a dominant, clear, colorless, tabular 4.5×4.5×0.6cm crystal with a subparallel stack of 2.5×2.5cm platy goshenite crystals. These beryls are overgrown by dark green elbaite prisms. The surfaces of the stacked beryl crystals are coated by an orange-to-orange-brown transparent film. Each of the goshenite crystals has a small triangular pyramid face at the intersection of a prism with the basal termination.

***Plate #5-12, Specimen #5273***
**BERYL, var. goshenite – Resplendor, Minas Gerais, Brazil. (MH)**

## BERYL, VAR. GOSHENITE
### PAKISTAN

**#5271 6.8×3.8×3.4cm Plate #5-13**

**Dassu, Shigar Valley, Pakistan**

Pakistan has recently become a major source of gem beryl crystals. This rare large goshenite measures 4.5×1.5cm and consists of a bright hexagonal prism with an unusual termination. Six nearly equally developed, lightly frosted pentagons join the corner of the hexagonal prism to a large bright basal face. A shallow depression on the basal faces shows a complex inner structure of frosted pyramidal faces truncated by the bright terminal face. The point of attachment of the crystal consists of two major terminated quartz crystals to 4.0×3.0cm, along with minor yellow-to-brown muscovite and tiny green elbaite. Although essentially colorless goshenite, the core of the crystal is a pale blue. The transition from the blue core to the goshenite shell is marked by numerous inclusions of quartz and small white feldspar crystals. This crystal is an outstanding specimen.

***Plate #5-13, Specimen #5271*** **BERYL, var. goshenite – Dassu, Shigar Valley, Pakistan. (MH)**

## BERYL, VAR. HELIODOR
### RUSSIA

**#1877 3.5×1.7cm**

**Westschinski, Ural Mountains, Russia**

This is an unusual healed heliodor crystal with a 120° open "V" cross-section. The upper half of the crystal contains a multitude of closely spaced, microscopic yellow and clear color zones, while the bottom half of the crystal is a pale yellow.

### UKRAINE

**#5101 18.5×7.0cm 1,530g**
**Plate #5-14 (left)**

**#5138 19.5×5.5cm 1,402g**
**Plate #5-14 (right)**

**Woldarsk, Wolysnky, Ukraine**

This is a magnificent pair of essentially flawless gem beryl crystals. The slightly smaller crystal is a uniformly greenish yellow heliodor, while the larger is a uniformly deep ocean-blue. Both have some minor flaws. Etching is a common feature in pocket beryls, and these crystals are no exception. They have a surface texture of sharp, square to rectangular hopper etch pits several millimeters wide and deep. These bright beryl crystals with contrasting colors make a spectacular pair.

***Plate #5-14, Specimen #5101*** **- left;** ***Specimen #5138*** **- right**
**BERYL, var. heliodor – Woldarsk, Wolysnky, Ukraine. (VP)**

*Plate #5-15, Specimen #1021* **BERYL, var. morganite – Corrego do Urucum, Minas Gerais, Brazil. (VP)**

## BERYL, VAR. MORGANITE

### BRAZIL

**#1021 10.5×9.5×4.0cm Plate #5-15**

**Corrego do Urucum, Minas Gerais, Brazil**

This deep pink, with orange tint, hexagonal morganite crystal has a subtle, more deeply colored 7.0×4.0cm phantom-zoned core. The morganite has a large six-sided basal termination and short prism faces that are etched and striated. The morganite rests on a matrix of somewhat rusty cleavelandite and blackish green elbaite needles to 2.0×0.2cm. Numerous similar elbaite crystals decorate the etched morganite prism. This is an aesthetic example from a famous find.

### CALIFORNIA

**#445 9.0×9.0cm Plate #5-16**

**Tourmaline Queen Mine, Pala, San Diego County, California**

The specimen contains three individual transparent pale cinnamon-pink morganite crystals to 4.3×3.0cm, jutting out of naturally recemented pocket debris of cleavelandite, lepidolite, and pink elbaite prisms. The morganites are typical of Tourmaline Queen Mine crystals, showing glassy basal terminations in combination with satiny etched small pyramids. The piece was self-collected in 1974.

**#2967 9.0×8.3cm**

**Little Three Mine, Ramona, San Diego County, California**

A pale pink 3.2×2.2cm morganite crystal and a nearly colorless, cloudy goshenite are both lightly to severely etched in patches. The matrix is a gem pocket lining consisting of cauliflower-like pale blue cleavelandite with several diamond-shaped lilac lepidolite crystals. A 4mm orange spessartine crystal validates this as being from the Little Three Mine.

*Plate #5-16, Specimen #445* **BERYL, var. morganite – Tourmaline Queen Mine, Pala, San Diego County, California. (MH)**

## BERYL, VAR. MORGANITE

**#5097 9.0×7.2cm Plate #5-17**

**Stewart Mine, Pala, San Diego County, California**

A peachy-pink, gemmy, 5.0×3.5×3.0cm morganite crystal is set in the center of a cleavelandite and lepidolite matrix. The termination of the morganite is unusual in that the basal face shows a slightly concave surface that is irregularly patterned by smooth curving growths and seven somewhat triangular sectors near the center of the termination.

***Plate #5-17, Specimen #5097***
**BERYL, var. morganite – Stewart Mine, Pala, San Diego County, California. (MH)**

## BERYL, VAR. RED

**Harris Mine, Wah Wah Mountains, Beaver County, Utah**

Red beryls have been a specialty in the Barlow Collection. (See Chapter 14, Harris Mine.) Although they were given the varietal name bixbite in the 19th century, the variety is known simply as red beryl today. The Barlow suite of specimens is the best assembled in the world to date, and all specimens described herein are from the Harris Mine (formerly the Violet Claims), Wah Wah Mountains. (Refer to Chapter 3, Plates #3-11 and #3-12. For more information refer to Chapter 14.)

### Pegmatite Minerals

The Barlow Collection is well endowed with the minerals that occur in pegmatites. The variety of species includes the common minerals that may constitute the bulk of the pegmatite: albite and microcline feldspars, quartz, and micas. Pegmatites are also famous for rare-element minerals that contain beryllium (chrysoberyl, euclase, hydroxyl-herderite, phenakite), boron (danburite, hambergite, jeremejevite), fluorine (fluorite, topaz), lithium (cookeite, lepidolite, montebrasite, spodumene, kunzite), niobium, tantalum, and titanium (betafite, columbite, manganocolumbite, stibiotantalite), phosphate (brazilianite, eosphorite, fluorapatite, hureaulite, wardite), tin (cassiterite, wodginite), zirconium, and uranium (uraninite, zircon).

***Plate #5-18, Specimen #5299***
**ALBITE – Shigar, Ratistan, Pakistan. (MH)**

## ALBITE (SODIUM ALUMINUM SILICATE)

**#5299 11.5×8.0×8.5cm Plate #5-18**

**Shigar, Ratistan, Pakistan**

Pegmatite gem pockets can have crystals clinging to their walls as well as loose crystals in the pocket-filling debris. This albite specimen is dominated by two white, interlocking, 7.0×3.0×2.5cm albite variety valencianite crystals. The albite surrounds a severely etched 3.5×3.0×2.0cm schorl crystal that is flanked by silvery-to-bronzy rhombic tabular muscovite crystals. The cluster is crowned by a partially gemmy, deep red 3.2×2.5×2.2cm garnet crystal with finely lined, stepped crystal faces.

## ALBITE, VAR. CLEAVELANDITE

**#2968 9.0×7.0cm**

**Little Three Mine, Ramona, San Diego County, California**

This pale blue rounded cleavelandite crystal is tipped by delicate orange and cream-colored fringes. The cleavelandite is associated with brilliant, black, multifaceted schorl crystals to 6.0×2.6cm and gemmy, flame-orange, etched 1.5cm spessartine crystals. There is a heavily etched 1.5cm microcline crystal at the base of the specimen.

## BETAFITE (CALCIUM TITANIUM OXIDE HYDROXIDE)

**#4730 13.7×11.5×5.7cm** **Refer to Chapter 12, Plate #12-32**

**Ambolotara, near Betafo, Madagascar**

This specimen is a cluster of four major octahedral etched betafite crystals to 9.0cm from the classic location Betafo, Madagascar. The dark olive crystal faces are highlighted by fine-grained yellow and tan minerals filling imperfections. These are reputed to be the largest crystals of the species. The piece came from the Sorbonne Collection, Paris through dealer Al McGuinness to Dr. Don Doell to Barlow.

***Plate #5-19, Specimen #855*** **BRAZILIANITE – Corrego Frio Mine, Conselheiro Peña, Minas Gerais, Brazil. (VP)**

## BRAZILIANITE (SODIUM ALUMINUM PHOSPHATE HYDROXIDE)

**#855 9.5×8.0cm Plate #5-19**

**Corrego Frio Mine, Conselheiro Peña, Minas Gerais, Brazil**

This outstanding gemmy yellow-green brazilianite crystal with some cloudy zones comes from a classic source. Several re-entry notches indicate a multiple growth effect with four faces on the termination variously sloping away from the lightly striated prism faces. The specimen was formerly in the collection of Peter Bancroft.

## CASSITERITE (TIN OXIDE)

**#2485 4.0×4.0×3.7cm**

**Araçuai, Minas Gerais, Brazil**

This unusual, etched black-to-coffee-brown multiple-growth, doubly terminated cassiterite twin comes from near Araçuai, Minas Gerais, Brazil. The upper termination consists of several pyramidal cassiterite summits forming a square surface that is partly overshadowed by an "elbow" cassiterite twin. The main cassiterite has an eight-sided cross-section, with alternating large and small prism faces. The lower termination consists of three major summits isolated by slightly notched valleys and minor summits. The crystal is approximately a rectangular solid with twinning represented by diagonal notches. The symmetry of this twinning is exceptional.

*Plate #5-20, Specimen #1548* **CHRYSOBERYL – Itaguacu, Espirito Santo, Minas Gerais, Brazil. (MH)**

## CHRYSOBERYL
**(BERYLLIUM ALUMINUM OXIDE)**

**#1548 4.5×3.8×3.0cm Plate #5-20**

**Itaguacu, Espirito Santo, Minas Gerais, Brazil**

This gemmy-to-cloudy nearly complete "sixling" has four strongly and symmetrically radiating chrysoberyl crystals and two short unterminated projections. The color is a deep yellow-green-to-olive with some brown accents.

## DANBURITE
**(CALCIUM BOROSILICATE)**

**#2396 4.5×4.0cm**

**Mount Bity, Madagascar**

This blocky tan-to-creamy-brown etched multiple growth 4.0×2.3×2.3cm danburite crystal is on a matrix of coarsely bladed cleavelandite. This mineral could be easily mistaken for topaz.

## EOSPHORITE
**(MAGNESIUM ALUMINUM PHOSPHATE HYDROXIDE HYDRATE)**

**#1003 5.5×4.5cm**

**Lavra de Ilha, Itinga District, Minas Gerais, Brazil**

This is an outstanding 5.0×4.0cm cluster of slightly reddish-to-amber-brown gemmy eosphorite crystals from the noteworthy island deposit of Lavra de Ilha. Some of the individual eosphorite groups range up to 8×2×2mm and show parallel growth, while other groups show a nearly radial sunburst. The matrix consists of stout quartz crystals to 3.5×2.0cm.

*Plate #5-21, Specimen #3232* **EOSPHORITE – Lavra do Criminoso, São José da Safira, Minas Gerais, Brazil. (MH)**

## EOSPHORITE

**#3232 10.0×9.0cm Plate #5-21**

**Lavra do Criminoso, São José da Safira, Minas Gerais, Brazil**

This specimen is a crystallized plate of gemmy brown-to-pinkish-brown 3.0×2.5×1.2cm eosphorite crystal cluster. The larger 3.0×2.0 crystal shows occasional skeletal thin overgrowths of brown eosphorite on more pinkish-brown core crystals.

## EUCLASE
**(BERYLLIUM ALUMINOSILICATE HYDROXIDE)**

**#782 5.5×2.5×1.3cm**

**Refer to Chapter 3, Plate #3-33**

**Saramenha, near Ouro Preto, Minas Gerais, Brazil**

This sharp, wedge-shaped, transparent, pale blue single euclase crystal is described and pictured in Chapter 3.

**#4725 3.3×2.5cm**

**7778 Pocket, Last Hope Mine, Karoi, Zimbabwe**

This medium-blue-to-deep-indigo, thick, "picket fence" arrangement of terminated euclase crystals to 1.6×1.3×1.2cm is tightly clustered with fragmented quartz crystals to 1.0×1.0cm. The euclase crystals, as is typical of the locality, are conspicuously sprinkled with regrown quartz shards to 1×1mm that resemble granulated salt. This ranks as one of the world's best.

*Plate #5-22, Specimen #973* **EUCLASE – Santa do Encoberto, Minas Gerais, Brazil. (VP)**

## EUCLASE

**#973 9.0×5.0cm Plate #5-22**

**Santa do Encoberto, Minas Gerais, Brazil**

This piece is a giant, clear-to-slightly-creamy, mirror-faced euclase crystal with several parallel growths of euclase terminations. One of the largest crystals of the species.

*Plate #5-23, Specimen #1503* **FLUORAPATITE – King Lithia Prospect, Keystone, Pennington County, South Dakota. (MH)**

## FLUORAPATITE

**(CALCIUM PHOSPHATE FLUORIDE)**

**#1503 6.0×6.0cm Plate #5-23**

**King Lithia Prospect, Keystone, Pennington County, South Dakota**

This is a cluster of six strongly color-zoned fluorapatite crystals to 1.2×0.5cm. The crystals are milky blue with fibrous pyramidal terminations on a matrix of drusy transparent quartz crystals on albite. Barlow self-collected this piece in 1977.

## FLUORAPATITE

**#2090 3.0×3.0cm**

**Rachid Mine, Itambacuri, Minas Gerais, Brazil**

This nearly equant, interlocking trilling of blue-to-milky-white zoned tabular fluorapatite crystals is terminated by low-angle pyramids and small basal faces. The blue color exists only on the terminations, while the milky white zone forms a central wafer of varying proportions. The specimen was a gift from the miner to Alice Barlow when the collector and his daughter were in Brazil.

## FLUORAPATITE

**#2863 8.5×7.0cm**

**Stak Nala, near Gilgit, Pakistan**

This complex, muted pink, 2.2×2.1×0.8cm, hexagonal fluorapatite crystal has a dominant basal termination and modified hexagonal prism. The association is dominated by strikingly color-zoned smoky-olive-green elbaite crystals, to 4.0cm, capped by thin pastel-pink-to-colorless zones. The order of growth of the various minerals on the specimen is clear. The elbaite forms triangular crystals that, in turn, are partly enclosed by a larger transparent, doubly terminated, 5.3×3.0cm quartz crystal. The quartz and elbaite intersect two 2.5cm white wedge-shaped albite, var. valencianite crystals that are coated by rosettes of white cleavelandite. The prominent pink fluorapatite is perched on this association of minerals. A thin druse of a tan-stained apatite, with minute muscovites, highlights the bottom edge of the pink fluorapatite as well as the cleavelandite. Finally, a half dozen or so colorless bladed crystals with chisel terminations that might be bertrandite are scattered over the specimen. This is an instructive, aesthetic specimen.

**#3099 2.0×1.0×1.0cm**
**Refer to Chapter 3, Plate #3-37**

**Pulsifer Quarry, Mount Apatite District, Auburn, Androscoggin County, Maine**

This reddish purple-to-light-royal-purple partially doubly terminated fluorapatite crystal comes from a classic locality in Maine.

**#4778 14.2×8.5cm**

**Otjua Mine, Karibib, Namibia**

The specimen consists of royal purple, hexagonal, tabular fluorapatite crystals to more than 1.0cm in a tight cluster measuring 3.7×2.0cm on coarse, pinkish, tan-stained cleavelandite. Light lemon-yellow, curved, 2mm cookeite mica crystals partially cover the cleavelandite. The specimen has a 9.0×5.0cm frosted dark smoky quartz crystal that, in turn, is selectively coated with royal purple fluorapatite crystals to more than 1.0cm.

**#4912 7.5×4.0cm**

**Inna Nala, Sighar Village, Lower Nagar, Pakistan**

This piece is an interlocking stack of five major light rose fluorapatite crystals without matrix. Individual crystals are up to 3.3×2.0cm. Several silvery muscovite crystals to 1.0cm dot the specimen.

## FLUORITE (CALCIUM FLUORIDE)

**#3047 2.7×2.6cm**

**Klein Spitzkopje, Namibia**

This is a rich cluster of unusually color-zoned and etched fluorite crystals to 0.8×0.8cm. The crystals have lime-green triangular corners with light peach-orange cores. They are closely grouped on crude, sooty-black, cloudy quartz crystals with minor microcline and siderite crystals.

**#5018 19.0×11.0cm**

**Chumar Bakhoor Nala, Nagar, Pakistan**

The specimen consists of six outstanding uniformly etched, pink, gemmy, dodecahedral fluorite crystals to 9.0×8.0cm, with mint-green cores. One minor transparent 2.3×2.2cm fluorite has a smooth satin finish yet is intergrown with fluorite that is roughly etched. The crystals are set on a plate of coarsely crystallized silvery muscovite. This is one of the best from the locality and one of the best pegmatite fluorites in the world.

**#5205 4.8×3.4×3.4cm**

**Klein Spitzkopje, Namibia**

This specimen is a slightly corroded and flawless fluorite cube. The faces contain linear and right-angled surface troughs that appear to be etched but are growth striations in the crystal faces. This is an important pegmatite fluorite specimen.

## HAMBERGITE
## (BERYLLIUM BORATE HYDROXIDE)

**#4779 5.7×6.4×1.9cm**
**Refer to Chapter 3, Plate #3-50**

**Turakuloma Range, East Pamirs, Tadzhikistan**

This rare, deeply etched, gemmy, colorless, blocky hambergite crystal shows several perfect cleavages. This mineral is rarely preserved because it decomposes when ground water enters the crystal pocket in which it forms.

## HUREAULITE
**(MANGANESE PHOSPHATE HYDRATE)**
**#2557 4.0×2.0cm**

**Lavra do Criminoso, São José da Safira, Minas Gerais, Brazil**

These attractive pink-to-pinkish-tan bundles of obliquely terminated hureaulite crystals reach 1.1×0.7cm. They form a 2.4×2.0cm cluster next to a sparkling and densely clustered 1.7×1.5cm group of rare, brilliant black, nearly equant, barbosalite crystals to 1mm. The matrix is massive quartz and apatite.

## HYDROXYLHERDERITE
**(CALCIUM BERYLLIUM PHOSPHATE HYDROXIDE)**
**#1411 14.0×10.0×4.5cm**
**Refer to Chapter #3, Plate #3-53**

**Xanda Mine, Virgem de Lapa, Minas Gerais, Brazil**

This famous Xanda Mine hydroxylherderite is certainly the best in the world for this species. It is described and pictured in Chapter 3, Plate #3-53. Pictured on the cover of *Mineralogical Record*, Volume 10, Number 1, 1979.

## JEREMEJEVITE
**#2470 5.5×3.5×1.0cm**
**Refer to Chapter 3, Plate #3-56**

This attractive jackstraw cluster is one of the best jeremejevites from the Mile 72 Pegmatite.

## LEPIDOLITE (POTASSIUM LITHIUM ALUMINOSILICATE FLUORIDE)
**#2993 9.5×5.3×4.0cm**

**São José de Pederneira, near Cruzero Mine, Governador Valadares District, Minas Gerais, Brazil**

The specimen consists of light purple stacks and plumes of small lepidolite crystals on the edge of a triangular patch of cloudy-gray elbaite crystals to 9.5×2.0cm, with a cluster of white cleavelandite crystals to 1.0cm. Numerous rounded tan cookeite botryoids to 2mm cover the back of this specimen.

*This unique specimen is a cross-section of a gem deposit.*

***Plate #5-24, Specimen #5389*** **LEPIDOLITE – Himalaya Mine, Mesa Grande, San Diego County, California. (MH)**

## LEPIDOLITE
**#5389 14.0×11.0×12.0cm Plate #5-24**

**Himalaya Mine, Mesa Grande, San Diego County, California**

This unique specimen is a cross-section of a gem deposit. The base consists of a thin 1.0cm zone of fine-grained country rock. The central zone of the matrix consists of normal pegmatite minerals with 2.0×0.5cm laths of green tourmaline between the layers of muscovite. The general crystal size increases to several centimeters, and the silvery white muscovite sheets show sharp crystal outline. The muscovite crystals have overgrowths, however, of beautiful purple lepidolite, which continued to increase in size to 5.0×2.0cm as they grew into a central gem-bearing cavity. The lepidolite crystals show two contrasting styles. The large purple lepidolite crystals project from the matrix on their "sides," exposing the prism faces. A second generation of gray-lilac-to-tan, diamond-shaped, 1–2mm lepidolite crystals forms a cushiony mass on the terminations of the large lepidolite crystals. Two clear-to-wispy 4.0×4.0cm quartz crystals are nestled on the lepidolite surface. This is an important paragenesis specimen and was a gift of friendship from Bill Larson to Barlow.

## MANGANOCOLUMBITE
(MANGANESE NIOBATE)
**#1893 8.0×4.5×7.5cm**

**Toca da Onca Mine, Virgem da Lapa, Minas Gerais, Brazil**

This black, parallel-growth, rectangular 5.0×3.4cm manganocolumbite crystal shows slight curvature as if healed after being subjected to gem pocket stresses. The specimen has a central open 4.0×3.5cm rosette cluster of gray-lilac, sutured lepidolite crystals accented by a cluster of white 1.5×1.0cm cleavelandite. The manganocolumbite rests on a bed of subparallel cleavelandite crystals sprinkled with several brilliant, black cassiterite crystals to 2mm. This is a particularly aesthetic specimen for the species.

## MICROCLINE (AMAZONITE, VAR. MICROCLINE)
(POTASSIUM ALUMINUM SILICATE)
**#937 8.7×7.3×5.3cm**
**Refer to Chapter 3, Plate #3-66**

**Park County, Colorado**

The blue amazonite variety of microcline has been known from Colorado pegmatites for over a century. This beautiful bright greenish-blue 5.0×5.0×3.5cm amazonite crystal is described and pictured in Chapter 3, Plate #3-66.

## MICROCLINE (AMAZONITE, VAR. MICROCLINE)
**#1005 10.5×6.0cm**

**Park County, Colorado**

This specimen consists of a partly skeletal, greenish-blue amazonite crystal with contrasting buff-white sandine variety of microcline in selective overgrowth highlights that cover entire crystal faces.

***Plate #5-25, Specimen #1704*** **MICROCLINE (AMAZONITE, var. microcline) – Crystal Peak, Teller County, Colorado. (MH)**

## MICROCLINE (AMAZONITE, VAR. MICROCLINE)
**#1704 5.5×5.0×3.8cm Plate #5-25**

**Crystal Peak, Teller County, Colorado**

This especially aesthetic miniature specimen of smoky quartz, amazonite, and cleavelandite albite has four sharp, tapered, brownish-gray smoky quartz crystals from 1.5-3.5cm tall rising above six vivid blue amazonite crystals from 1.0-2.0cm across. Minor bladed white cleavelandite partially fills the depressions between the amazonite crystals and between the amazonite and the smoky quartz crystals. Medium-grained granite forms the base of the specimen. This is one of the finest miniatures from this classic locality.

## MONTEBRASITE (LITHIUM ALUMINUM PHOSPHATE FLUORIDE HYDROXIDE)
(PROBABLY AMBLYGONITE)
**#943 4.3×4.0×2.0cm** **Refer to Chapter 3, Plate #3-68**

**Linopolis, Minas Gerais, Brazil**

This exceptional pair of pale yellow montebrasite crystals is described and pictured in Chapter 3, Plate #3-68.

## MUSCOVITE (POTASSIUM ALUMINOSILICATE HYDROXIDE)
**#4734 13.0×8.0×8.0cm**

**Aldeia do Eme Mine, Minas Gerais, Brazil**

Beautiful dark rum-colored, pseudohexagonal muscovite crystals to 4.5×4.5cm are stacked in a cluster, studded with numerous dark green, hexagonal columns of fluorapatite crystals to 1.0cm.

## ORTHOCLASE

**#5021 9.0×4.5×3.0cm**

**Refer to Chapter 12, Plate #12-44**

**Itrongay, Madagascar**

This is one of the great feldspar crystals from an old-time location and was in the collections of the Sorbonne and the Los Angeles County Museum. The specimen is described and pictured in Chapter 12, Plate #12-44.

## PHENAKITE (BERYLLIUM SILICATE)

**#1009 4.1×3.8cm**

**Caraca Hill, Sao Miguel de Piracicaba, Minas Gerais, Brazil**

This specimen is a clear-to-cloudy, tabular, low-angle rhombohedron with complex faces. A partly doubly terminated, clear-to-rusty, included, 2.4×0.9cm quartz crystal has grown flat on a terminal phenakite face.

***Plate #5-26, Specimen #5023*** **PHENAKITE – Anjanabonoina, Madagascar. (VP)**

## PHENAKITE

**#5023 2.9×0.9×0.9cm Plate #5-26**

**Anjanabonoina, Madagascar**

This asymmetrically and obliquely terminated, gemmy and facetable pale yellow phenakite crystal has hexagonal prism and small modifying forms and originates from the classic Madagascar locality. Fine phenakites from other classic localities are also included in the collection.

## QUARTZ (SILICON DIOXIDE)

**#2479 6.5×5.0×3.0cm**

**Mile 72 Pegmatite, near Cape Cross, Namibia**

This is a superb tricolored, skeletal, and distorted doubly terminated quartz crystal. The crystal is zoned approximately in thirds. The upper termination is a bright raspberry-amethyst. The color is most intense at crystal corners, less intense on crystal edges, and still less intense on the deeply skeletal faces. The essentially colorless central prism zone is deeply hoppered over one half or more of each face. The lower, multiple-growth, termination is a coffee-colored smoky quartz with a small patch of amethyst. While not a true scepter, the whole crystal is supported by a gnarled smoky quartz "root."

## QUARTZ

**#5206 8.5×7.0cm**

**Jonas Mine, Itatiaia, Minas Gerais, Brazil**

This astounding enhydro-included (movable water bubbles) smoky quartz crystal has a healed 4.7×1.8cm "shard crystal" that has grown through the termination, resulting in a branching of the single termination. The unusually large, 2.1×1.1×~0.3cm, fluid inclusion contains a movable bubble occupying about half of the inclusion's volume. It can be seen under one of the pyramidal faces. Another movable bubble, 3.6×0.7×0.3cm, occupies about half of its chamber under the prism surface. Several additional dry and movable bubble inclusions are also easily visible. This piece is a spectacular teaching tool used by Barlow when college students visit his museum.

*This specimen is pre-eminent among the skeletal quartz crystals of the world.*

***Plate #5-27, Specimen #2748*** **QUARTZ – Mina La Limeña, Artigas, Uruguay. (MH)**

## QUARTZ

**#2748 22.0×15.0×11.0cm Plate #5-27**

**Mina La Limeña, Artigas, Uruguay**

This is a remarkable doubly terminated skeletal and sceptered quartz crystal. This specimen is pre-eminent among the skeletal quartz crystals of the world. The crystal has a seven-peaked multiple growth termination with a central chasm that extends through to the base of the crystal. The pyramidal terminations, as well as the prism, are individually cavernous, sometimes from the very edge of the face to its center. The outer zones of the crystal contain numerous interrupted planes where opaque tan-to-mottled, rusty-orange clay is included in the overgrowth of quartz, resulting in distinctive phantoms. The frontal prism edge is uniformly etched and skeletal. The outside edges of the midsection of the crystal are etched and cavernous. Finally, the few outer transparent parts of the quartz show thin amethyst tints on corners and edges of the upper pyramidal faces, while the lower pyramids show a slightly greater amount of smoky color with some intermixed amethyst. An etched milky quartz shaft underlies the skeletal quartz. This is one of the great crystals of any species. The mine, part of a relatively flat cattle ranch, is named in honor of the miner's wife, who was from Lima, Peru.

## SPODUMENE (LITHIUM ALUMINUM SILICATE)

Several varietal names exist for the gem varieties of spodumene. Kunzite is used to classify pink-to-violet gem spodumene colored by manganese. This material yields virtually all of the important crystallized specimens. Hiddenite is a green spodumene, colored by chromium, first discovered in North Carolina.

***Plate #5-28, Specimen #1007*** **SPODUMENE, var. hiddenite – Adams Property, Hiddenite Mine, Alexander County, North Carolina. (VP)**

## SPODUMENE, VAR. HIDDENITE

**#1007 3.3×2.7cm Plate #5-28**

**Adams Property, Hiddenite-Emerald Mine, Hiddenite, Alexander County, North Carolina**

This specimen, from the classic locality for hiddenites, is a transparent-to-translucent, etched, but sharply terminated, emerald-green 1.7×1.7cm hiddenite crystal on mottled black and gray gneissic matrix. The hiddenite is deeply striated along its length and has an etched "picket fence" termination. Some minor etched and rounded hiddenite crystals to 1.2cm are associated on the matrix.

*Plate #5-29, Specimen #1606* **SPODUMENE, var. kunzite – Stewart Mine, Pala, San Diego County, California. (MH)**

*Plate #5-30, Specimen #2861* **SPODUMENE, var. kunzite – Kabul, probably Mawi, Laghman Province, Afghanistan. (JS)**

## SPODUMENE, VAR. KUNZITE

**#1606 13.0×13.0cm Plate #5-29**

**Stewart Mine, Pala, San Diego County, California**

This is an extraordinary transparent, etched and twinned spodumene crystal sitting in a mixture of brecciated rubellite-bearing gem pocket debris cemented by cream-colored montmorillonite clay. The barely pink-tinted spodumene is typically rounded and deeply striated, perhaps more extensively due to etching.

**#1243 3.5×3.0cm**

**Kabul (probably Mawi), Laghman Province, Afghanistan**

This specimen is a sharp and symmetrical, gemmy, pale pink kunzite crystal. The tabular crystal has lightly etched faces with a satiny luster. Like all kunzites, it shows its deepest color when viewed down the long axis.

**#2408 9.3×5.9cm**

A large number of kunzite specimens have come from the gem-bearing pegmatites of Nuristan, Afghanistan, but they do not generally show sharp crystal forms as this one does, for many crystals have succumbed to the etching effects of late-stage corrosive fluids. This one is a well-formed, single, tabular, delicate magenta kunzite crystal with slightly etched "picket fence" terminations. The front view of the transparent crystal is pale magenta, while the side view reveals a vivid magenta color.

**#2861 12.0×9.5cm Plate #5-30**

**Kabul, probably Mawi, Laghman Province, Afghanistan**

Matrix kunzites are a rarity. This specimen is a bright, deep-to-light-magenta, well-terminated kunzite crystal, 10.5×4.3cm, flat lying on a rosette of white cleavelandite, 5.0×4.0cm, in turn guarded by a cloudy, well-terminated, skeletally capped quartz crystal, 11.7×5.6cm. The kunzite shows the usual differential color intensity — darker along the elongation and lighter through the minimum dimension — a feature well shown by the species. This piece is a most aesthetic representative of the species. Pictured on the cover of *Rocks and Minerals*, Volume 62, Number 2, 1987.

**#1541 8.5×5.0×4.5cm**

**Governador Valedares District, Minas Gerais, Brazil**

This blocky, marvelously etched, essentially flawless gem crystal with a delicate cinnamon-pink (lilac along the elongation) color is an uncommonly fine kunzite. The crystal shows scalloped and broadly terraced etch planes. Some edges of the crystal are rounded and eroded while others are sharp and straight. One side of the crystal shows four-sided hoppered pyramidal etch pits to 7×5×4mm. This is one of the superb kunzites in the collection.

## SPODUMENE, VAR. KUNZITE

**#2160 7.3×3.8×1.6cm**

**Nuristan District, Afghanistan**

This highly etched, pale-to-dark-magenta kunzite crystal section is typical of the Nuristan District's gem rough crystals. Etching on the prism tends to be streaked to corrugated, while the etched termination frequently shows closely spaced pyramidal growth hillocks.

## STIBIOTANTALITE
### (ANTIMONY TANTALATE)

**#3712 2.1×2.0cm**

**Himalaya Mine, Mesa Grande, San Diego County, California**

This doubly terminated, tabular, variegated dark-amber-to-reddish amber stibiotantalite crystal is 1.4×1.2×0.2cm. It is a rare specimen and has a small etched microcline fragment at the base.

## TOPAZ
### (ALUMINUM SILICATE FLUORIDE HYDROXIDE)

**#1236 6.0×5.3cm**

**Virgem da Lapa, Minas Gerais, Brazil**

This specimen is a gemmy, nearly flawless, pale blue, steep wedge-shaped topaz crystal. The dominant "wedge" faces are bright but microscopically pitted while the side prismatic faces show etching.

**#1546 5.0×4.7×4.5cm**

**Mursinka, Ekaterinburg (Sverdlovsk), Russia**

Mursinka is one of the classic sources for topaz. This outstanding transparent, blocky, lightly veiled, light sky-blue gem topaz crystal of classic proportions with prism, prominent basal termination, moderately sized domes, and small modifying pyramid faces comes from this famous locality.

***Plate #5-31, Specimen #1019*** **TOPAZ – Virgem da Lapa, Minas Gerais, Brazil. (MH)**

**#1019 12.6×5.6×5.5cm Plate #5-31**

**Virgem da Lapa, Minas Gerais, Brazil**

Virgem da Lapa is known for fine topaz like this transparent, nearly flawless, medium blue crystal. It is nearly equally apportioned by glassy crystal faces, forming a steep pair of faces in contrast to the deeply corroded side of the crystal. A beautiful specimen with an exciting past, this is one of several fine topaz specimens from Virgem da Lapa in the collection.

**#2323 6.5×4.0cm**

**Pech, Afghanistan**

A major new source of fine topaz crystals is the Pech area of Afghanistan. This colorless, gemmy, blocky, 4.7×4.5×3.3cm, crystal is a fine example. It shows a form reminiscent of Russian topaz with dominant prisms, large basal termination, and small domes and pyramids. The topaz is set on a coarse 4.5×3.5cm rosette of white cleavelandite blades with several lilac lepidolite crystals to over 1.0cm, and a pink elbaite crystal, 2.6×0.5cm. The matrix and the back of the topaz are coated by a thick-to-sparse druse of cream-colored muscovite crystals generally 1.0mm and smaller.

## TOPAZ

**#5043 5.2×3.3cm**

**Klein Spitzkopje, Namibia**

This colorless and essentially flawless blocky topaz crystal, 2.8×1.4cm is in canted subparallel contact with a lightly frosted light smoky quartz crystal, 5.2×1.6cm. Both are sprinkled with black tourmaline needles to 7.0×1.0mm.

*Plate #5-32, Specimen #3360* **URANINITE – Swamp #1 Quarry, Topsham, Sagadahoc County, Maine. (MH)**

## URANINITE (URANIUM OXIDE)

**#3360 1.8×1.5×1.1cm Plate #5-32**

**Swamp #1 Quarry, Topsham, Sagadahoc County, Maine**

This is a rare, brilliant, sharp, parallel growth, 1.8×1.5cm cluster of two black unoxidized cubo-octahedral uraninite crystals. This world-class specimen was formerly in the Trebilcock Collection.

## WARDITE (SODIUM ALUMINUM PHOSPHATE HYDROXIDE HYDRATE)

**#1023 9.5×4.5×6.0cm**

**Lavra da Ilha, Itinga District, Minas Gerais, Brazil**

Lavra da Ilha is known for exceptional phosphate minerals. This wardite is among the best from that site. It consists of extraordinary milky-white, bipyramidal crystals to 1.7cm, with basal terminations flecked with minute pyrite crystals on a skeletal quartz crystal, 9.5×5.5cm. Also present are other phosphate minerals, including interspersed dark green zanazziite botryoids to 4mm with a radial fibrous texture as well as some associated gemmy brown eosphorite crystals to 8mm. There are also some drusy rose quartz crystals. This is one of the largest wardite crystals from Brazil.

*Plate #5-33, Specimen #1316* **WODGINITE – Lavora Jabuti, Sao Geraldo do Baixio, Galileia, Minas Gerais, Brazil. (MH)**

## WODGINITE (MANGANESE TIN TANTALATE)

**#1316 3.7×2.3cm Plate #5-33**

**Lavora Jabuti, Sao Geraldo do Baixio, Galileia, Minas Gerais, Brazil**

This rare element compound from Galileia is a brilliant, wedge-shaped, 3.7×0.7cm wodginite crystal overgrown on both sides of a brighter black spine of cassiterite. The cassiterite shows a small twin notch while the central wodginite does not. This is a world-class specimen of wodginite.

## ZIRCON (ZIRCONIUM SILICATE)

**#2276 5.5×3.0cm**

**Seiland, Norway**

A doubly terminated, reddish-to-dark-coffee-brown, partly gemmy, zircon crystal, 1.7×0.8cm, extends out of pearly ashen-gray oligoclase feldspar.

## ZIRCON

**#2381 2.8×2.6×2.4cm**

**Miask, Ural Mountains, Russia**

This is a fine single, low-angle bipyramidal and horizontally striated lustrous dark brown zircon crystal.

***Plate #5-34a, Specimen #5844*** **BERYL GEM POCKET (MH)**

***Plate #5-34b, Specimen #5844*** **BERYL GEM POCKET (MH)**

## BERYL GEM POCKET

**#5844 30.6×8.7×10.6cm Plate #5-34**

This work of art was produced by the meticulous craftsmanship of John Jefferson, Chemung, New York.

It is made from black walnut. The locks are made of purpleheartwood, and the white pegs are from an unidentified wood. The pocket matrix is from Wausau, Wisconsin and consists of quartz and feldspar with several rhombohedral pseudomorphs of goethite after siderite. There is one phenakite crystal located at the base of the large multiple quartz crystal hanging from the ceiling. At the floor is the aquamarine is from Afghanistan, and the schorl is from Pakistan.

This is a major work of craftsmanship from a mineral collector and mineralogist.

## REFERENCE

Shepard, C. U. (1844) *A Treatise on Mineralogy*, 2nd Ed., A. H. Maltby, New Haven, Connecticut, 188 pp.

Chapter 6

# Gemstones and Rare and Important Minerals

*by F. John Barlow and Peter J. Juneau*

The determination of whether a mineral specimen or cut gem is considered a rare or important specimen results from a complex interplay of many factors. The mode of occurrence, the availability, and the quality, size, and/or uniqueness of the specimen all influence the decision. Another factor is whether the mineral occurs in the size and quality necessary for faceting into a cut gem.

Less than 100 of the approximately 3,600 described minerals commonly occur in the earth's crust. Feldspars (plagioclase, microcline, and orthoclase), quartz, and ferromagnesium minerals (olivine, hornblende, and biotite) are very common igneous rock-forming minerals. However, favorable geologic conditions in restricted locations have resulted in superb mineral specimens that could be considered rare and important, be it from crystal habit, quality, or uniqueness. The remaining 3,500 mineral species vary from uncommon to extremely rare. Some of the mineral species, or specimens of one species that are of superb quality, are limited to only one discovery or occurrence. For example, red beryl (Chapter 14) is found in several locations, but specimens of gem quality come from only one mine in Utah.

The scarcity of rare mineral species limits the possibilities of acquisition by a collector. However, included in the Barlow Collection are specimens of some of the world's important rare minerals and gemstones. Of the 481 different mineral species cataloged in the collection, approximately 200 species are considered very rare. A few of these mineral specimens are unique, and some are the best of their kind, dramatically overshadowing even the second ranked specimen of that species.

*Of the 481 different mineral species cataloged in the collection, approximately 200 species are considered very rare.*

The cut gems described in this chapter are representative of the total holdings of the collection and, admittedly, along with the mineral specimens, represent a subjective selection by Barlow based on his interpretation of what constitutes rare and important.

Although everyone may not agree with all of the selections, we hope you will enjoy a look at some rare and important mineral specimens and cut gems from the Barlow Collection.

## ARMSTRONGITE
**(CALCIUM ZIRCONIUM SILICATE HYDRATE)**
**#2832 5.2×3.5×1.0cm**

**Khan-Bogdinskii Massif, Mongolia**

Armstrongite is a rose to rose-brown massive mineral in an undistinguished amphibole-bearing rock. While the rationale for obtaining rare minerals has been discussed, this one was acquired, without pursuit, while the collector was in Russia. It was added to the collection for a unique reason.

Dr. Leo Bulgak, Dr. Godovikov's secretary, of the Fersman Museum handed Barlow the specimen and asked with a smile, "Do you know what this is?" Barlow said, "No." Leo said, "I'm surprised. This is armstrongite." By the puzzled look on Barlow's face, Leo knew he had him. After a moment of silence, Leo said that the mineral was named after "your astronaut who was the first man on the moon." Thus, a rare species entered the Barlow Collection.

## BERYL, VAR. AQUAMARINE
**(BERYLLIUM ALUMINUM SILICATE)**

Beryls are described in Chapter 5, but several additional specimens are included here. A matched pair of spectacular gem aquamarines is included. Another beryl deserving of "significant" status and, therefore, included in this chapter is referred to as "trapiche" emerald. The trapiche emerald is not a pure beryl but an intergrowth of a carbonaceous material with emerald in hexagonal-shaped skeletal crystals. Trapiche emeralds, found in the emerald mines of Muzo, Colombia, grow in carbon-rich sediments, and as they grow, include carbonaceous material in their structure. When well developed, trapiche emeralds exhibit a strong hexagonal pattern — six emerald-green, wedge-shaped segments radiating outward from a black core and separated from each other by very thin lines of black inclusions.

**#2092 8.2×2.2×1.9cm**
**Plate #6-2 (bottom)**

**#2093 7.0×2.5×2.0cm Plate #6-2 (top)**

**Jaqueto Mine, Minas Gerais, Brazil**

A superb matched pair of highly etched, gemmy aquamarine crystals with an excellent pale blue color. Both crystals are tapered due to etching and have tubular etch pits on the terminations. Etching on the prism faces produced prominent grooves parallel to the length of the crystals. Most of the grooves have striations perpendicular to their long dimension. The more severely etched tapered ends of the crystals have a somewhat frosted appearance. This is an outstanding pair of gem crystals.

***Plate #6-2, Specimens #2092 and 2093*** **BERYL, var. aquamarine – Jaqueto Mine, Minas Gerais, Brazil. (MH)**

## BABINGTONITE
**(CALCIUM IRON SILICATE HYDROXIDE)**
**#1369 6.5×5.5×3.5cm Plate #6-1**

**Westfield, Middlesex County, Massachusetts**

Babingtonite is the state mineral of Massachusetts. On this piece, two bright black multifaceted babingtonite crystals (to 3.5×2.1×1.5cm) just intersect on a bed of pale yellow datolite and dark lime-green prehnite. Westfield is a classic American babingtonite locality, long since closed to collecting.

***Plate #6-1, Specimen #1369*** **BABINGTONITE – Westfield, Middlesex County, Massachusetts. (MH)**

*Plate #6-3, Specimen #2351* **BERYL, var. emerald, trapiche – Muzo, Colombia. (MH)**

*Plate #6-4, Specimen #B1* **BERYL, var. emerald, trapiche – Muzo, Colombia. (MH)**

## BERYL, VAR. EMERALD, TRAPICHE

**#2351 4.4×3.5×3.5cm Plate #6-3**

**Muzo, Colombia**

This is an excellent matrix specimen of the rare trapiche emerald from the classic Colombian locality. The rather fractured, 1.5×1.8 gemmy green crystal has the six rays of black carbonaceous inclusions parallel to the *c*-axis, characteristic of trapiche emeralds. The crystal is partially enclosed in a brecciated black carbonaceous shale. White calcite and scattered pyrite crystals cement the matrix breccia. One of two known matrix specimens.

## BERYL, VAR. EMERALD, TRAPICHE

**#B1 2.5×1.7cm Plate #6-4**

**Muzo, Colombia**

This roughly hexagonal prism section of a crystal shows somewhat rounded wedge-shaped units of gemmy green emerald separated by black carbonaceous walls. One end of the crystal section has a hexagonal-shaped black core; the other end has a thin, black, six-rayed "star" but contains no black core. The outer surface of the crystal section is mostly coated with carbon. Both ends of the sample have been polished to better show the typical cross-section of trapiche emeralds.

## TRAPICHE EMERALD, LOOSE GEM

**1.5×1.0×0.5cm 3.17ct Plate #6-3**

Cut: Pear shaped cabochon.

Color: Medium light, slightly bluish green.

Clarity: Very slightly included.

## BUSTAMITE

### (MANGANESE CALCIUM SILICATE)

**#1138 5.5×5.5×2.5cm Plate #6-5**

**Broken Hill, New South Wales, Australia**

Gemmy, very well terminated, 5.0×2.0cm orange crystal in galena. The specimen came from the Albert Chapman Collection to Peter Bancroft to the Barlow Collection.

***Plate #6-5, Specimen #1138*** **BUSTAMITE – Broken Hill, New South Wales, Australia. (MH)**

***Plate #6-6, Specimen #4902*** **CAFARSITE – Mount Cervandonne, Piedmont, Italy. (MH)**

## CAFARSITE (CALCIUM IRON TITANIUM ARSENITE HYDRATE)

**#4902 4.2×3.6×2.6cm Plate #6-6**

**Mount Cervandonne, Piedmont, Italy**

Cafarsite is named for some of its constituent elements. The Barlow specimen has a large (3.2×2.6×2.4cm) dark brown octahedron with cube faces perched on a nearly equally sized, selectively etched but gemmy, smoky quartz crystal.

***Plate #6-7, Specimen #2237*** **CALCITE – Florita Limestone Quarry, near LeCanto, Hernando County, Florida. (MH)**

## CALCITE (CALCIUM CARBONATE)

**#2237 8.5×7.0×6.5cm Plate #6-7**

**Florita Limestone Quarry, near LeCanto, Hernando County, Florida**

Although calcite ranks among the commonest of minerals, this calcite is astonishing. It consists of gemmy to translucent, complex, pale yellow calcite crystals to 2.0cm grown in columns arranged at the corners of a square. At the apex of the specimen, and at a midlevel, the crystallized columns widen broadly like blossoms on a vine. Each calcite blossom consists of a hollow cuplike growth around the crystallized columns. Each ends at the same level with its neighbors. A hollow tube extends along the core of each column where a palm tree root started the calcite crystallization.

*Plate #6-8, Specimen #1941* **CASSITERITE – Cornwall, England. (MH)**

*Plate #6-9, Specimen #5066* **CHALCOPYRITE – Mine à Giraud, La Gardette, Bourg d'Oisans, Isère, France. (MH)**

### CASSITERITE (TIN OXIDE)

**#1941 6.8×3.5×3.0cm Plate #6-8**

**Cornwall, England**

Rarely does a collector encounter as aesthetic a specimen as this one. A commonly referred to, but seldom seen, symmetrical twin has historically been called the "tin beak" in allusion to its resemblance to a baby bird with its mouth open for food. This specimen appears to be such an unusual twin. In spite of the abundance of cassiterite mined in Cornwall, fine crystal specimens like this are decidedly uncommon. From the Richard Kelly Collection.

American mining has its roots in Cornwall. Mining districts throughout the world benefited by the "Cousin Jacks" from Cornwall who supplied the hard rock mining expertise to develop properties, be they silver, gold, or lead-zinc deposits.

As a major tin deposit, the Cornwall mines should have flooded the world with good cassiterites, the only commonly crystallized tin mineral. Yet, fine cassiterites from here are decidedly uncommon. So, the Barlow cassiterite is much less common than one would normally assume. *R.W.J.*

### CHALCOPYRITE (COPPER IRON SULFIDE)

**#5066 9.0×8.0×6.0cm Plate #6-9**

**Mine à Giraud, La Gardette, Bourg d'Oisans, Isère, France**

Outstanding chalcopyrite crystals are rare, even though the mineral constitutes a primary ore of copper. Found about 1900, this is a group of four, large, interpenetrating wedge-shaped sphenoidal crystals recovered during the reworking of the mine under the direction of A. Bordeaux, noted mining engineer.

### CLINOHUMITE (MAGNESIUM IRON SILICATE FLUORIDE HYDROXIDE)

Clinohumite is a member of the humite group of minerals norbergite, humite, clinohumite, and chondrodite, which generally occur as small granular crystals in limestone or dolomite in contact with igneous intrusions. They are often highly fractured and commonly filled with inclusions. As a result, cut gems of the humite group of minerals are almost always small (1-3ct), and generally only chondrodite has crystals the quality and size necessary to yield facetable gem-quality material. Therefore, the clinohumite described here is significant in that it is a faceted gem, one of only a few known.

*Plate #6-10, Specimens #B2 and #3721* **CLINOHUMITE – The Pamirs, Kazakhstan. (MH)**

## CLINOHUMITE, LOOSE GEM

**#B2 4.95×4.92×3.82mm .71ct**
**Plate #6-10 (left)**

Cut: Square barion.

Color: Intense, medium dark, slightly brownish yellow.

Clarity: Slightly included.

## CLINOHUMITE, NATURAL CRYSTAL

**#3721 1.1×0.9×0.7cm Plate #6-10 (right)**

**The Pamirs, Kazakhstan**

As rock-forming minerals, all members of the humite group tend to be murky to opaque. Clinohumite, a rarer member of the group, is no exception. This nearly flawless single crystal is one of the best existing specimens.The mineral is intense golden-orange and is bounded by innumerable crystal faces.A natural jewel.

*Plate #6-11, Specimen #2698* **CREEDITE – Santa Eulalia, Chihuahua, Mexico. (MH)**

## CREEDITE (CALCIUM ALUMINUM SULFATE FLUORIDE HYDRATE)

**#2698 4.0×3.0×3.0cm Plate #6-11**

**Santa Eulalia, Chihuahua, Mexico**

The Barlow Collection probably has the world's largest selection of creedite specimens — virtually all of which are of high quality up to "top of the line." One specimen is described in Chapter 3, Plate #3-28, and five others are described in Chapter 13, but one more specimen should be mentioned. Specimen #2698 consists of two very large single crystals arranged in a "V" shaped group (2.0×2.0cm) on the edge of a shelf of matrix partially coated with much smaller creedite crystals. The creedite has a light amethystine hue and is nearly transparent. The unusual feature of this group is that the individual terminations mimic a steep pyramid.

## DIAMOND (CARBON)

Diamond has been a significant mineral for thousands of years. World-class deposits have been found in Siberia, South America (Brazil, Venezuela), India, Australia, southern and central Africa (especially South Africa, which is the most productive source of gem-quality diamonds at the present time), and the Northwest Territories, Canada (where the area surrounding Yellowknife appears to represent the next premier source of gem-quality diamonds well into the 21st century). Diamonds have been found in Arkansas, Michigan, and Wisconsin where geological exploration is presently being conducted.

Diamond, which forms under extremely high temperatures and pressures similar to those conditions that occur approximately 200 miles below the earth's surface, are generally found in a rock called kimberlite. Weak zones in the earth's crust, commonly along old plate boundaries, are favored areas for the diamond-bearing rock to "explode" to the earth's surface, ripping off clasts of existing rock as the material rises. Nearer the earth's surface, the fissures become more circular and are known as kimberlite pipes. It is in these kimberlite pipes, and in the alluvial deposits derived from the pipes during weathering and erosion, where diamonds are found. Not all kimberlite pipes contain diamonds, and when diamonds are found, they are generally too small or contain too many inclusions to be considered gem quality.

Diamonds crystallize in the isometric (cubic) system and occur in octahedra, dodecahedra, and combinations of other crystal forms. Occasionally, spinel twinning during crystal growth results in a flattened octahedron; such flattened diamonds are called "macles." The crystal faces of a diamond are often rounded and may contain trigons, believed to have formed during the crystal growth process. Diamonds may be colorless, but they generally occur in a rainbow of shades and colors — yellow, pink, red, green, brown, blue, and black. Diamonds may contain many inclusions. Bort is one such example where the inclusions are graphite particles rendering the diamond useless except for industrial purposes.

***Plate #6-12*** **DIAMOND**
***Specimen #5016*** **(left) – Kimberley, South Africa.**
***Specimen #D2Q*** **(center) – Kinasha Mine, Zaire.**
***Specimen #5782*** **(right) – Mimy Region, Yakutia, Russia. (MH)**

### DIAMOND

**#5016 1.5×1.5×1.2cm 212ct Plate #6-12 (left)**
**Kimberley, South Africa**

A spectacular diamond crystal composed of a smoky-gray portion and white portion. Approximately 75% of the crystal is the gray bort; the remainder of the crystal is white. The crystal has prominent triangular striations that are continuous from the bort into the white portion of the crystal. Some contacts between the gray and the white portions are smooth planar surfaces suggesting crystallographic faces. Elsewhere the contact is highly irregular; in these areas the striations do not continue across the contact. This is a most unique specimen.

**#D2Q 2.0×1.3×1.2cm 24.08ct Plate #6-12 (center)**
**Kinasha Mine, Zaire**

A unique specimen composed of two intergrown 1.1cm cubes with a butterscotch-yellow color. The surface of the cubes is very pitted with numerous tiny lustrous faces. There is a general reflection in the octahedral positions, suggesting that the cubes are produced by intergrown octahedral crystals. A very interesting specimen.

**#5782 1.3×1.0×1.0cm 7.12ct Plate #6-12 (right)**
**Mimy Region, Yakutia, Russia**

A spectacular smoky-gray-to-black octahedron, 8.0mm on an edge, with minor black inclusions. The sharp octahedral faces have prominently striated triangular areas in the center of the faces with relatively smooth corners of the octahedra.

## DIAMOND

**#T228 1.2×1.2cm 6.4ct Plate #6-13**

**Tongoma, Sierra Leone**

A superb, flawless, colorless crystal with sharp octahedral faces to 1.2cm. The faces show no growth lines. This is an exquisite, brilliantly lustered crystal. Acquired from Bill Larson in 1975.

***Plate #6-13, Specimen #T228***
**DIAMOND – Tongoma, Sierra Leone. (MH)**

***Plate #6-14*** **DIAMOND**
***Specimen #3419*** **(left) – Mir pipe, Siberia, Russia.**
***Specimen #3385*** **(center) – Mir pipe, Siberia, Russia.**
***Specimen #3386*** **(right) – Mir pipe, Siberia, Russia. (MH)**

**#3419 3.0×2.5×1.1cm**
**Plate #6-14 (left)**

**Mir pipe, Siberia, Russia**

A brilliant 7.0mm octahedron on greenish gray matrix. Four octahedron faces are exposed. Each face has concentric, steplike, triangular faces. These three matrix specimens are stream-rounded pebbles of the unaltered diamond-bearing kimberlite.

**#3385 3.5×2.5×2.0cm Plate #6-14 (center)**

**Mir pipe, Siberia, Russia**

A superb, colorless, octahedral diamond crystal 8mm on an edge in a greenish gray kimberlite matrix. The diamond has prominent triangular growth lines and pits on the crystal faces.

**#3386 3.0×2.5×1.5cm Plate #6-14 (right)**

**Mir pipe, Siberia, Russia**

A brilliantly lustered, colorless, twinned octahedron (macle) diamond crystal 6.0mm on an edge. One triangular octahedron face is exposed in the greenish gray matrix. Fine triangular growth lines are evident on the crystal face.

## DIAMOND

**#B-3 6.47ct**
**Plate #6-15 (upper)**

**Sierra Leone, Africa**

A slightly light yellowish green crystal with prominently curved faces. The major crystal form is a twinned dodecahedron with curved faces producing a rather rounded crystal. Most of the faces are smooth, but several faces that cross the plane of the spinel twin are striated.

**#B-4 .87ct**
**Plate #6-15 (bottom left)**

**Sierra Leone, Africa**

A medium dark yellow crystal with curved octahedral faces. This is a superb example of a spinel twin (macle) with a sharply defined twin plane.

***Plate #6-15*** **DIAMOND**
***Specimen #B-3*** **(upper) – Sierra Leone, Africa.**
***Specimen #B-4*** **(bottom left) – Sierra Leone, Africa.**
***Specimen #B-5*** **(bottom center) – Kimberley, South Africa.**
***Specimen #B-6*** **(bottom right) – Kimberley, South Africa. (MH)**

**#B-5 1.44ct Plate #6-15 (bottom center)**

**Kimberley, South Africa**

A slightly orange-brown crystal with prominently curved (bulging) faces. The major crystal shape is a somewhat flattened, rounded octahedron modified by rounded cubic faces around the girdle.

**#B-6 .59ct Plate #6-15 (bottom right)**

**Kimberley, South Africa**

A superb, clear, octahedral crystal with slightly blue-green patches that appear to be confined to the outer "skin" of the crystal. The crystal has prominently striated octahedral faces. A very striking specimen.

***Plate #6-16, Specimen #1852*** **DIAMOND – Kimberley, South Africa. (MH)**

**#1852 4.0×4.0×3.0cm Plate #6-16**

**Kimberley, South Africa**

A very fine, 7.0mm, slightly yellowish white, cloudy octahedral diamond on a greenish gray matrix of rather fresh kimberlite. The diamond crystal shows prominently striated triangular octahedral faces. The striations and somewhat rounded edges may have been formed by etching of the crystal during its transport toward the earth's surface. The matrix contains olivine, enstatite, reddish pyrope, green chrome diopside, phlogopite, and chromite. This is an excellent matrix specimen from this famous diamond-producing area.

## DIASPORE (ALUMINUM OXIDE HYDROXIDE)

**#5072 5.7×4.0×2.0cm Plate #6-17**

**Mugula Province, Turkey**

Diaspore is usually an inconspicuous component in rock, but at least five localities in the world have yielded good crystals. A small prospect in Turkey has produced unquestionably the best specimens. Barlow #5072 is exceptional not for its size, but because it consists of an etched, light-to-pale-brown, gem-quality "V" twin. The longer arm of the twin has an etched termination.

***Plate #6-17, Specimen #5072*** **DIASPORE – Mugula Province, Turkey. (MH)**

## FLUORITE (CALCIUM FLUORIDE)

**#896 9.5×7.5×6.0cm Plate #6-18**

**Weardale, Durham, Cumbria, England**

Fluorite is a very "collectible" species with literally hundreds of localities having produced good specimens. Yet, no locality can compete with Weardale's fluorescent fluorite. Even the term "fluorescence," as well as the element fluorine, owe their origins to the name of the mineral — fluorite. As a crystal specimen, this one can "hold its head high" as it consists of two major interlocking crystals. Each crystal in turn is geometrically twinned by a slightly smaller crystal. The cluster is on a matrix of smaller but sharply developed fluorites. The rare feature of this specimen is its fluorescence due to a trace of rare earth elements. Under incandescent light, the specimen shows an amethyst color that on the crystal edges has a faint rosy-purple glow. However, when the specimen is in direct sunlight, it seems to leap from your hands with a bright, eerie indigo-to-purple glow that even in light shade fails to release the crystals from its excitement. A choice specimen showing an extraordinary phenomenon.

***Plate #6-18, Specimen #896*** **FLUORITE – Weardale, Durham, Cumbria, England. (MH)**

The rare feature of this specimen is its fluorescence due to a trace of rare earth elements. Under incandescent light, the specimen shows an amethyst color that on the crystal edges has a faint rosy-purple glow. However, when the specimen is in direct sunlight, it seems to leap from your hands with a bright, eerie indigo-to-purple glow that even in light shade fails to release the crystals from its excitement.

## KÖTTIGITE (ZINC ARSENATE HYDROXIDE)

**#3340 10.0×5.0×5.0cm** **Refer to Chapter 13, Plate #13-6**

**Mina Ojuela, Mapimi, Durango, Mexico**

Köttigite was found sparingly at this locality as radiating sprays of metallic-looking blue-gray needlelike crystals. Crystals to 6.0cm have been found, but most sprays reach only 2.0cm. This specimen has a well-colored thick spray of crystals to 2.2cm with a spread of 2.0cm. The spray rests in an open cavity within the typical hard brown matrix rock. Many köttigites collected here have a dull coating that probably resulted from repeated undulations of ground water as wet and dry seasons occur. The Barlow specimen, however, is pristine in luster and hue.

## KYANITE (ALUMINUM SILICATE)

**#5027 10.0×9.0×6.0cm Plate #6-19**

**Barra do Salinas, Minas Gerais, Brazil**

An aggregate of large kyanite crystals in glassy quartz matrix. The largest kyanite crystal measures 8.0×1.0×0.5cm. Crystals exhibit a striking deep blue color with a gemmy appearance where relatively free of fractures. Numerous <0.1cm, resinous, reddish brown grains at one end of the specimen could be gemmy staurolite.

***Plate #6-19, Specimen #5027*** **KYANITE – Barra do Salinas, Minas Gerais, Brazil. (MH)**

## KYANITE, LOOSE GEM

**#514 35×20×12mm 105.58ct**
**Plate #6-20 (left)**

**Barra do Salinas, Minas Gerais, Brazil**

Cut: Rectangular step cut.

Color: Medium light, slightly greenish blue with dark blue "line" running through center.

Clarity: Very slightly included.

**#510 11.20×20.09×6.34mm**
**16.97ct Plate #6-20 (bottom)**

**Barra do Salinas, Minas Gerais, Brazil**

Cut: Rectangular step cut.

Color: Slightly intense, medium dark blue (irregular color patches).

Clarity: Slightly included.

**#513 20×15×10mm 37.28ct**
**Plate #6-20 (right)**

**Barra do Salinas, Minas Gerais, Brazil**

Cut: Rectangular step cut.

Color: Medium intense, medium dark, slightly greenish blue with dark blue "line" through center.

***Plate #6-20*** **KYANITE**
***Specimen #514*** **(left) – Barra do Salinas, Minas Gerais, Brazil.**
***Specimen #510*** **(bottom) – Barra do Salinas, Minas Gerais, Brazil.**
***Specimen #513*** **(right) – Barra do Salinas, Minas Gerais, Brazil. (MH)**

## MANGANITE (MANGANESE OXIDE HYDROXIDE)

**#789 8.0×5.5×3.0cm Plate #6-21**

**Ilfeld, Harz, Germany**

Ilfeld manganites are commonly regarded as being among the classic minerals a display collection should have. This piece consists of brilliant black tubular manganite crystals (to 2.3×1.5×1.4cm) in a cluster with nicely contrasting plates of white barite. This specimen is unusual because of the three-dimensionally open exposure of its main crystals. While the collection has larger, very fine specimens, this one is the collector's favorites. Acquired from David Wilber in 1979.

***Plate #6-21, Specimen #789*** **MANGANITE – Ilfeld, Harz, Germany. (MH)**

## NAMBULITE (LITHIUM SODIUM MANGANESE SILICATE HYDROXIDE)

**#3138 5.0×4.0×3.0cm**

**Refer to Chapter 3, Plate #3-69**

**Kombat Mine, Namibia**

Nambulite is an extremely rare lithium manganese silicate mineral, being found in only two locations associated with manganese oxide ores — the Funakozawa Mine in northeastern Japan, where it occurs in crystals up to 8mm long, and at the Kombat Mine in Namibia, where crystals up to 3cm long are found. Arem (1977) stated that as of 1977, no gems had been cut from nambulite, but gems up to approximately 10ct could possibly be cut from some of the South African material. The Barlow Collection contains one such faceted nambulite.

## NAMBULITE, LOOSE GEM

**#3138-1 3.35×6.74×2.70mm .58ct Plate #6-22**

**Kombat Mine, Namibia**

Cut: Rectangular octagon faceted by Art Grant.

Color: Intense, medium dark, slightly pinkish orange.

Clarity: Very slightly included.

***Plate #6-22, Specimen #3138-1*** **NAMBULITE, loose gem – Kombat Mine, Namibia. (MH)**

## NICKEL-SKUTTERUDITE (NICKEL ARSENIDE)

**#5342 9.0×7.0×5.0cm Plate #6-23**

**Pohle Mine, Erzgebirge, Saxony, Germany**

Chloanthite variety of nickel-skutterudite is a well-known ore mineral, but some specimens deserve attention. This specimen is composed of sharply crystallized dark metallic-gray cubic crystals with triangular octahedral faces (to 3mm), grouped in conical dendritic 1.5-2.0cm clusters, profusely covering the specimen, reminiscent of a beautiful Christmas tree.

***Plate #6-23, Specimen #5342*** **NICKEL-SKUTTERUDITE – Pohle Mine, Erzgebirge, Saxony, Germany. (MH)**

## PARISITE-(Ce) (calcium cerium lanthanum fluorocarbonate)

Parisite-(Ce), although found in many places, is generally known in quantities that would serve only academic interests, except perhaps at the Snowbird Mine in Montana. Parisite-(Ce) has also been found in veinlets and pockets within the carbonaceous shales hosting the emerald deposits of the Muzo district, Colombia. Cut gems are very rare.

### PARISITE

**#2746 3.1×3.0×2.0cm Plate #6-24 (left)**

**El Indio Pit, Muzo Mine, Quipama, Colombia**

This specimen is among the largest well-formed crystals known. The crystal is translucent brown with numerous gemmy areas. It tapers from a broad base to a truncated termination. It is also corrugated as many of its kind typically are. When viewed down the vertical axis, the crystal shows a strong coppery-orange color and pearly luster. A thin columnar crystal has grown parallel with the main crystal, but the *c*-axis of the smaller crystal is oriented obliquely to the main crystal.

**#5741 6.0×6.0×4.0cm Plate #6-24 (right)**

**El Indio Pit, Muzo Mine, Quipama, Colombia**

A 1.5cm sharply terminated crystal on a carbonaceous shale with quartz matrix.

***Plate #6-24, Specimen #2746 Left, Specimen #5741 Right***
**PARISITE – El Indio Pit, Muzo Mine, Quipama, Colombia. (MH)**

***Plate #6-25, Specimen #5741-1*** **PARISITE, loose gem – Muzo, Columbia. (MH)**

### PARISITE, loose gem

**#5741-1 6.68×6.00mm 1.38ct Plate #6-25**

Cut: Modified round cut.

Color: Medium intense brown with slight orange overtones.

Clarity: Very slightly included.

This is the largest cut parisite gemstone known to the authors to date.

## PHOSGENITE (LEAD CARBONATE CHLORIDE)

**#T389 2.6×2.2×2.0cm**

**Cromford, Matlock, Derbyshire, England**

Phosgenite is one of the classic rare minerals known best from Sardinia, where it was found in commercial quantities in the 19th century. Phosgenite from Matlock is rare, known in very few clusters. This cluster of colorless, gemmy, complex cylindrical crystals is outstanding. The grouping consists of five main crystals of various lengths, all in parallel growth with the terminations offset. Several kinds of etch pits decorate all faces, with some resembling arcuate dendritic hoarfrost. Acquired from the Richard Kelly Collection.

## PHOSPHOPHYLLITE (ZINC IRON MANGANESE PHOSPHATE HYDRATE)

**#T379 2.0×1.5×1.3cm Plate #6-26**

**Potosi, Bolivia**

The crystal is a superb, gemmy, blue-green, fish-tail twinned thumbnail specimen, repaired. The two elements of the twin are almost equally developed, producing a very attractive specimen. This fine example of a rare species is from the classic locality of Potosi, Bolivia, where it was formed by hydrothermal fluids that produced the famous Bolivian tin deposits.

***Plate #6-26, Specimen #T379*** **PHOSPHOPHYLLITE – Potosi, Bolivia. (MH)**

## PYRARGYRITE, LOOSE GEM

**5.58×7.42mm 27.29ct**

Cut: Modified fan cut.

Color: Dark gray (reflected light) intense red (transmitted light). Faceted by Mike Gray.

Clarity: Slightly included.

## QUARTZ (SILICON DIOXIDE)

**#1547 7.0×7.0×6.5cm Plate #6-27**

**Yamanashi Ken, Honshu, Japan**

Quartz is not a rare mineral, but high-quality visible Japan law twins of quartz are uncommon. This specimen consists of bright transparent crystals variously growing out of the matrix with an exceptional “right angle” twin, 4.5×3.5cm, dominating the well-proportioned specimen. Such twins were originally named for an important source in La Gardette, France, but a major find in Japan in the 19th century generated the now familiar term, Japan law twin. Purchased from John Patrick. The specimen was brought from Japan by John Patrick's wife, Masako.

***Plate #6-27, Specimen #1547*** **QUARTZ – Yamanashi Ken, Honshu, Japan. (MH)**

## QUARTZ

**#1875 5.3×2.3×2.0cm Plate #6-28**

**Jefferson County, Montana**

Amethyst and smoky quartz are not strangers; they often can be found together. However, intensely colored amethyst and deep smoky quartz, together in a scepter combination, is very worthy of notice. This amethyst is doubly terminated with very interesting arcuate, terraced growth on the termination. It has a deep rose-purple axial thread that grades into the generally rose-amethyst-tinted crystal. The shaft crystal is impervious smoky-black, a combination rarely matched in the world. From the Collman Collection, Northland College, Ashland, Wisconsin.

***Plate #6-28, Specimen #1875*** **QUARTZ – Jefferson County, Montana. (MH)**

## RHODIZITE

**(CESIUM ALUMINUM BERYLLIUM BORATE)**

**#5355 3.2×2.1×1.7cm Plate #6-29**

**Anjanabonoina, Madagascar**

Rhodizite is a mineral with an unusual chemical composition and is rarely found in granite pegmatites. This enormous, complexly terminated, greenish yellow crys tal has both octahedral and dodecahedral forms.

***Plate #6-29, Specimen #5355*** **RHODIZITE – Anjanabonoina, Madagascar. (MH)**

***Plate #6-30, Specimen #5578*** **SEAMANITE – Chicagoan Mine, Iron County, Michigan. (MH)**

## SEAMANITE

**(MANGANESE PHOSPHATE BORON HYDROXIDE)**

**#5578 4.5×3.5×2.0cm Plate #6-30**

**Chicagoan Mine, Iron County, Michigan**

Abundant lustrous pale pink, needlelike seamanite crystals to 7.0mm are randomly oriented on the surface of somewhat brecciated and porous black hematite. A fine pinkish material, possibly also seamanite, cements some of the hematite fragments together. Several 1.5cm areas are covered with a "jackstraw" arrangement of lustrous seamanite crystals. An outstanding specimen of this very rare mineral from the only reported locality. Label signed by A.E. Seaman after whom the mineral, as well as the Michigan Technological University Museum, was named. The handwriting on the label was verified by Seaman's great granddaughter, Donna Cole.

## SEMSEYITE (LEAD ANTIMONY SULFIDE)

**#5441 5.4×4.0×3.0cm**

**Herja, Maremures, Transylvania, Romania**

Semseyite is not only a rare mineral, it is an unusual mineral. Its chemistry is not particularly odd, but its crystal groups are. This specimen is typical except for the large size of the crystals (to 1.4×1.2×0.1cm). The clusters consist of bright silvery gray semseyite crystals in subparallel arcuate growth. Individual crystals typically are elongated platy hexagons joined in their largest faces but with a serpentine axis rather than linear axis. Some of the crystal groups have a slight radial development. The specimen (5.5×4.0×3.0cm) has two side-by-side clusters. Minor black sphalerite, gray galena, and botryoidal tan siderite are associated.

***Plate #6-31, Specimen #2414*** **STANNITE – Potosi, Bolivia. (MH)**

***Plate #6-32, Specimen #1940*** **TORBERNITE – Old Gunnislake Mine, Gunnislake, Cornwall, England. (MH)**

## STANNITE (COPPER IRON TIN SULFIDE)

**#2414 6.0×4.0×3.5cm Plate #6-31**

**Potosi, Bolivia**

Even at tin mines, crystallized stannite is seldom encountered and the mineral is indistinct. This specimen consists of an unusually large, 3.0×3.0×2.5cm, conical grouping of black multitwinned crystals associated with a few bright cassiterite crystals and some drusy quartz. From the collection of Dick Jones, noted Arizona field collector and mineral dealer, now deceased.

## TORBERNITE (COPPER URANYL PHOSPHATE HYDRATE)

**#1940 5.0×4.5×3.0cm Plate #6-32**

**Old Gunnislake Mine, Gunnislake, Cornwall, England**

Torbernite from the Old Gunnislake Mine used to be the standard of quality for the species before the crystals from Zaire were known. This antique and classic specimen is unique for its exceptional placement of bright, square, tabular green crystals arranged in a "house of cards" intergrowth. The torbernite is on a typical matrix of milky-white-to-gray phantomed quartz crystals. Rare and aesthetic. Acquired from the Richard Kelly Collection.

## VESZELYITE (COPPER ZINC PHOSPHATE HYDROXIDE HYDRATE)

**#3093 2.7×1.5×2.0cm Plate #6-33**

**Black Pine Mine, Philipsburg, Granite County, Montana**

As has been the case with many rare minerals in this chapter, one locality is paramount in producing specimens. The Black Pine Mine took on this distinction for veszelyite. While larger crystals are known, this specimen is one of the aesthetic matrix specimens. The bright veszelyite, 1.5×1.0cm, is a midnight blue in color and occurs in a subparallel crystal cluster canted forward with respect to a translucent greenish yellow group of quartz crystals.

***Plate #6-33, Specimen #3093*** **VESZELYITE – Black Pine Mine, Philipsburg, Granite County, Montana. (MH)**

## SPECIAL JEWELRY

The Barlow gem and mineral collection has many very fine jewelry and artifact specimens. The following specimens have a special meaning to Barlow.

**#SJ1 Red Beryl Drop and Pin Plate #6-34**

Two very fine modern pieces of jewelry in the Barlow Collection are unique. A brooch combining a natural specimen of Wah Wah Mountain red beryl crystals is set in an angular 18-karat gold frame, 3.0×4.0cm. This is complemented by two 18-karat gold flowers with faceted red beryl centers and five diamonds on 18-karat gold stems arching over the red beryl crystal specimen.

***Plate #6-34, Specimen #SJ1*** **RED BERYL DROP AND PIN. (MH)**

***Plate #6-35, Specimen #SJ2*** **RED BERYL RING. (MH)**

**#SJ2 Ring Plate #6-35**

The second outstanding piece is a 18-karat ring composed of a spiral band set with small diamonds climaxing in a brilliant-cut 1.3ct red beryl centerstone with two emerald-cut red beryls on a subsidiary band. Barlow commissioned Martha Gilchrist of Dallas, Texas to make these two very special jewelry pieces in 1980.

*Plate #6-36, Specimen #SJ3* **CHRISTMAS BERYL RING – Wah Wah Red Beryl and Muzo Emerald. (MH)**

*Plate #6-37, Specimen #SJ4* **INITIALED EMERALD RING – Muzo Emerald. (MH)**

### #SJ3 Christmas Beryl Ring Plate #6-36

### Wah Wah Red Beryl and Muzo Emerald

Two emerald-cut red beryls and one emerald-cut emerald set in 18-karat gold. Barlow commissioned a goldsmith in Los Angeles to produce this "Christmas beryl ring" in 1979.

### #SJ4 Initialed Emerald Ring Plate #6-37

### Muzo Emerald

This very special 18-karat gold ring is set with one cabochon-cut emerald. The emerald is a carved intaglio with gold-filled initials "JB." It was a gift to Barlow in 1984 from Gerhardt Becker and Hans Jacob Klein of Idar-Oberstein to repay the favor of flying Becker and Klein from Tucson for a bird's eye view of the Grand Canyon.

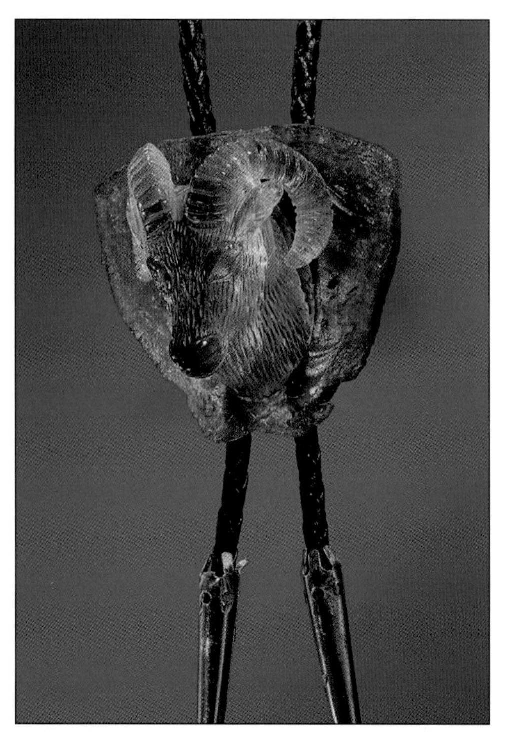

*Plate #6-38, Specimen #SJ5* **TOURMALINE CARVING – Idar-Oberstein. (MH)**

### #SJ5 Tourmaline Carving Plate #6-38

### Idar-Oberstein

### 5.0×4.5×5.0cm

In Barlow's words, "As a big game hunter, I climbed mountain peaks from Alaska to Baja, Mexico, stalking the mighty ram (the four species inhabiting the North American continent) known as the Grand Slam. I bagged each one with one shot. It took 18 years to accomplish this goal to become #46 on the grand slam list and the first Wisconsin hunter to be a grand slammer."

Scaling mountain peaks and dry river beds picking up pretty rocks was the beginning of the Barlow mineral collection. Because tourmaline is one of his favorite minerals, Barlow decided his bolo tie would be a gem tourmaline carved ram and must be a piece of gem rough from his collection. The carving had to be the finest. In 1986, Barlow sent a model to Gerhardt Becker of Idar-Oberstein with the rough, and the result was spectacular. Many have admired it, and many have tried to purchase or trade for it. Dave Wilber has requested that when Barlow dies, the bolo tie will go to Dave Wilber. Barlow's comment, "Good luck, Dave."

### Is It Bola or Bolo?

In spite of the error promulgated by certain individuals in Arizona to have the state legislature pass a bill in 1971 making the "bola" tie the state official neckwear, it is correctly a *bolo tie*.

The bolo tie did not originate in Arizona. It was worn by Native Americans from both North and South America long before Arizona became a state.

In *Webster's New World Dictionary, bola* is defined as "a weapon made of a long cord or throng with heavy balls at the end used for throwing at and entangling cattle, etc."

*Bolo tie* in *Webster's New World Dictionary* is defined: "from its resemblance to the bola, a man's string tie held together with an ornamented slide device."

Also refer to *The Random House Dictionary of the English Language,* 2nd edition, "Bolo tie, a necktie of thin cord fastened in front with an ornamental clasp or other device."

***Plate #6-39, Specimen #SJ6*** **JADEITE – Palmilla River area, Guatemala. (MH)**

***Plate #6-40, Specimen #SJ6*** **JADEITE – Palmilla River area, Guatemala. (Upside down.) (MH)**

## JADEITE

**#SJ6 3.7×3.1×2.5cm Plates #6-39 and #6-40**

**Palmilla River area, Guatemala**

The prehistoric jade carving was worn as a neck drop many centuries ago. Dr. Peter Keller writes from the Bowers Museum of Cultural Art, "There is no question in my mind that the material is typical of jadeite from Palmilla River area in Guatemala. This is the material that was studied by Foshag in the mid 1950s."

"As far as the material as an artifact goes, our curator believes that it is authentic and is typical of Mayan objects from Southern Mexico. Try looking at the piece oriented as a human face, and then turning it upside down. You should see a bird."

## REFERENCE

Arem, J.E. (1977) *Color Encyclopedia of Gemstones*, Van Nostrand Reinhold, New York, 149 pp.

# *PART IV*

## *THE NATIVE ELEMENT SUITE*

### *Introduction*

The native elements, in this case metals, have long fascinated mankind and are among the first of earth's minerals to be collected and preserved. Gold, silver, and platinum have been elevated to the status of valued media of exchange and objects of personal adornment. Their compounds, as expressed in Chapter 10, not only may be important ores, but are much sought by collectors due to their rarity and beauty.

The native metals lend themselves to a separate grouping, particularly since they all form in the isometric or cubic system of crystallography. Descriptions of the Barlow native metals offered by authors include crystallographic references more easily understood with reference to crystal drawings. The editor recommends that the reader refer to:

Klein, C. and C. Hurlbut (1993) *Manual of Mineralogy* (after James D. Dana), 21st ed., John Wiley and Sons, New York, 681 pp.

Berry, L.G., B. Mason, and R.V. Dietrich (1983) *Mineralogy*, 2nd ed., W.H. Freeman and Co., San Franciso, California, 561 pp.

> *Many an amateur, far from improving his fortune by gathering together a fine lot of artistic treasures, has actually maneuvered himself into comparative poverty in the process.*
>
> *-Douglas and Elizabeth Rigby*

***Plate #7-1*** **GOLD SPECIMENS (MH)**

# Chapter 7

# Gold

*by Richard Thomssen*

Over a period of 20 years, F. John Barlow has put together a remarkable collection of gold specimens from localities around the globe. Most of the specimens have been obtained from contemporary sources, with some localities represented by numerous pieces. The acquisition of several private collections early in Barlow's efforts to build his collection broadened the representation of world gold localities. This is especially true of those localities that have yielded specimens of well-crystallized gold. Gold won from placer sources has received special attention in the collection, particularly if those pieces show crystal form. The resulting collection of 1,250 specimens provides abundant evidence of Barlow's interest in gold's form and style, as well as its being a good worldwide reference. Space permits only a representative selection of specimens with photographs and descriptions from this large collection.

*The resulting collection of specimens provides abundant evidence of Barlow's interest in gold's form and style, as well as its being a good worldwide reference.*

In February of 1974 Barlow acquired his first gold specimen. This first example of gold from the 16 to 1 Mine in California's famous Alleghany District added fuel to the collecting fire. A month later in March, Barlow purchased the first of many specimens from Ernest Butler, a gold dealer from Oregon, at the Tucson Gem and Mineral Show. During the next two years significant specimens, including some superb crystallized golds from Hungary, California, Colorado, Africa, Venezuela, Washington, Canada, and Romania were added to the collection.

The congregation of dealers from around the world at the 1975 and 1976 Tucson Shows was particularly helpful to John in acquiring pieces from unusual locations. The wide variety of locations and style of gold acquired indicates that John was undergoing a program, albeit informal, of familiarization and education into the nature of gold. This process has continued up to the present and is responsible for the depth and eclectic nature of Barlow's gold collection.

## The Right Place at the Right Time

An incident at the California Federation of Mineralogical Societies' Show at the Cow Palace in San Francisco in July 1976 must have appealed to John's trophy hunting instincts. A miner appeared at this show with a handful of superbly crystallized gold specimens. These, complete with dirt and iron oxides, were offered to a dealer who referred the miner to other potential buyers, and on the second offering of what became a feeding frenzy, several of the better pieces were purchased by a more knowledgeable dealer. With great dispatch, the dealer sought John as a possible customer for an immediate resale of the cream of these gold specimens. Upon consummating a purchase of the best, John carefully wrapped the specimens, and, with a certain finality, put them in his pocket. These specimens remain in Barlow's collection as superb representatives of Colorado Quartz Mine crystallized gold (Plates #7-42 and #7-43) (see also Kampf and Keller, 1982).

*John, not to let opportunity pass untested, inquired of the locals about gold specimens and was soon rewarded for his efforts through contact with an Eskimo who had some placer gold.*

While on an Alaskan hunting trip in February of 1979, John reached McGrath in the valley of the Kuskokwim River. This settlement, typical of central Alaskan towns, is accessible during the summer only by plane or boat from the Bering Sea, some 480 miles away, and in winter by plane or dog sled. It is a truly isolated place, but it is the rallying point for gold prospectors from the central portion of the Kuskokwim Mountains, hosting several gold districts. John, not to let opportunity pass untested, inquired of the locals about gold specimens and was soon rewarded for his efforts through contact with an Eskimo who had some placer gold. Several pieces had been christened by the Eskimo. Now, "Big Foot" (#4516) is pictured in Plate #7-25.

Barlow's long association with Ernest Butler, during which he acquired numerous gold specimens from western North America, culminated in November 1979 when John purchased the Butler Collection. This acquisition added a significant number of fine crystallized golds to his collection. Three localities are extensively represented in the Butler suite: Ace of Diamonds Mine, Liberty, Kittitas County, Washington, where gold in brilliant wirelike crystals was found in aggregates up to 186g within calcite seams at the junction of a shear zone and a calcareous layer interbedded in the enclosing shales; 16 to 1 Mine, Alleghany District, Sierra County, California in irregular concentrations within ribbon quartz veins ten or more meters in thickness; and Mariposa County, California (possibly the Diltz Mine), where crystallized gold occurred in fractured areas of massive quartz that had been recemented by calcite (Julihn and Horton, 1940).

Starting in February of 1975 at the Chicago Show, Barlow began an association, which was to span nearly two decades, with Theo and Sharon Manos of St. Troy Consolidated Mines, Ltd. Over the next seven years, John acquired from St. Troy a small number of exceptional gold specimens from Alaska and, more importantly, from various localities in the Siberian portion of Russia. These included some of the best single crystals in John's collection. Though the specimens showed some abrasion from stream action, as is the case with most of the Siberian golds, the thumbnail-sized cube and cluster of cubic crystals are among the best known examples of this uncommon crystal form.

In April 1982, after acquiring a small collection of Siberian golds from St. Troy at the Tucson Show, John traveled with Manos to Moscow to deal for additional specimens. During the ensuing two-and-one-half years and two more trips to Moscow, Barlow acquired an extensive collection of beautifully sculptured Russian gold nuggets, as well as many fine crystallized pieces. Numerically, golds from Siberia represent the largest single source in the Barlow gold collection. This fact alone makes the collection exceptional.

Nearly half a world away, superb, water-worn, single crystals of gold were being liberated from the placers of southern Venezuela. Two entrepreneurs, John Carlson and Roger La Rochelle, were marketing these crystals in the early 1980s as jewelry, complete with soldered-on jump rings. (See "What's New in Gold," *Mineralogical Record*, Volume 18, 1987, for further details of this discovery.) Upon learning of the specimen value of their gold, Carlson contacted John in February 1981 and, after prolonged negotiations, some of the best crystals ended up in the Barlow Collection (Plate #7-57). Barlow acquired additional specimens in February and October of 1981 from Lawrence Conklin, a noted mineral dealer in New York City, who was largely responsible for educating Carlson and La Rochelle about the specimen value of their golds. In October, 1983, John acquired two of the best Venezuelan single crystals (#3730) from Larry Conklin, one of which is the one-inch hopper-growth octahedron pictured in Plate #7-57, second row, center. The second figure (#3731), Plate #7-57, first row, right, was acquired from Allan Caplan.

The locality for the modern Venezuelan golds has been given as "La Gran Sabana, 40 kilometers west of Santa Elena," a nebulous area in the far southeastern corner of the state of Bolivar, near the Brazilian border. Feverish activity is going on in this area now that the Venezuelan army has exported thousands of Brazilian garimperos (itinerant diggers) south across the international border. The Kilometer 88 District, more or less equivalent to "40 kilometers west of Santa Elena" is the label given to this area, which is considered to be one of the largest gold discoveries of the 20th century. Although over the past decade local miners have recovered an estimated 1.6 million ounces of gold, there remains a probable reserve of up to 30 million ounces. The opportunity for obtaining additional monstrous crystals of gold from both placer and hardrock deposits is without equal elsewhere on the globe; John already has his contacts searching.

In January 1981, John purchased his first specimen from a locality near Forest Hill in Placer County, California. The golds from the DeMaria Mine (other names applied to specimens from the immediate area of this mine include the Eagle's Nest and Michigan Bluff Mines) are extremely bright and well crystallized in branching, flattened, twinned octahedra. They are mined from flat veins only a few inches thick, occurring in the medial portion, lying on quartz crystals. Extraction requires double mining with a drift opened above the vein, then the vein is taken up without blasting. The quality of the specimens recovered attests the benefit of using this mining system and to the care with which mining is conducted.

The challenge of obtaining specimens from this locality, without dealing with the avowed sole source, appealed to John's nature and he set out to test market availability. In January 1985, John was led to Auburn, California where he found the "Gold Lady" selling material from the De Maria Mine. Recognizing a significant alternative source for the high-quality golds from this mine, John proceeded to acquire some fine to superb pieces over the next eighteen months from Lee Petretti, the "Gold Lady." In 1988, he acquired additional specimens from J. Gordon McGinley, a well-known mineral prospector, trader, and dealer. Today, John's suite of specimens from the DeMaria Mine certainly rivals any in the world.

To rise above the average, a collection must contain more variety than that obtainable on the current market, no matter how spectacular and numerous the specimens. Classic localities must be represented — and by exceptional examples.

The Monts Apuseni District in Romania (western Transylvania Erzgebirge or Siebenburgen) is the principal gold-producing district of Europe, excluding Russia. Several mining areas within this 30×50km area have produced fine crystallized gold specimens. Among these are the region around Brad, in the southwestern part of the district, and mines in the vicinity of Rosia Montana, formerly known by its Hungarian name, Verespatak.

According to Lieber (1982), mining has been carried out in the region around Brad for more than 2,000 years, with some of the earliest tunnels having been excavated by the Romans. Mining reached its peak during the 18th century and documents date back to 1774. The miners then were usually Romanian farmers with a technical staff composed of Austrians, Germans, and Hungarians. A mining school was founded at Sacaramb (formerly known by its Hungarian name, Nagyag) in 1835, with German as the official teaching language.

The veins and lodes in the Monts Apuseni District are associated with late Tertiary volcanoes and volcanic necks, which cut early Tertiary sediments and volcanic rocks and older sedimentary and metamorphic rocks. Individual veins seldom exceed one meter in

*During the ensuing two and one half years and two more trips to Moscow, Barlow acquired an extensive collection of beautifully sculptured Russian gold nuggets, as well as many fine crystallized pieces. Numerically, golds from Siberia represent the largest single source in the Barlow gold collection. This fact alone makes the collection exceptional.*

width but in places are closely spaced, forming fissure zones and stockworks. Gold of around 700 fineness (70% gold) occurs free or with the rare tellurides sylvanite, nagyagite, and petzite. The element tellurium was first discovered in ore from Sacaramb (Nagyag). Associated sulfides include pyrite, sphalerite, stibnite, tetrahedrite, bournonite, galena, arsenopyrite, and locally, argentite, pyrargyrite-proustite, stephanite, marcasite, and cinnabar. Gangue minerals are quartz and carbonates, including rhodochrosite. Wall rock alteration is of the Comstock-type with zoned propylitization, kaolinization, sericitization, and silicification from fresh wall rock toward the vein.

*To rise above the average, a collection must contain more variety than that obtainable on the current market, no matter how spectacular and numerous the specimens. Classic localities also must be represented — and by exceptional specimens.*

Significant specimens from Rosia Montana in the collection include a fine 9.0cm leaf with numerous growth trigons and local coatings of millimeter-size, tarnished marcasite crystals (refer to Plate #7-16). This piece was formerly in the Alfred M. Buranek Collection. Additional pieces showing fine crystals have been acquired through mineral dealers from collections of the American Museum of Natural History, Lazard Cahn, Sir W. Sergeants, and Carl Bosch (USNM #B31). From the Musariu (also spelled Muszari) Mine near Brad comes a small but choice specimen with bladed crystals showing some growth trigons. It was originally acquired from the Harvard Collection, which, in turn, had acquired it from the L. C. Wyman Collection. An interesting aspect of the provenance of these Romanian golds is that the recycling of old collections is virtually the only means of gathering together specimens from classic localities such as this.

*Gold specimens of unusual mineral associations or occurrence are also included in the Barlow Collection.*

In the United States, gold in payable quantities was first discovered in the southern Appalachian states in placer deposits in Cabarrus County, North Carolina in 1802. The first lode deposit was opened in 1825 in North Carolina's Stanly County.

Specimens dating from this early period of development are rare indeed. In the Barlow Collection, a slightly worn, flattened and twinned, one-inch octahedron from the Reed Gold Mine, Cabarrus County, North Carolina (#T334, Plate #7-52) was found in 1828. This piece was in the Abtheilung Museum Collection (of Humboldt University, Berlin), ultimately finding its way into the Alfred M. Buranek Collection in Salt Lake City. The piece was then traded by Buranek to Bill Larson of Pala Properties who, in turn, sold it to Barlow.

Two gold specimens from Georgia, one from the Dahlonega area and the other from a White County placer, and an early specimen from Virginia are also in the collection. The Georgia specimens were both purchased from dealers who acquired them from private or institutional collections. The Virginia specimen (#1959, Plate #7-53) came out of the Lawrence University Collection, through Jack Baragwanath/Leech, when it was sold to raise funds for the university. The provenance of these historically valuable golds again emphasizes the part that recycling of old collections plays in the building and broadening of modern collections.

Gold specimens of unusual mineral associations or occurrence are also included in the Barlow Collection. Pieces of quartz monzonite porphyry with round spots of bright gold to 5.0mm associated with paratacamite, a rhombohedral copper chloride, from Mina La Farola near Copiapo, Chile were acquired by John from a German dealer, Helmut Weidner. The better than 970 fineness of the gold in these specimens is remarkable and supports the observation prompted by the association with paratacamite that the gold is secondary, having been redistributed by near-surface oxidizing solutions.

## Artifacts Formerly in the Collection

In 1967, John went on an African safari with three friends. John befriended the safari's purser, Abdul Suleman, who offered to obtain anything John wanted in Dar Es Salaam. Abdul's efforts resulted in John's acquiring five pieces of gold lion and leopard claw jewelry and antique ivory carvings. These later were part of the treasure traded for some superb Indian gems that ultimately went to the Smithsonian.

Three years later John went on a tiger hunt in India. After his African experience in buying fine jewelry, he went prepared to buy old Indian gems and jewelry. This trip was very important to John as it encouraged his tremendous interest in minerals and gems. Additionally, it gave him the opportunity to acquire more trading material. During the trip, while staying at the Rambow Palace in Jaipur, John met the maharaja's jeweler, Mr. Lunia, and became friendly with him. Mr. Lunia asked John if he would help his two sons start a gem and jewelry business in America and John promised to try. He also purchased a number of cut gems and some fine jewelry from Mr. Lunia. Barlow later traded these back to one of Lunia's sons in America.

*In February, 1973, Lunia arrived in Appleton with a 1,111 carat carved emerald which was cut in about 1860 for an Indian Prince, Raju Khusal Singhji, of Jaipur, India.*

About a year later John was contacted by one of the sons, J.P. (Raj) Lunia, who traveled to Appleton with a gift of a 20ct ruby. John purchased a number of cut emeralds that, along with the ruby, he later traded back to Lunia. Over the next two years he met with Lunia about twice a year and became good friends with him while learning a good deal about Indian culture, gems, and jewelry.

In February, 1973, Lunia arrived in Appleton with a 1,111ct carved emerald which was cut in about 1860 for an Indian Prince, Raja Khusal Singhji, of Jaipur, India. After a night of excited dealing, John acquired the emerald for cash and trade. Subsequently, he acquired four additional pieces of exceptional quality, including a 650ct carved emerald inlaid with 22-karat gold piping to form an oval around a six outer-petaled and six inner-petaled lotus flower (pictured on page 174-K of *Emerald and Other Beryls* by John Sinkankas). In the center is a mine-cut diamond. On the girdle, a long stem with four leaves curves upward, with a smaller leaf holding a spray set with five small diamonds on one side and four on the other. The bottom is carved with another lotus with a diamond in its center. Another piece is a 455ct low-domed ruby, carved with leaf motif, and three squarish ovals. These are engirdled with 22-karat gold piping, studded with 18 diamonds alternating with small emerald cabochons, with a kite-shaped diamond in the center. This completes the top side. The underside is carved with crossed leaves within a circle. This piece is from a maharaja's collection. Two necklaces completed this acquisition, one of emerald containing 56 beads in two strands measuring from 10 to 25mm and weighing 671.5ct, and the other of ruby with 44 beads measuring from 10 to 25mm each and weighing 1,900ct. These pieces were a part of a display entitled "Jewels of India," which appeared at the James Hunt Barker Galleries in Palm Beach in 1973. John donated the five superb emerald and ruby pieces to the Smithsonian in 1977.

Among other pieces acquired by John through Raj Lunia in 1974 from the exhibit at the James Hunt Barker Galleries were nine exceptional 19th century objets d'art in the Mogul-style of gold overlaid with enamel in the champlévé manner. Mogulware is a term reserved for wares of the 16th, 17th, and 18th centuries and refers to the Moguls who invaded India in the 16th century and brought with them the art of enamel decoration on gold, an art that has nearly been lost. John gave these remarkable pieces to the Smithsonian in 1979 after he concluded that the Mogulware deserved a larger audience.

## Artifacts Currently in the Collection

Currently in the Barlow Collection there are pieces of Indian gold jewelry and objets d'art. One is a superb complete ceremonial temple necklace with twenty Rudraksh beads dating from 1820 to 1840. The necklace is approximately 61.0cm long with a temple medallion 8.7cm high and 9.3cm wide embellished with nine ruby cabochons, two pearls, and one emerald. Hanging below the tem-

***Plate #7-2*** **This necklace once belonged to the Maharaja of Mysore and was worn on religious and festive occasions. (MH)**

ple is a two-piece incense burner 11.5cm high and 8.2cm wide decorated with 34 ruby cabochons and one diamond. The ornate clasp is 4.3cm high and 7.6cm wide with seven ruby cabochons (one is missing). The temple medallion and incense burner consist of heavy gold leaf applied over carved resin, producing a prominent three-dimensional aspect. There are three seated figures in front of the temple. This necklace once belonged to the Maharaja of Mysore and was worn on religious and festive occasions (Plate #7-2).

A rare gold pendant with a detailed history complements the temple necklace. The pendant may have been a gift to a Shiva temple as an ornament to the main deity, Lord Shiva or Goddess Parvati. The pendant is embossed with three temple arches (Gopurams). In the center, Uma Marheshwara (Shiva and Parvati) are seated on the holy bull, Nandi. The elephant-headed god Lord Ganesha on his mouse, and his younger brother Katikya on a peacock, complete the principal figures. Between the Gopurams there are two devotees, the temple attendants, Devadasi standing above the mouth of Makaharas (Mythical Alligator) (Plate #7-3).

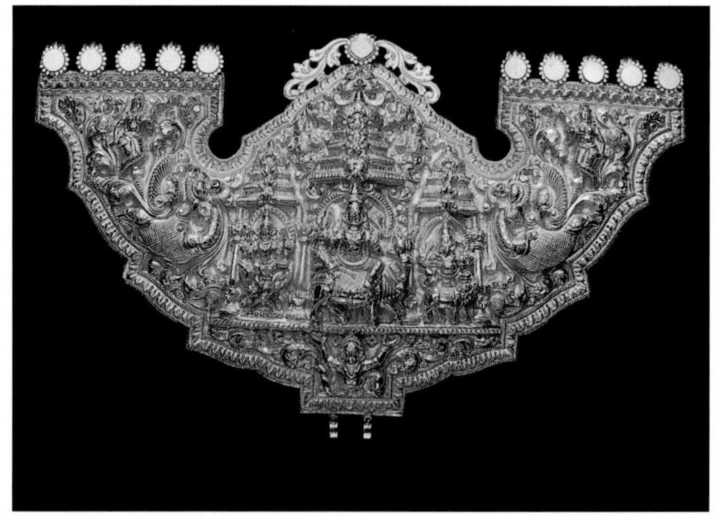

***Plate #7-3*** **A rare gold pendant, possibly a gift from a donor to a Shiva Temple as an ornament to the main deity, Lord Shiva or Goddess Parvati. (MH)**

Several Mogulware gold boxes (Plate #7-4) are among the objets d'art in the collection. A description of one will suffice to illustrate the exquisite character of these pieces. A small box, 4.0cm by 3.3cm by 1.7cm high has a decorated lid and sides with vivid colored enamels executed in the champlévé style. Set in the lid are nine diamonds in a circular pattern. This particular gold box is lined, bottom and sides, with thin plates of ivory. Such boxes were used to hold sweetmeats or pandans, small rolls of betel leaf (Plate #7-4).

***Plate #7-4*** **Several Mogulware gold boxes are among the objets d'art in the Barlow collection. (MH)**

*Gold objets d'art from other cultures and times are also represented in the Barlow Collection. These include a number of pieces, probably from a pre-Christian Grecian burial.*

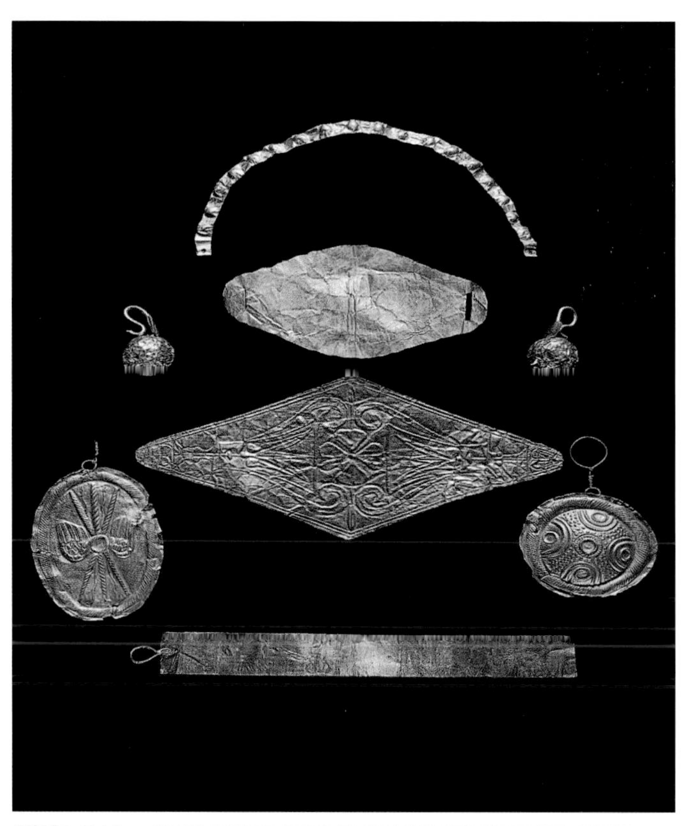

*Plate #7-5* Gold objets d'art from Grecian sources. (MH)

Gold objets d'art from other cultures and times are also represented in the collection. These include a number of pieces, probably from a pre-Christian Grecian burial. Among these are thin spangles in simple geometric forms, either rectangular or diamond shaped, to which are attached jump rings, which indicate their use as earrings. A larger piece has slots cut near its perimeter, suggesting attachment by ribbons to clothing. They are all elegant in their simplicity, in keeping with the usual dress of the Greeks.

In contrast to the essentially two-dimensional aspect of these pieces are a pair of hollow crinkled gold ball earrings, apparently formed from foil crushed over a ball of beeswax, some of which is still present in the hole through which the earclips are attached. These earrings are large, 1.6cm in diameter, with the gold earclips hidden behind 8.0mm round gold shields. The overall effect is quite unusual and would be a welcome addition to any discerning modern woman's jewelry (Plate #7-5).

Throughout the ages, gold has been used extensively in jewelry making to adorn humanity. It was used especially by royalty as a symbol of their station in society. Gold has also been used for centuries as a medium of exchange. Gold coins have a long history of use throughout civilization.

The collection contains a large selection of gold coins from a variety of sources. The coins illustrated in Plates #7-6 and #7-7 are but two of the collection with an interesting history. Much of the gold recovered from the New World by the Spaniards was sent to Spain in the form of coinage referred to as *cobs.* Pieces of gold or silver of the approximate size of the intended denomination were chopped from the ends of flat bars of refined bullion. They were trimmed to the exact weight, heated, and hand hammered between crudely engraved dies. The resulting coins were somewhat irregular as they were crudely struck. The Spanish authorities were not concerned with the shape or the date on the coins, but they were very concerned with the identification of the mint and the assayer's mark.

***Plate #7-6*** **View of obverse side of the two-Escudo coin (left) and the eight-Escudo coin (right). (MH)**

**_Plate #7-7_ View of reverse side of the two-escudo coin (left) and the eight-escudo coin (right). (MH)**

*The coins shown were recovered from a Spanish galleon that sank during a hurricane on July 31, 1715 off the east coast of Florida.*

Gold coins are called *escudos*; silver coins are called *reales*. During the reign of King Philip V in the early 18th century, the escudos and reales were struck in denominations of eight, four, two, and one in both the Bogota and Lima mints (Plate #7-6 and Plate #7-7).

The coins shown were recovered from a Spanish galleon that sank during a hurricane on July 31, 1715 off the east coast of Florida. The coins originated in the Bogota and Lima mints, and had been part of the Spanish Plate Fleet. The plates show the obverse and reverse sides of the two-escudo and eight-escudo coins. These coins were recovered by the Mel Fisher team and awarded to Salvors Inc. 265 years after the flotilla was wrecked.

While gold has been used throughout history for personal adornment or as a means of exchange, it also had been used for strictly practical purposes well before the advent of gold lettering and electronic applications of the 20th century. A gold fishhook (refer to Plate #7-11) in the collection from Colombia, dating from the 16th century (1540), demonstrates the practicality of the post-Spanish Colombian Indians. The hook is fashioned out of 0.5mm gold wire and is 17mm long with a 5mm opening. There are no barbs within the curve of the hook. This piece was acquired by John from the Lawrence University Collection when the collection was sold. The piece was part of the Jack Baragwanath/Leech Collection before that.

In addition to jewelry and coinage, gold is sometimes used in the creation of fine works of art. Plate #7-8 shows a particularly pleasing rendition in native gold on black fabric. The 19×24cm creation depicts a royal elk walking past a cabin in a forested mountain setting beneath a full moon. The scene has a striking three-dimensional effect — a testament to the ability of the artist. The scene was created by Irene Butler. John purchased the artwork along with the Butler Collection in 1979.

**_Plate #7-8_ Artistic rendition in native gold by Irene Butler. (MH)**

## GOLD (NATIVE ELEMENT)

### AUSTRALIA

**#2620  5.0×4.3×4.0cm  Plate #7-9**

**Ballarat, Victoria, Australia**

Frosted, platy crystals to 1.2cm rest in quartz with some included schist and iron oxide stain from this famous mining center.

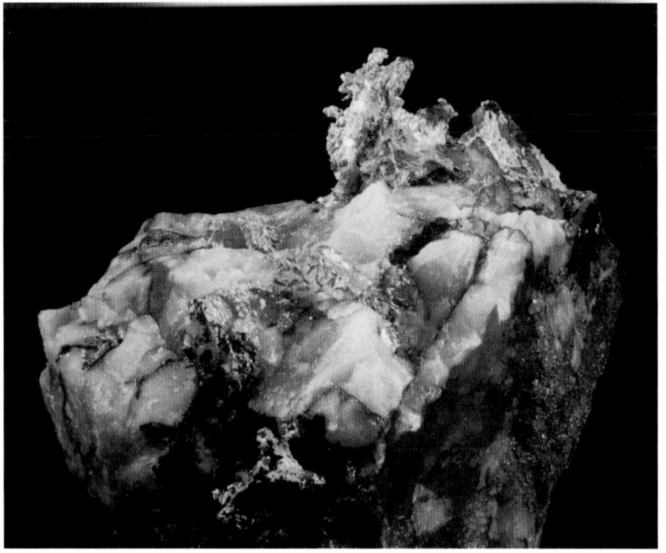

*Plate #7-9, Specimen #2620* **GOLD – Ballarat, Victoria, Australia. (MH)**

**#5036  4.0×3.5×2.5cm**

**Meekathara, Western Australia**

This specimen is coarsely crystallized with an individual 6mm cubo-octahedron on quartz with a crude 9mm crystal on one side of the piece. The specimen has a total weight of 50.22g.

### CANADA

**#995  8.6×7.5×1.8cm**

**Northern British Columbia, Canada**

This very attractive nugget is well abraded, with an estimated 50% quartz by volume (approximately 8% by weight).The piece weighs 382g.

### CHILE

**#2567, #3443, #3449**

**Mina La Farola, Copiapo, Chile**

These three specimens show small rosettes to 5mm of bright, richly colored gold with, on, or under a green copper mineral that proved to be paratacamite, basic copper chloride, when it was chemically analyzed with an SEM/EDS.The crystal morphology was resolved as rhombohedral to distinguish it from monoclinic atacamite.The occurrence on dark gray, weakly metamorphosed volcanic rock is certainly secondary, indicating substantial redistribution of gold, perhaps by chloride solutions.The complete lack of admixed silver within the limits of detection of the SEM/EDS also points to an unusual origin for this gold.

**#4708  3.3×3.0×2.4cm  Plate #7-10**

**Cuevitas, Chile**

This excellent large group of twinned, cubo-octahedral crystals shows numerous step faces with three crystals to 8mm on a matrix of quartz and silicified rock. The specimen has a total weight of 57.97g.

*Plate #7-10, Specimen #4708* **Cuevitas, Chile. (MH)**

## COLOMBIA

**#4675 2.7×1.3×0.7cm Plate #7-11 (left)**

This unusual, moderately abraded, gold wire weighs 12.049g. The wire is fluted, indicating that it is composed of radially twinned and elongated crystals.

**#4694 3.8×1.5×0.6cm Plate #7-11 (center)**

This specimen is a thick, fluted, "young ram's horn," formed in the same manner as #4675.

**#4640 1.7cm Plate #7-11 (right)**

This fishhook from the Chibcha Indians, circa 1540, passed from Jack Baragwanath/Leech to Lawrence University. It was acquired by Barlow from Lawrence University in June 1979.

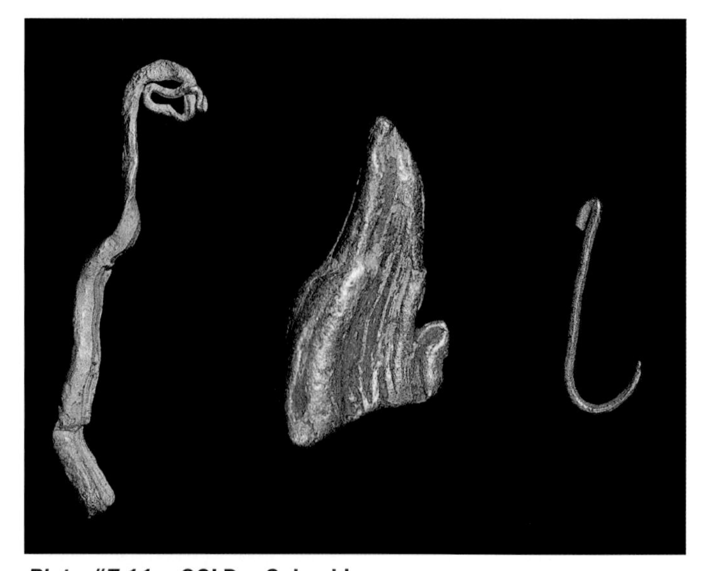

*Plate #7-11* **GOLD – Colombia.**
**Left:** ***Specimen #4675***
**Center:** ***Specimen #4694***
**Right:** ***Specimen #4640*** **(MH)**

*Plate #7-12, Specimen #1294* **GOLD – Gabon. (MH)**

## GABON

**#1294 9.2×6.4×1.5cm Plate #7-12**

This slightly water-worn spongy mass has negative casts after small (1–2mm) quartz crystals. It weighs 311g.

## GREAT BRITAIN

**#3020 3.9×2.8cm Plate #7-13**

**Hope's Nose, Torquay, Devon, Great Britain**

This reticulated cluster is composed of elongated and twinned crystals to 1.0cm with a peculiar brownish gold color reportedly due to a minor palladium content. The gold is on a limestone matrix with a brownish orange iron oxide stain. The gold from Hope's Nose is found in calcite veins in limestone discovered in 1922. Pictured in *Mineralogical Record*, Volume 18, Number 1, 1987, page 87, figure 4.

*Plate #7-13, Specimen #3020* **GOLD – Hope's Nose, Torquay, Devon, Great Britain. (MH)**

*Plate #7-14, Specimen #5705* **GOLD – San Julian Mine, Ramos, Chihuahua, Mexico. (MH)**

## MEXICO

**#5705 3.0×3.0×0.6cm Plate #7-14**

**San Julian Mine, Ramos, Chihuahua, Mexico**

This roughly square plate weighing 5.5g is composed of radiating fans of finely reticulated dendritic elongate octahedral gold crystals. A fine satiny luster is imparted by parallel growth lines of the dendrites. This piece was clearly removed from selvage within an epithermal vein. Tiny euhedral quartz crystals are locally lodged among the gold crystals. This old classic gold was in the Bally Collection and traded out of a German museum. It carries an old acquisition number (2838) and an exceptional catalog card handwritten in German indicating location, weight, and a general description. The card indicates the specimen was originally obtained in 1914 for 50 currency units (marks?). Although the Ramos District is not a name currently in use, the name "San Julian" and the dates of the specimen match well with the now defunct San Julian high-grade vein mining district that lies almost on the Chihuahua-Durango border southeast of the more famous Guadalupe y Calvo District. (See Chapter 13, page 347.) *P.K.M.M*

*Plate #7-15, Specimen #4744* **GOLD – Mount Kare, near Porgera, Enge Province, Papua New Guinea. (MH)**

## NEW GUINEA

**#4744 4.6×3.5cm Plate #7-15**

**Mount Kare, near Porgera, Enge Province, Papua New Guinea**

This very fine, bright specimen consists of sharp, flattened, twinned octahedra with step faces, locally with second generation of tiny gold crystals scattered on both sides. The piece weighs 12.46g.

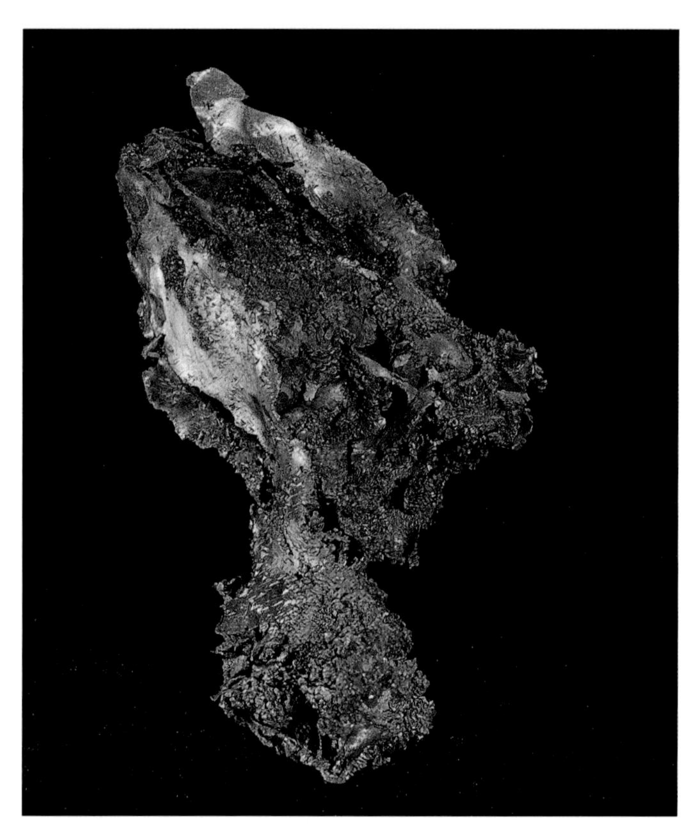

*Plate #7-16, Specimen #1388* **GOLD – Verespatak, Siebenburgen, Romania. (MH)**

## ROMANIA

**#1388 8.8×4.8cm Plate #7-16**

**Verespatak, Siebenburgen, Romania**

Pale, locally tarnished, large crystal plates to 6.5cm lie in curved, subparallel position with clusters of modified octahedra on one end of this piece. Superb growth trigons, for which this locality is famous, adorn the surfaces of the plates. Local coatings of sharp, tarnished brassy crystals were chemically analyzed by the SEM/EDS as iron sulfide and were identified by composition and crystal morphology as marcasite. This superb piece has a total weight of 123.35g and was formerly in the collection of Al Buranek.

**#1621 4.0×4.5cm**

**Verespatak, Siebenburgen, Romania**

A 3.6cm leaf with a remarkable reddish brown color is perched on a surface of drusy quartz coating an altered volcanic rock. It is sealed in a plastic dome with a wooden base.

*Plate #7-17, Specimen #G130* **GOLD – Verespatak, Siebenburgen, Romania. (MH)**

**#G130 9×6mm Plate #7-17**

**Verespatak, Siebenburgen, Romania**

Pale, silvery-yellow, sharp, twinned, and slightly flattened cubo-octahedra to 5mm are without obvious re-entrant faces denoting a twin plane. This specimen was originally in the collection of Carl Bosch, a noted German collector, and was acquired from the U.S. National Museum (#B31) in 1981.

*Plate #7-18, Specimen #2607* **GOLD – Berezovsk (Beresov), Ural Mountains, Russia. (MH)**

## RUSSIA

**#2607 5.0×3.5×3.0cm Plate #7-18**

**Berezovsk (Beresov), Ural Mountains, Russia**

This piece is an exceptional combination of bright, complex, twinned clusters of gold crystals in and on quartz crystals to 3.0cm. From the I. Philip Scalisi Collection.

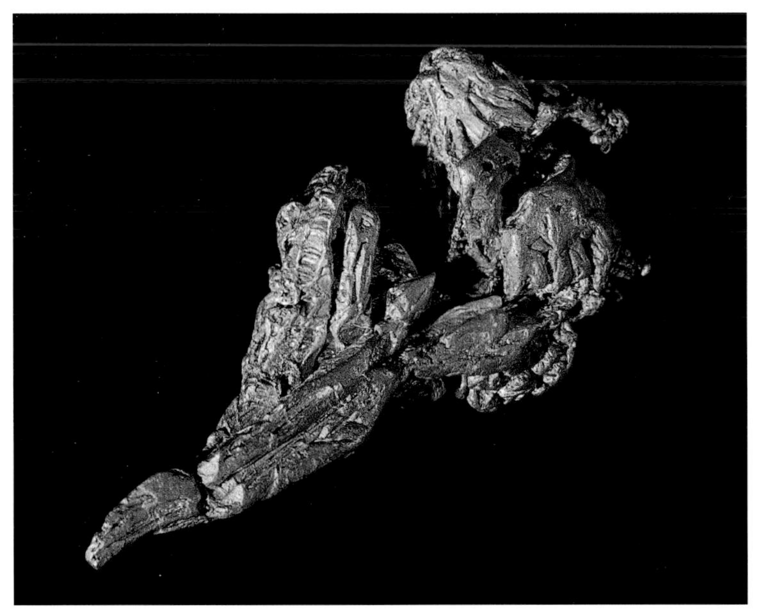

Plate #7-19, Specimen #3930 GOLD – Siberia, Russia. (MH)

**#3930 6.5×3.2×1.3cm Plate #7-19**

**Siberia, Russia**

These elongated, multiply twinned, octahedra to 4.9cm form five out of six sections of a complete 360° twinned crystal. This is the best example of this type of twin in the collection. The specimen weighs 54.93g. Attached to the twin are rounded and elongate incomplete octahedra, hoppered and ribbed. Pictured in *Mineralogical Record*, Volume 13, Number 6, page 367, 1982.

**#3727 5.0×3.5×1.8cm Plate #7-20**

**Western Siberia, Russia**

This tabular, composite cluster with cubic crystals on both sides weighs 87.93g. The largest crystal is 1.0×0.7cm. Pictured in *Mineralogical Record*, Volume 13, Number 6, 1982, page 367

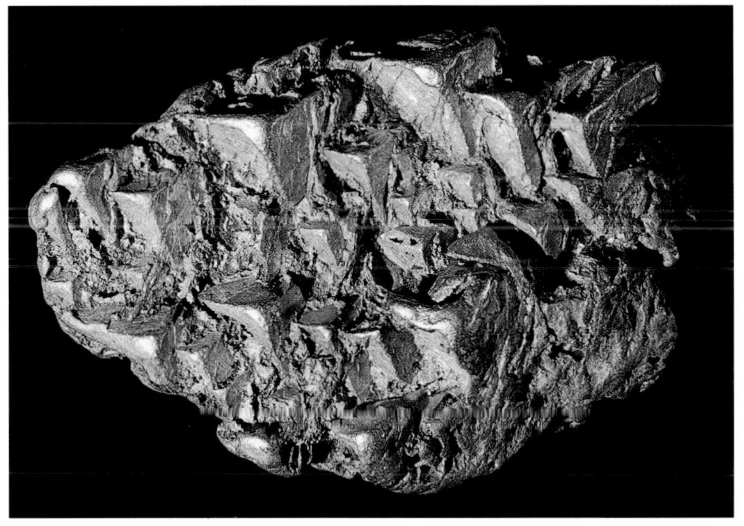

*Plate #7-20, Specimen #3727* **GOLD – Western Siberia, Russia. (MH)**

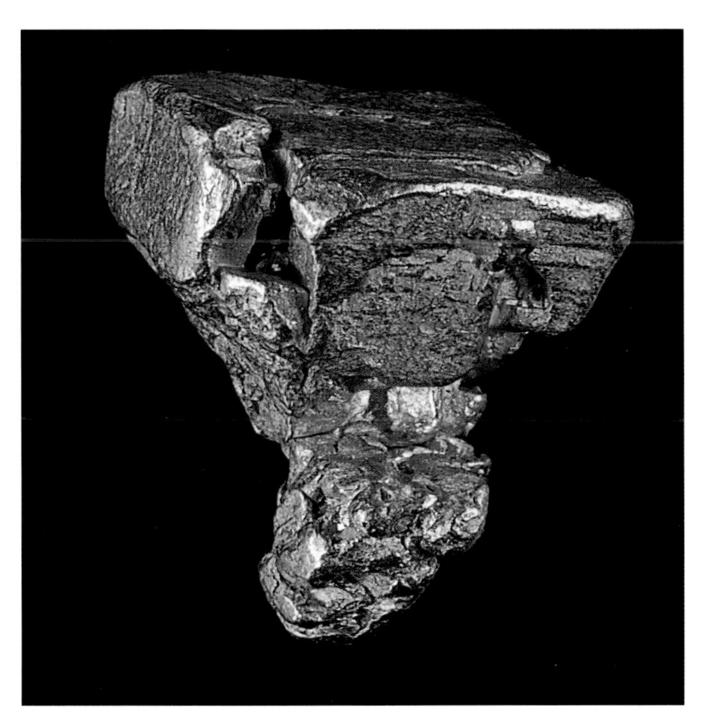

**#3757 2.0×1.2cm Plate #7-21**

**Lena River, Yakutsk, Siberia, Russia**

This spectacular cube (more than 50% complete) is slightly hopper formed, tapering down to a small cluster of modified crystals. A cube face measures 1.0×1.2cm. The specimen is thumbnail size and weighs 17.23g.

*Plate #7-21, Specimen #3757* **GOLD – Lena River, Yakutsk, Siberia, Russia. (MH)**

**#3756 3.7×3.0×1.5cm Plate #7-22**

**Lena River, Yakutsk, Siberia, Russia**

This is an exceptional cluster of cubic crystals with the largest measuring 1.6×1.4cm. The central cube is very large and is nested in other cubes ranging from sharp to rounded. The crystals are slightly water worn on edges and corners with some development of hoppers on some cube faces. The specimen weighs 49.63g. (For another view, see Chapter 3, page 66, Plate #3-49.)

***Plate #7-22, Specimen #3756*** **GOLD – Lena River, Yakutsk, Siberia, Russia. (MH)**

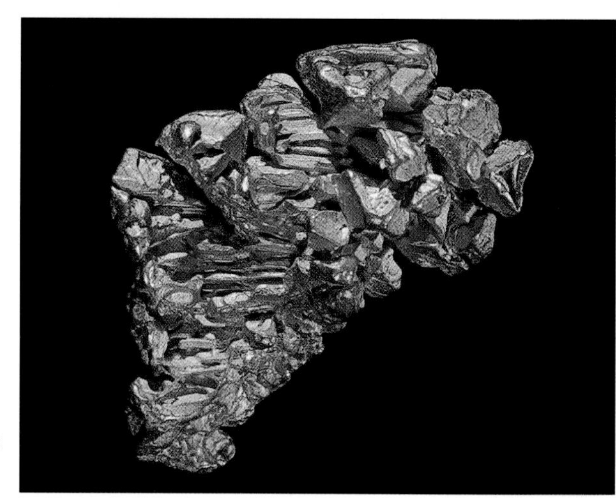

**#3728 3.4×2.0×1.7cm Plate #7-23**

**Northern Siberia, Russia**

This very fine "floater" cluster of parallel skeletal or hopper-shaped octahedral crystals has a rich gold color. The piece weighs 39.91g.

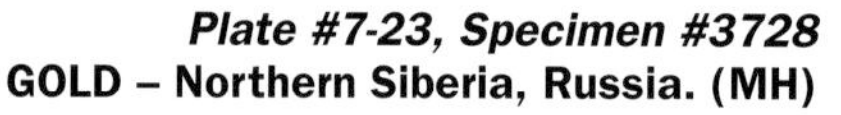

***Plate #7-23, Specimen #3728*** **GOLD – Northern Siberia, Russia. (MH)**

***Plate #7-24, Specimen #3740*** **GOLD – Siberia, Russia. (MH)**

**#3740 2.8×1.5×1.2cm Plate #7-24**

**Siberia, Russia**

This specimen consists of lightly abraded, arborescent, twinned crystals with dominant cube corners and edges to 8.5mm in quartz.The piece weighs 11.11g.

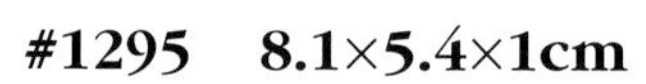

**#1295 8.1×5.4×1cm**

**Berezovsk (Beresov), Ural Mountains, Russia**

Multiple plates of flattened and twinned octahedra rest on quartz.The piece has been repaired.

**#3755 3.4×3.0×1.0cm**

**Yakutsk, Siberia, Russia**

This piece is a dendritic mass of deeply fluted, twinned, crystals.The crystals are slightly rusty on abraded surfaces. The piece weighs 34.03g.

**#1897 6.8×2.5×1.6cm**

**Lena River, Yakutsk, Siberia, Russia**

This abraded, somewhat curved, plate shows numerous growth trigons on its protected concave side. Minor sand and clay fill the crevices.The piece weighs 74.26g.

## UNITED STATES

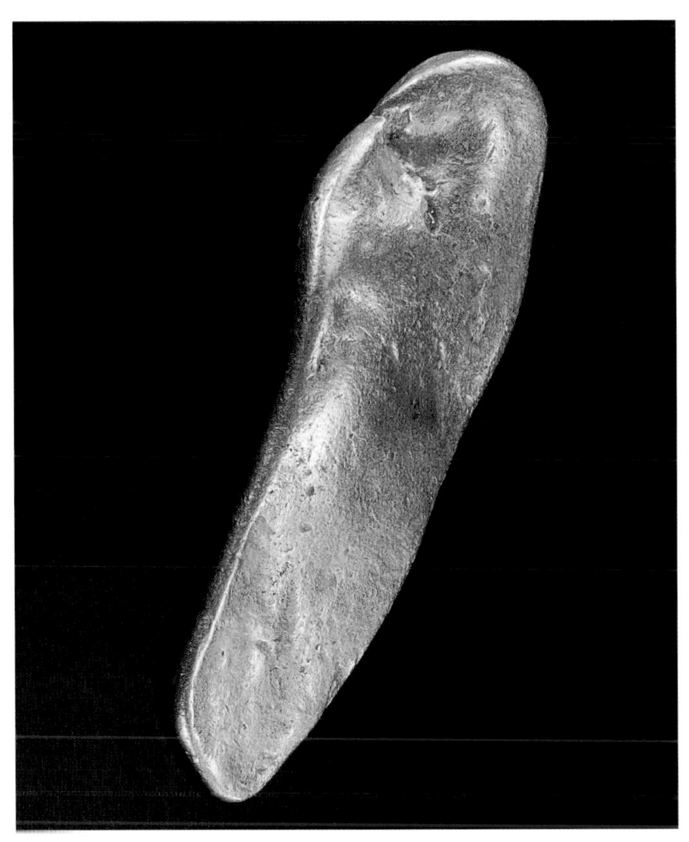

*Plate #7-25, Specimen #4516* **GOLD – Kuskokwim River, Alaska. (MH)**

### ALASKA

**#4516 6.5×2.3×0.4cm Plate #7-25**

**Kuskokwim River, Alaska**

This bright nugget is in the shape of a footprint and therefore was christened "Big Foot" by its Eskimo discoverer. It was acquired in McGrath on the shore of Alaska's second longest river during a hunting trip. The nugget possesses a slight abrasion ridge on one edge. The piece weighs 50.6g.

**#G238 5.7×5.0×2.5cm**

**Klondike Region, Alaska**

This nugget from a classic locality is approximately 60% quartz by volume with dark brownish iron oxide stain. It weighs 233.29g. (Note: specimens labeled Klondike Region may well be from the Canadian side of the border.)

**#3880 1.3×1.0cm Plate #7-26**

**Talkeetna, Susitna River, Alaska**

This specimen consists of several bright, sharp, intergrown dodecahedral crystals. One is a well-defined 0.9×0.9cm dodecahedral crystal with some hopper faces perched on several smaller 3-5mm dodecahedral crystals. The specimen shows unusual crystal forms for gold. The piece weighs 6.4g.

*The specimen shows unusual crystal forms for gold.*

*Plate #7-26, Specimen #3880* **GOLD – Talkeetna, Susitna River, Alaska. (MH)**

**#G137 6.9×4.6×2.4cm**

**Klondike Region, Alaska**

This very attractive nugget consists of 50% quartz by volume with some iron oxide stain. The piece weighs 232.67g.

**#3724 7.2×5.0×1.0cm**

**Klondike Region, Alaska**

From a classic locality, this exceptionally flat nugget with some quartz weighs 295.50g. The piece still contains some fine sand in crevices.

*The specimen was sold at an auction in Boston, Massachusetts in the 1860s to a German buyer and ended up in a German university where in September 1992 it was traded for meteorites and returned to the United States.*

## CALIFORNIA

**#5060 4.5×2.5×1.6cm** **Refer to Chapter 3, Plate #3-44**

**Coloma, El Dorado County, California**

This extraordinary crystal, weighing 94.44g and collected in 1860, represents a portion of a huge octahedron. The crystal is slightly water worn with hopper-shaped octahedral faces and growth ribs parallel to the edges of the octahedron. Its luster is typical of water-worn gold, the edges rounded but distinct. Minor sand grains and clay in the ribs attest to the transient placer origin of this piece. The specimen was sold at an auction in Boston, Massachusetts in the 1860s to a German buyer and ended up in a German university, where in September 1992 it was traded for meteorites and returned to the United States. This piece is the largest crystal of gold known from an American locality.

**#G119 9.0×8.0×3.5cm**
**Refer to Chapter 3, Plate #3-45**

**Spanish Dry Diggings,
El Dorado County, California**

This is a lovely gold. A beautiful "nest" of countless crystals in the center of the piece is surrounded by quartz crystals and spongy gold. Weighing 338.7g, this specimen is an outstanding small cabinet-size piece. The specimen was formerly in the collection of the American Museum of Natural History.

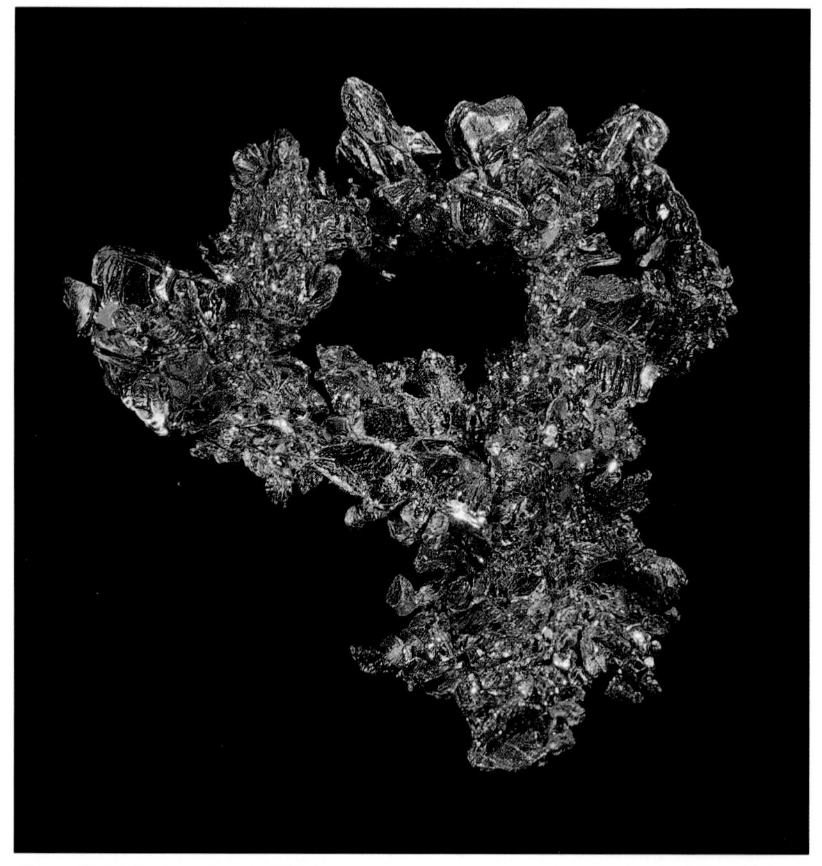

***Plate #7-27, Specimen #3668*** **GOLD – DeMaria Mine, Forest Hill, Placer County, California. (MH)**

**#3668 4.8×4.3×1.3cm Plate #7-27**

**DeMaria Mine, Forest Hill,
Placer County, California**

This exceptional piece is composed of numerous flattened and twinned octahedra, some to 7mm. Weighing 42.32g, this bright and aesthetic specimen is representative of the best from this locality.

*An exceptional piece...*

***Plate #7-28, Specimen #5179*** **GOLD – Diltz Mine, Mariposa County, California. (MH)**

**#5179 4.0×3.6×2.5cm Plate #7-28**

**Diltz Mine, Mariposa County, California**

This coarse, spongy aggregate of complexly overgrown octahedra in arborescent groups weighs 124.37g. The lightly frosted texture is typical of this locality in the East Belt adjacent to the southern Mother Lode. This specimen was collected about 1949.

## The Diltz Mine

The Diltz Mine is about six miles northwest of Mariposa, the county seat of Mariposa County, California. The vein strikes N 10° E and dips 38–40° to the east and has an average width of less than one meter, though it widens to a maximum of 3 meters. The hanging wall of the vein consists of greenstone (metamorphosed volcanic rock) and the footwall of dark granitic rock and a basic dike rock. The ore contains arsenopyrite, chalcopyrite, pyrite, tetrahedrite, galena, and gold in matrix of ribbon quartz, frequently accompanied by calcite and manganese oxides. The ore occurs in well-defined shoots in fractured zones of massive quartz that have been recemented by calcite. It has been concluded that most of the gold was deposited with the calcite. The feature has contributed to the development of pockets of beautifully crystallized gold. A magnificent specimen weighing 52 troy pounds was mined on the 100-foot level north on May 12, 1932. Unfortunately, this piece was processed, yielding 43 pounds 2oz. of fine gold. Another specimen, 9 inches long and 6 inches high, weighed 20 pounds 2oz. Although pockets containing 5–25oz. are common, the average gold content of the ore, excluding high-grade pockets, is about 0.2oz. per ton.

**#3753 3.0×2.1×0.5cm Plate #7-29**

**DeMaria Mine, Forest Hill, Placer County, California**

This piece with flattened and twinned hopper-shaped crystals to 2.1cm shows some tetrahexahedral faces along the edges of the thick flattened plates. It weighs 14.66g.

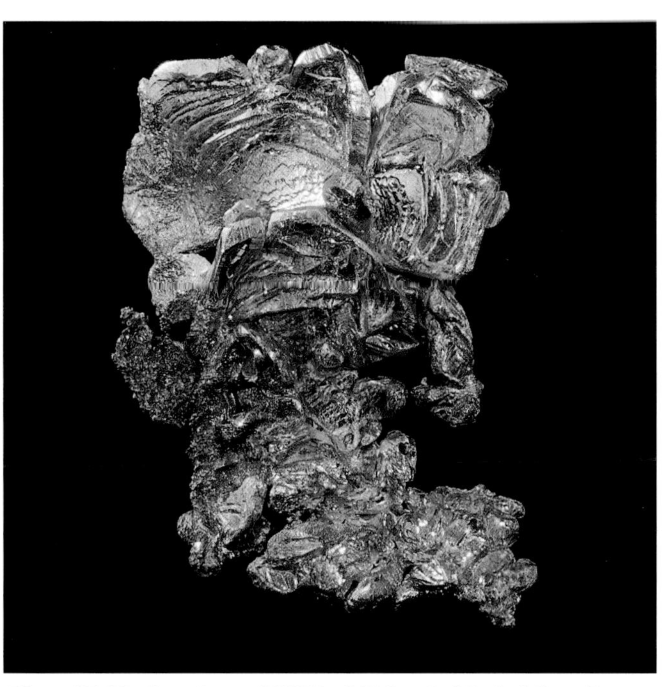

***Plate #7-29, Specimen #3753*** **GOLD – DeMaria Mine, Forest Hill, Placer County, California. (MH)**

### #3785 4.6×3.0×1.2cm Plate #7-30

**DeMaria Mine, Forest Hill, Placer County, California**

This showy piece weighs 13.69g and has flattened and twinned octahedra to 8mm in parallel growth in a common plane.

### #3030 5.5×3.4×3.8cm

**DeMaria Mine, Forest Hill, Placer County, California**

This particularly bright combination of sharp, flattened, and twinned octahedra to 1.0cm has arborescent twinned crystals in parallel arrangement on and in quartz.

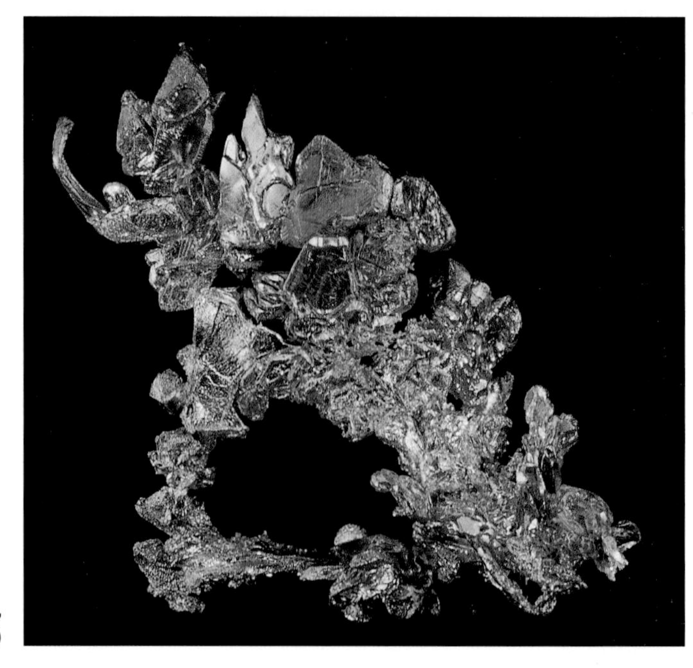

***Plate #7-30, Specimen #3785*** **GOLD – DeMaria Mine, Forest Hill, Placer County, California. (MH)**

### #4707 8.5×4.2×1.5cm Plate #7-31

**Bodie (a ghost town, now a state park), Mono County, California**

Large plates of finely reticulated crystals with a reddish brown tarnished surface rest on a calcite matrix. This fine piece comes from a rare locality.

### #5290 4.5×2.0×0.5cm

**Garden Valley District, El Dorado County, California**

This exceptional piece from the northern Mother Lode consists of flattened, twinned octahedra from 1.0 to 1.5cm in an arborescent fan. Its color is normal with minor tarnish on exterior crystal faces.

***Plate #7-31, Specimen #4707*** **GOLD – Bodie, Mono County, California. (MH)**

### #G166 6.8×7.5×2.5cm Plate #7-32

**Mariposa County, California**

This specimen, acquired from the Ernest Butler Collection, consists of two plates of gold (3.0×4.0cm and 1.5×2.0cm) with arborescent rims standing on a quartz-ankerite base. This spectacular matrix piece displays beautifully.

***Plate #7-32, Specimen #G166*** **GOLD – Mariposa County, California. (MH)**

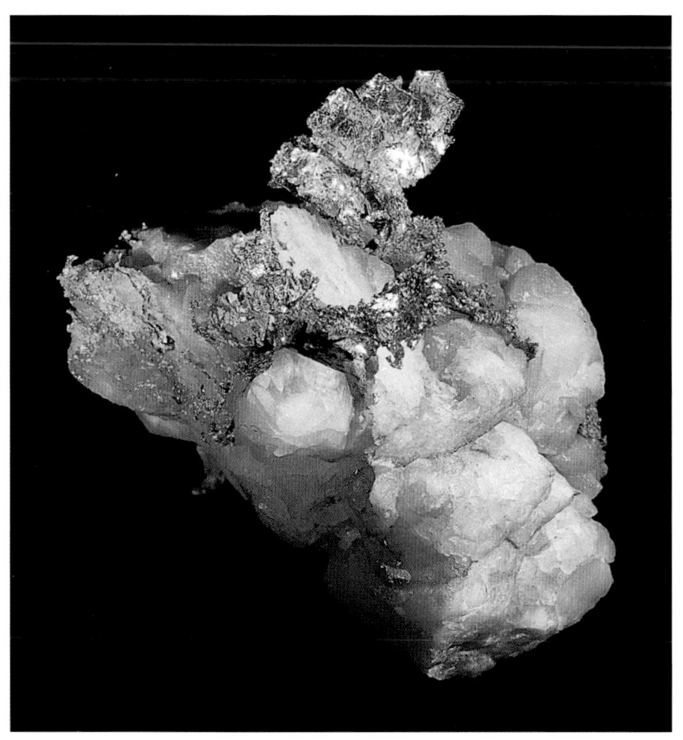

*Plate #7-33, Specimen #G167* **GOLD – Mariposa County, California. (MH)**

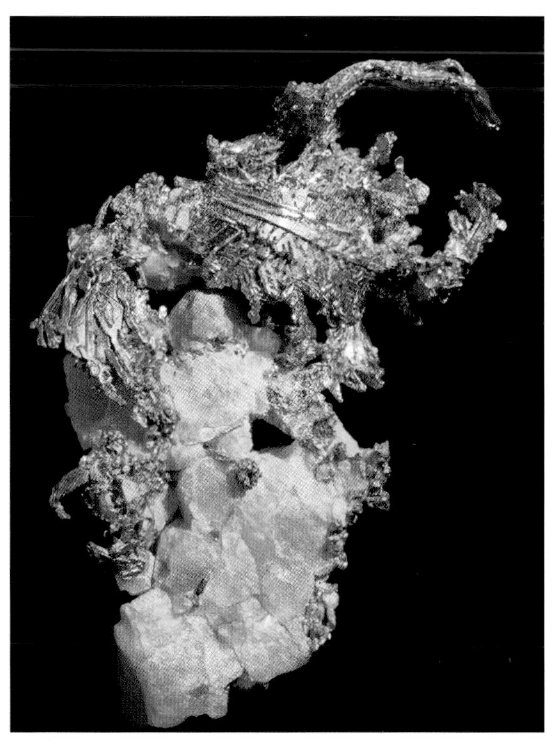

*Plate #7-34, Specimen #1841* **GOLD – Forest Hill/Michigan Bluff area, Placer County, California. (MH)**

**#G167 6.7×7.5×5.8cm Plate #7-33**

**Mariposa County, California**

Acquired as part of the Ernest Butler Collection, this piece displays beautifully, with two arborescent plates of gold (2.5×2.5cm and 2.0×1.5cm) standing on a slightly iron-stained quartz base. Additional gold plates and crystal clusters are situated below "eye-level" on this specimen.

**#1841 7.5×4.3×1.6cm Plate #7-34**

**Probably Forest Hill/Michigan Bluff area, Placer County, California**

This specimen is an aggregate of brilliant, arborescent, multiple-twinned, flattened, wirelike octahedral crystals on quartz weighing 66.71g. Mined in early 1920s, it was obtained from Jim Dines in a trade for meteorites.

*Plate #7-35, Specimen #3667* **GOLD – Forest Hill, Placer County, California. (WW)**

**#3667 6.2×2.5×1.3cm Plate #7-35**

**Forest Hill, Placer County, California**

This specimen is pictured on the inside back cover of *Mineralogical Record*, Volume 18, Number 1, 1987 in the Mockingbird Gold Mining Company advertisement. Weighing 31.63g, this bright aggregate of flattened, twinned, octahedra to 9mm and crystal plates is a superb example of material from this locality.

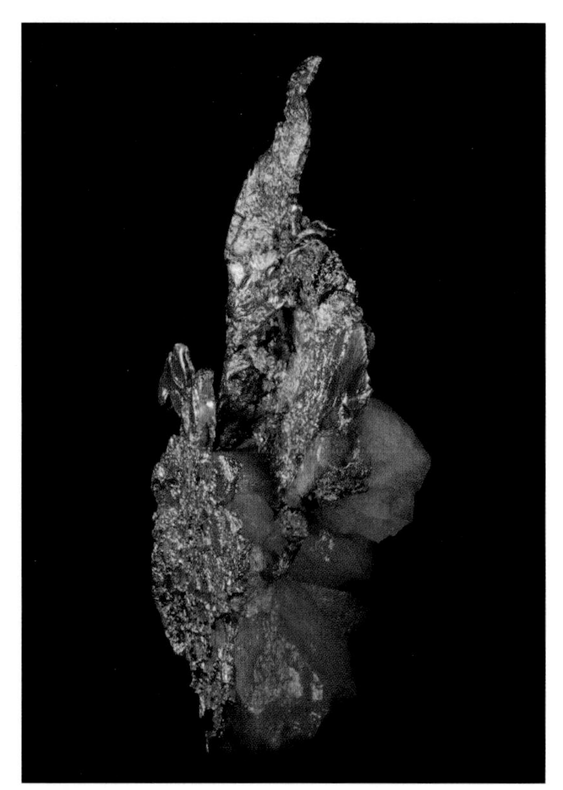

*Plate #7-36, Specimen #4512* **GOLD – Tuolumne County, California. (MH)**

**#4512 7.0×4.5×2.0cm Plate #7-36**

**Tuolumne County, California**

This excellent example of flattened and twinned octahedra to 9mm in parallel plates separated by quartz is a rich gold color. It weighs 60.43g, including the quartz. The specimen was formerly in the collection of the American Museum of Natural History.

**#3754 4.2×2.9×1.3cm**

**Forest Hill, Placer County, California**

This bright gold with minor iron oxide tarnish is composed of flattened, elongated, twinned crystals to 8.0mm with one elongated crystal an exceptional 2.3cm. Some development of hopper growth on the octahedral faces is unusual for this locality. The specimen weighs 23.83g.

**#3665 5.5×3.2×2.0cm**

**DeMaria Mine, Forest Hill, Placer County, California**

This large, twinned, arborescent crystal, 3.0cm across, on quartz is an excellent example of this particular habit from this locality. The piece weighs 36.03g.

**#3794 3.5×1.5×0.7cm**

**DeMaria Mine, Forest Hill, Placer County, California**

A flattened and twinned octahedron, 7.0mm, across sits at one end of a group of platy twinned crystals. The specimen weighs 6.21g.

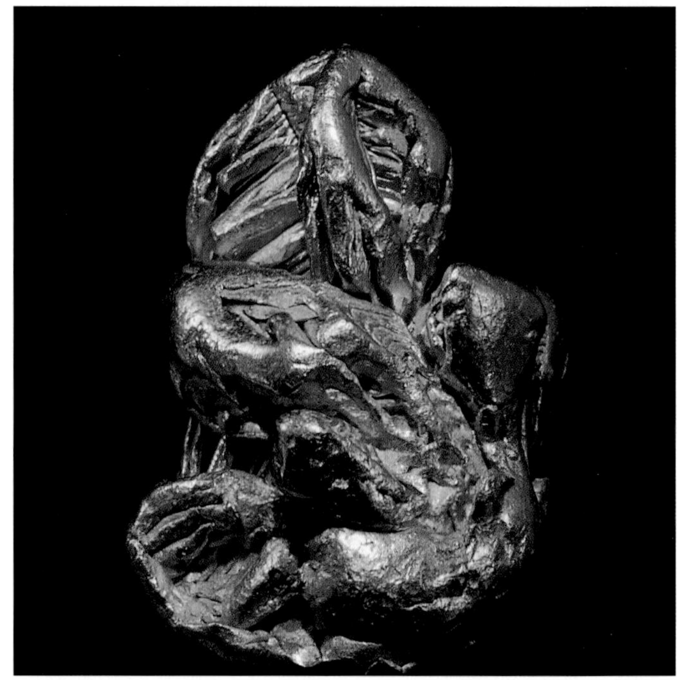

*Plate #7-37, Specimen #3739* **GOLD – Middle Fork of the American River, Placer County, California. (MH)**

**#3739 2.6×1.5×1.3cm Plate #7-37**

**Middle Fork River (probably Middle Fork of the American River), Placer County, California**

This slightly water-worn cluster of skeletal or hopper-shaped octahedral crystals shows pronounced ribs within the cavities substituting for the octahedral faces. It has crystals to 1.5×1.2cm. This superb specimen weighs 25.84g.

**#3752 5.2×2.4×0.7cm**

**DeMaria Mine, Forest Hill, Placer County, California**

This exceptionally brilliant, flattened, twinned, arborescent crystal plate has octahedra modified by the trapezohedron perched on the edges. The piece weighs 12.08g.

**#G121 8.7×6.3×3.1cm**

**New York Hill Mine, Grass Valley, Nevada County, California**

This specimen consists of bright curved plates of twinned crystals with octahedra to 2mm among the plates. Gold occupies the medial portion of a quartz vein approximately 4.0cm thick. The specimen weighs 194.16g, including the quartz. The specimen was formerly American Museum of Natural History specimen #G18190.

**#3751 3.3×3.2×1.0cm**

**DeMaria Mine, Forest Hill, Placer County, California**

Bright, flattened, twinned octahedra to 4mm are stacked on quartz. The specimen weighs 12.00g, including the quartz.

***Plate #7-38, Specimen #4705*** **GOLD – Red Ink Maid Mine, Placer County, California. (VP)**

**#5209 5.5×1.7×2.7cm**

**German Bar, Nevada County, California**

This cluster of pale yellow, complex crystals to 7mm has minor adhering quartz.

**#4705 7.2×4.9×1.5cm Plate #7-38**

**Red Ink Maid Mine, Placer County, California**

This very showy 66.59g piece consists of a spongy mass of arborescent, flattened, twinned octahedra with vicinal faces (faces on a crystal that approximate or take the place of fundamental planes) on edges between faces. The octahedral crystal measures 0.7×0.4cm.

**#G126 4.0×3.8×1.9cm**

**Placer County, California**

This piece consists of an aggregate of brassy crumpled plates on and in gray, greasy-appearing, quartz. Originally in the Norman Spang and American Museum of Natural History Collections, it weighs 40.04g, including quartz.

**#4704 6.1×3.4×1.3cm Plate #7-39**

**El Dorado County, California**

This cluster of arborescent, distorted, twinned crystals and plates to 2.0cm shows step growth patterns on the plate surfaces. The piece weighs 47.67g.

***Plate #7-39, Specimen #4704*** **GOLD – El Dorado County, California. (VP)**

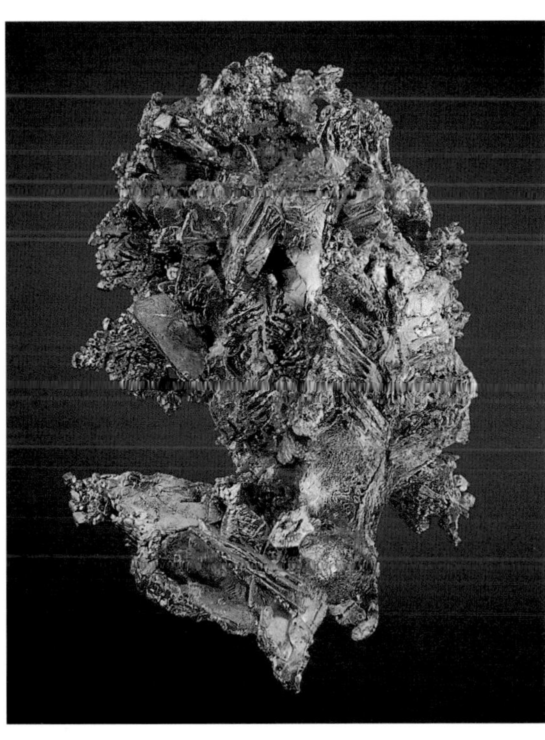

**#3031 3.0cm**

**DeMaria Mine, Forest Hill, Placer County, California**

These bright, dendritic, elongated crystals have 2mm twinned crystals projecting out of quartz matrix in two clusters. The specimen has associated limonite after pyrite.

**#1960 4.2×1.7×0.8cm Plate #7-40**

**Shaw Mine, Tuolumne County, California**

These showy, flattened, and twinned octahedra from 0.5 to 1.0cm across are associated with minor limonite and quartz. The piece weighs 15.57g.

***Plate #7-40, Specimen #1960*** **GOLD – Shaw Mine, Tuolumne County, California. (MH)**

***Plate #7-41, Specimen #T435*** **GOLD – Grass Valley, Nevada County, California. (MH)**

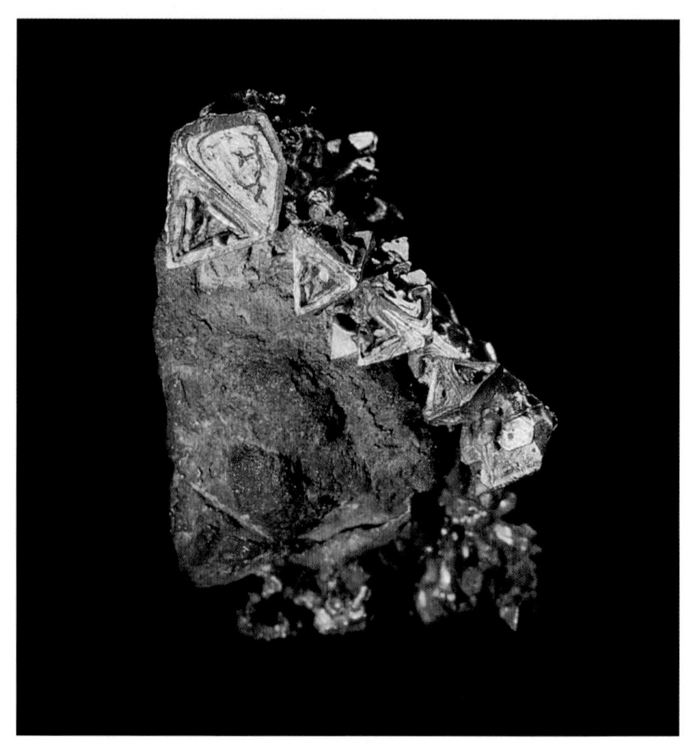

***Plate #7-42, Specimen #T325*** **GOLD – Colorado Mine (Colorado Quartz Mine), near Midpines, Mariposa County, California. (MH)**

**#T435 2.0×2.0cm Plate #7-41**

**Grass Valley, Nevada County, California**

These slightly rusty, sharp crystals with one 9mm distorted and twinned octahedron rest on quartz.The specimen weighs 9.79g.

**#T325 2.5×1.4cm Plate #7-42**

**Colorado Mine (Colorado Quartz Mine), near Midpines, Mariposa County, California**

These five unusually fine hopper-shaped octahedra, to 0.5cm, rest on the edge of a dendritic crystal perched on a fragment of an argillite matrix. Refer to this chapter, page 173,"The Right Place at the Right Time." The specimen weighs 9.26g and is pictured in *Mineralogical Record*, November-December, 1982, page 366.

## Colorado Quartz Mine, Mariposa County, California

This mine was originally called the Colorado Mine. Only in recent years has it been called the Colorado Quartz Mine.

Superbly crystallized gold has been found for many years in pockets within quartz veins at the Colorado Quartz Mine, near Midpines, Mariposa County, California. The mine is situated in the East Belt, which is some 8 to 15 miles east of and parallel to the Mother Lode. Specimens from this mine are generally small, measuring 1 to (exceptionally) 10cm across, but are beautifully crystallized in arborescent or fan-shaped plates and clusters.

The gold-bearing pockets at the Colorado Quartz Mine are localized at the intersection of quartz veins of different ages with one set cutting and brecciating the other. These veins, which occur within a foliated andesitic dike, are referred to as "floor" and "cutter" veins. The greater porosity that occurs at the vein intersections in the breccia developed in the earlier (floor) quartz veins apparently represents a favorable site for gold deposition. Some pockets also have been discovered at the margins of the dike where the quartz veins appear to terminate against the schistose country rock.

Red and black mud occurs with the crystallized gold and probably represents completely altered ankerite deposited on and around the gold. This is a common, though by no means universal, feature of Mother Lode and East Belt gold deposits.

**#5675 7.0×5.5×5.0cm Plate #7-43**

**Colorado Mine (Colorado Quartz Mine), near Midpines, Mariposa County, California**

The matrix of altered ankerite has two gold clusters aesthetically positioned with 0.3×0.3cm dodecahedral crystals. This is a classic specimen.

**#4252 1.8×1.2×0.9cm**

**Eureka Mine, Tuolumne County, California**

This specimen consists of exceptionally sharp 0.5cm octahedra of gold crystals and quartz. The specimen weighs 3.80g, including quartz, and was obtained from a miner at the 1979 Tucson Show.

***Plate #7-43, Specimen #5675***
**GOLD – Colorado Mine (Colorado Quartz Mine), near Midpines, Mariposa County, California. (MH)**

***Plate #7-44, Specimen #4172*** **GOLD – Printer Boy Hill, Leadville, Lake County, Colorado. (MH)**

***Plate #7-45, Specimen #G182*** **GOLD – Dixie Mine, Need County, Colorado (MH)**

## COLORADO

**#4172 4.3×2.9×1.2cm Plate #7-44**

**Printer Boy Hill, Leadville, Lake County, Colorado**

This showy specimen with unusual yellowish color consists of an arborescent cluster of complexly twinned crystals to 2mm. The piece weighs 25.31g.

**#G182 2.5×2.0×1.1cm Plate #7-45**

**Dixie Mine, Need County, Colorado**

This specimen is a showy, rare plate crystal form from Colorado. It weighs 7.1g.

## Farncomb Hill, Breckenridge, Summit County, Colorado

Some of the most remarkable gold specimens ever unearthed have come from Farncomb Hill, near Breckenridge. The veins here are mostly calcite, and the gold crystallized in seams and joints. In many cases the gold specimens show a perfect rhombic outline, apparently having formed in the cleavage joints of the host calcite.

Later solutions apparently caused the growth of perfect octahedra of gold, to several millimeters, along the inner surfaces of these rhombic septa.

In other cases, the leaves of gold are a myriad of interlocking octahedral gold crystals, none very large, but usually equidimensional, sharp, and brilliant.

The Denver Museum Collection is largely the collection of one of the early investors in the Farncomb Hill properties, a man named Campion. He also served on the museum board decades ago, which probably accounts for the collection's ending up in the museum. It is interesting to note that the collection was not always on display in the museum. There is abundant information suggesting that the museum lost track of the gold for a time. When a collection inventory was ordered, a local bank notified the museum that the bank had a couple of crates in their vault that belonged to the museum. These were subsequently opened, revealing much of the missing Campion Collection, plus two huge nuggets that when joined, turned out to be another missing gold specimen called "Tom's Baby," also a Breckenridge trophy.

Finally, one should also note the abundant tangled wire golds found at Breckenridge. One mine, called the Wire Patch, yielded multi-pound tangled masses of hairlike wires of gold, some of which are now displayed at the Denver Museum, a place well worth visiting.

*R.W.J.*

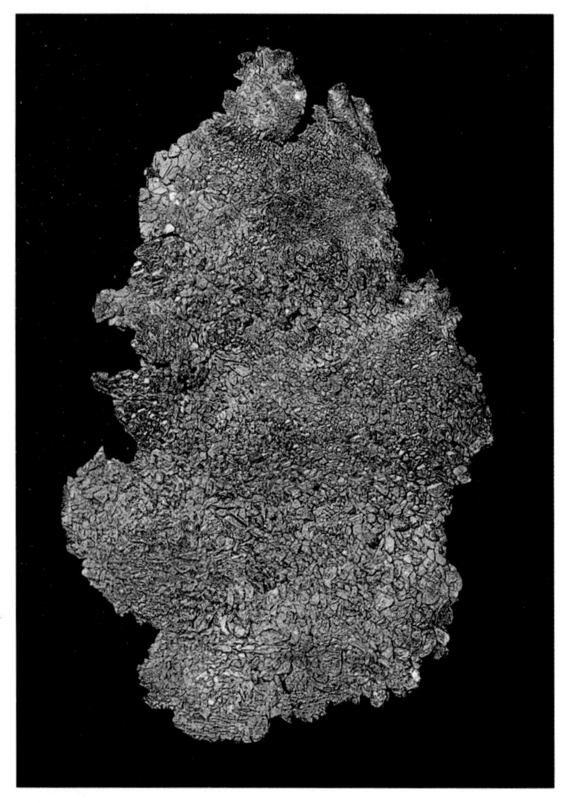

***Plate #7-46, Specimen #G125*** **GOLD – Gold Flake Vein, Farncomb Hill, Breckenridge, Summit County, Colorado. (MH)**

**#G125 7.0×4.0×0.2cm Plate #7-46**

**Gold Flake Vein, Farncomb Hill, Breckenridge, Summit County, Colorado**

This large single crystal from a classic locality is in the form of a flattened, twinned octahedron with a cuneiform growth pattern on the large faces. The piece weighs 17.72g and was formerly in the collection of the American Museum of Natural History (#G41945).

**#4277 3.5×0.3cm**

**Ibex Mine, Greece Hill, Leadville, Lake County, Colorado**

This piece is a sturdy, bright 3.5cm wire obtained from the American Museum of Natural History.

**#4195 2.0×2.0×1.2cm Plate #7-47**

**Ground Hog Mine, Gilman, Need County, Colorado**

This example of heavy wire gold to 1.2cm rests on a quartz matrix.

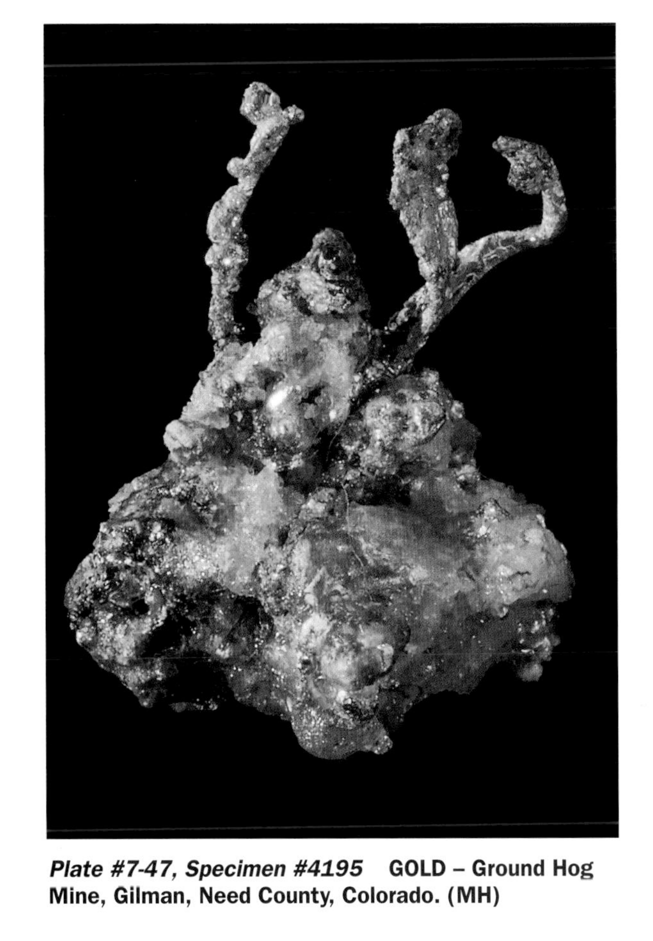

*Plate #7-47, Specimen #4195* GOLD – Ground Hog Mine, Gilman, Need County, Colorado. (MH)

**#1252 4.7×3.7×2.0cm**

**Leadville, Lake County, Colorado**

This specimen consists of a spongy mass of light-colored, elongated, twinned, indistinct crystals to 2mm. Growth of the crystals appears to have been impeded by quartz and possibly calcite, resulting in the lack of sharp crystal faces.The piece weighs 66.56g.

*Plate #7-48, Specimen #G-28* GOLD – Smuggler Mine, Marshall Basin, Telluride, Need County, Colorado. (MH)

**#G-28 4.0×3.0cm Plate #7-48**

**Smuggler Mine, Marshall Basin, Telluride, Need County, Colorado**

This major section of a 4.0cm flattened octahedral crystal shows many octahedral crystals on its surface. Pictured in *Mineralogical Record*, Volume 13, Number 6, page 368.

## GEORGIA

**#4263 3.0×1.2×1.0cm Plate #7-49**

**Dugas Mine, White County, Georgia**

These bright, slightly worn dendritic crystals are a superb example from this location. The specimen was obtained from the Buranek Collection.

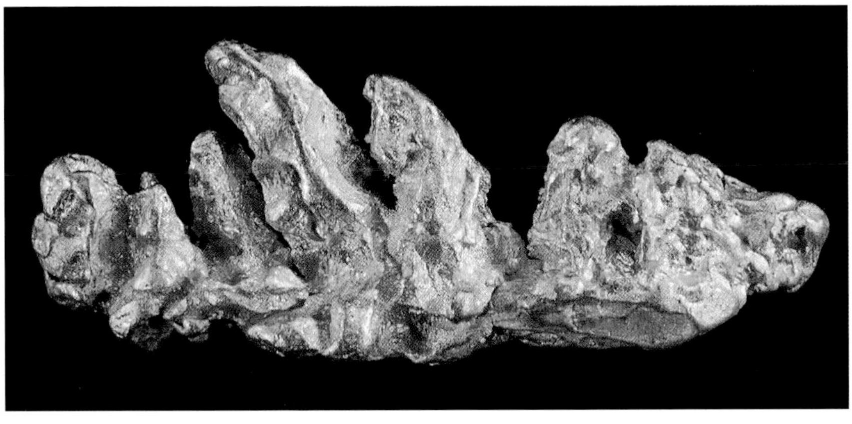

*Plate #7-49, Specimen #4263* GOLD – Dugas Mine, White County, Georgia. (MH)

## MONTANA

**#1133 6.0×4.5×2.0cm**

**Butte, Silver Bow County, Montana**

This flat elliptical nugget shows some spongy crystal aggregates with a dark, slightly brassy, color. The piece, from a famous mining district, weighs 400.01g.

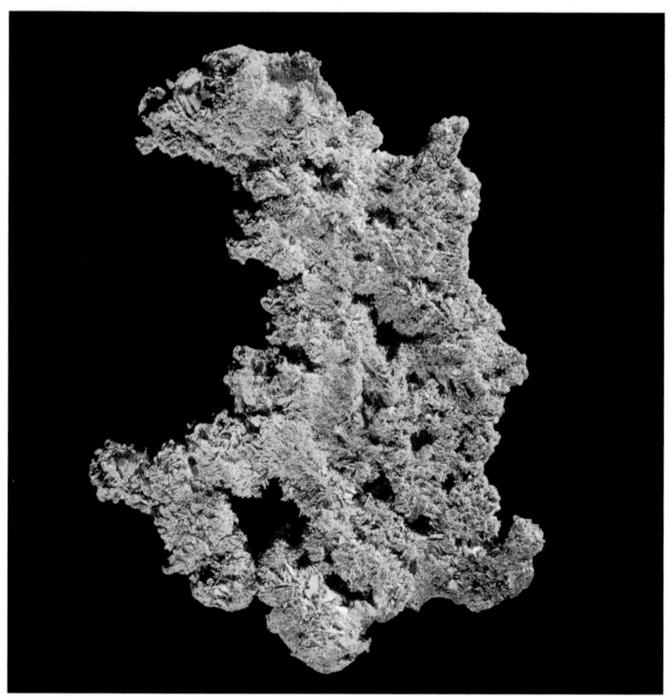

***Plate #7-50, Specimen #1018*** **GOLD – Winnemucca, Humboldt County, Nevada. (MH)**

## NEVADA

**#1018 12.8×8.5×1.5cm Plate #7-50**

**Winnemucca, Humboldt County, Nevada**

This piece, from either the Winnemucca or adjoining Ten Mile District, consists of a spongy mass of small, arborescent plates and "wire" or spicular crystals that have been etched from a large quartz matrix. The specimen weighs 191.11g.

***Plate #7-51, Specimen #5030*** **GOLD – Nightingale District, Pershing County, Nevada. (MH)**

**#5030 4.0×3.4cm Plate #7-51**

**Nightingale District, Pershing County, Nevada**

This superb specimen consists of many spinel-twinned wires.

## NEW MEXICO

**#3854 3.0×2.5×2.0cm**

**San Pedro Mine, Santa Fe County, New Mexico**

This is an attractive combination of fine spicular crystals and platy bands in calcite, with individual crystals to 1.7cm.

## NORTH CAROLINA

**#T334 3.4×2.3×1.4cm Plate #7-52**

**Reed Gold Mine, Cabarrus County, North Carolina**

This exceptional, richly colored and slightly water-worn crystalline mass has a 2.4cm edge on a flattened, twinned, octahedron. The piece was mined in 1828 from the first gold mine in the United States. It was acquired in 1978 with a Humboldt University Museum label when the Al Buranek Collection was broken up. The piece weighs 34.97g and is pictured in *Mineralogical Record*, Volume 18, Number 1, 1987, page 76.

***Plate #7-52, Specimen #T334*** **GOLD – Reed Gold Mine, Cabarrus County, North Carolina. (MH)**

## UTAH

**#G135 3.5×0.6×0.5cm**

These bright, elongated, deeply fluted, complexly twinned crystals to 2.8cm have especially sharp faces. The specimen weighs 7.81g. It was originally in the DeLaBouglise Collection and was acquired from the Harvard Museum (#100364E) in 1981.

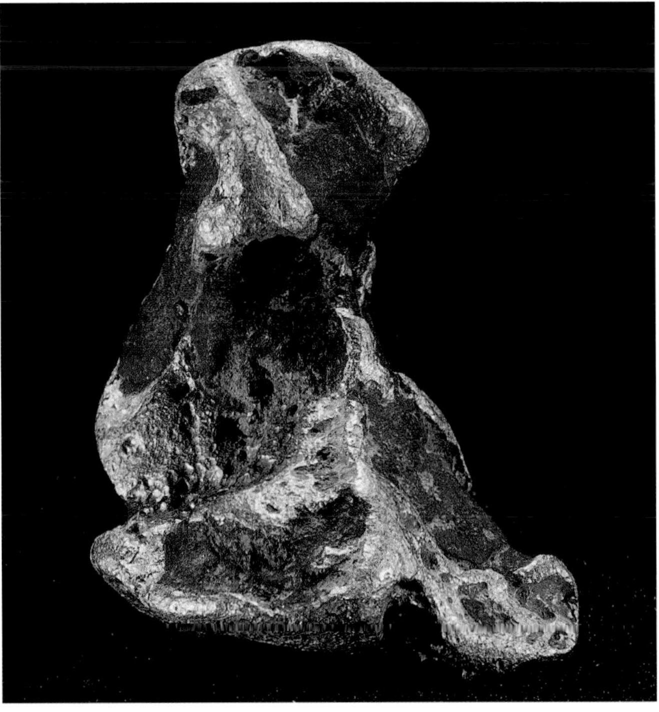

***Plate #7-53, Specimen #1959*** **GOLD – Old Classic Mine, Virginia. (MH)**

## VIRGINIA

**#1959 5.0×3.0×2.2cm Plate #7-53**

**Old Classic Mine, Virginia**

This rare Old Classic vein gold weighing 121.3g was formerly in the collection of geologist Jack Baragwanath, husband of the famous Neysa McMein of the Algonquin Roundtable, whose daughter, Joan Leech, is a friend of Barlow's.

## WASHINGTON

**#G206 10.3×6.0×1.5cm** **Refer to Chapter 3, Plate #3-46**

**Ace of Diamonds Mine, Liberty, Kittitas County, Washington**

This famous specimen, weighing 99.64g, consists of intergrown bright wires and brilliant arborescent crystals. It is one of the two finest golds recovered from this classic American locality. Barlow acquired the specimen from Ernest Butler, who handled the best pieces recovered from the mine. The piece was pictured on the cover of *Lapidary Journal*, September, 1971.

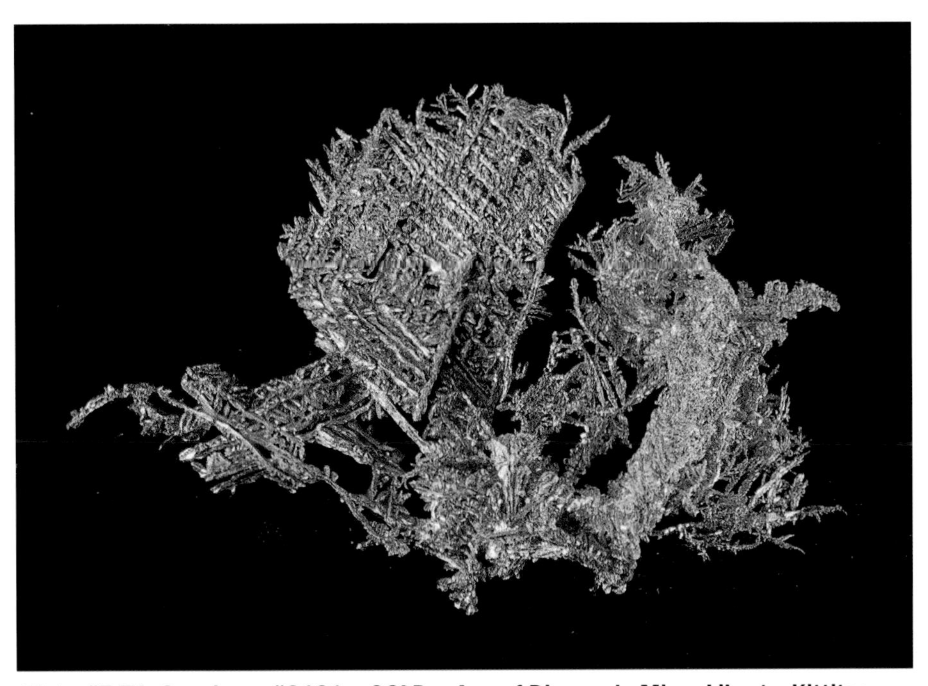

***Plate #7-54, Specimen #G191*** **GOLD – Ace of Diamonds Mine, Liberty, Kittitas County, Washington. (MH)**

**#G191 4.7×3.8cm**
**Plate #7-54**

**Ace of Diamonds Mine, Liberty, Kittitas County, Washington**

This particularly spectacular specimen acquired as part of the Ernest Butler Collection is composed of brilliant, pale yellow, reticulated crystals in flat sprays.

*Plate #7-55, Specimen #G202* **GOLD – Ace of Diamonds Mine, Liberty, Kittitas County, Washington. (MH)**

**#G202 7.5×5.0×1.0cm Plate #7-55**

**Ace of Diamonds Mine, Liberty, Kittitas County, Washington**

This specimen consists of a fine spongy mass of bright, arborescent crystals and plates, some of which are individual crystallographically continuous units to 2.0cm. Leached out of the enclosing calcite, the piece weighs 52.17g. This piece is the best of the Butler Collection.

*This piece is the best of the Butler Collection.*

**#3726 6.7×4.3×1.5cm**

**Ace of Diamonds Mine, Liberty, Kittitas County, Washington**

This brilliant aggregate of arborescent crystals has individual multiply twinned crystal units to 1.5cm. The piece, leached out of enclosing calcite, weighs 38.71g.

**#3840 4.5×3.5cm**

**Liberty Mine, Liberty, Kittitas County, Washington**

This exquisite, fragile, specimen consists of brilliant elongated and twinned octahedra in wires or spicules and arborescent crystal groups.

**#3840A 6.9×3.4×1.5cm**

**Liberty Mine, Liberty, Kittitas County, Washington**

A mate to #3840, this very fragile specimen similarly consists of brilliant elongated and twinned octahedra in wires or spicules and arborescent crystal groups, leached out of enclosing calcite.

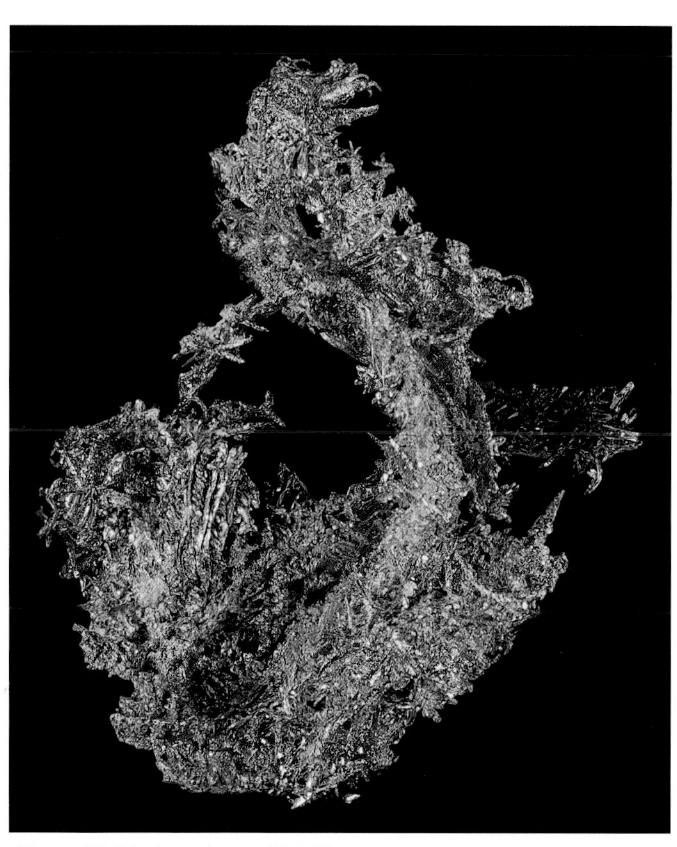

***Plate #7-56, Specimen #G199*** **GOLD – Ace of Diamonds Mine, Liberty, Kittitas County, Washington. (MH)**

**#3839 4.0×2.8×1.8cm**

**Ace of Diamonds Mine, Liberty, Kittitas County, Washington**

This specimen consists of a pale silvery-yellow aggregate of sharp, dendritic, elongated, and twinned crystals in and on quartz. The piece is mounted on a wooden base.

**#1017 8.2×4.6×2.0cm**

**Ace of Diamonds Mine, Liberty, Kittitas County, Washington**

This bright and showy, shell-like piece consists of platy mats of finely twinned and elongated crystals 1-2mm in length in arborescent clusters. Probably leached out of enclosing calcite, the piece weighs 107.53g.

**#3666 5.5×4.9×3.4cm**

**Liberty Mine, Liberty, Kittitas County, Washington**

This piece consists of showy clusters of bright, rich yellow, platy to equant crystals in quartz. The specimen, more massive than is typical from this locality, weighs 7.3oz.

**#G175 3.8×1.8cm**

**Ace of Diamonds Mine, Liberty, Kittitas County, Washington**

This unusual combination of brilliant, pale yellow, reticulated crystals and platy crystal clusters was acquired as part of the Ernest Butler Collection. The piece is mounted in a plastic box.

**#G199 5.0×3.5cm Plate #7-56**

**Ace of Diamonds Mine, Liberty, Kittitas County, Washington**

This spectacular specimen is composed of brilliant yellow reticulated crystals in a swirling spray. An aesthetic showy piece. It weighs 9.8g.

*An aesthetic showy spray.*

***Plate #7–57*** **GOLD – Tetrahexahedral forms.**
**Top row left to right: *#3733* (1.9cm); *#3757* (1.0cm); *#3731* (2.3cm)**
**Middle row left to right: *#3738* (9.0cm); *#3730* (3.8cm); *#3734* (1.2cm)**
**Bottom row left to right: *#3732* (2.2cm); *#3743* (1.0cm); *#3742* (1.3cm)**

**Crystals center left and bottom right are twinned tetrahexahedra. Crystals at top center and bottom center are untwinned tetrahexahedra. The intergrown group at top right is composed entirely of tetrahexahedral faces. (WW)**

## VENEZUELA

**#3730 2.5cm Plate #7-57**
**Refer to Chapter 3, Plate #3-48**

**Kilometer 88, Tiouiou, Zapata Field near Santa Elena, La Gran Sabana, Venezuela**

This spectacular, bright, single octahedron is 2.5cm high with a *c*-axis vertical. This world-class "floater" crystal, weighing 25.79g, is approximately 50% complete, with hopper-shaped octahedral faces exhibiting numerous ribs. The crystal edges are sharp, the luster is fine, and the color is excellent.

**#3743 10.0×6.0×6.0mm Plate #7-57**

**Kilometer 88 District, west of Santa Elena, La Gran Sabana, Venezuela**

This specimen consists of untwinned tetrahexahedra approximately 40% complete, with hoppers at positions of modifying octahedral faces. It weighs 3.20g.

**#3731 2.1×2.5cm Plate #7-57**
**Refer to Chapter 3, Plate #3-47**

**Kilometer 88, Tiouiou, Zapata Field near Santa Elena, La Gran Sabana, Venezuela**

This extraordinarily fine 1.8cm crystal is composed entirely of tetrahexahedral faces and is one of the world's choice thumbnail-size gold specimens. This specimen (pictured in *Mineralogical Record*, Volume 18, Number 1, 1987, page 90) passed through the hands of the noted gem and mineral dealer Allan Caplan on its way to the Barlow Collection.

**#3741 2.5×1.6×1.4cm**

**Kilometer 88 District, west of Santa Elena, La Gran Sabana, Venezuela**

This fine hopper-shaped octahedral crystal has pronounced ribs within the hoppers. It weighs 23.21g.

**#3733 1.9×1.4×1.0cm Plate #7-57**

**Kilometer 88 District, west of Santa Elena, La Gran Sabana, Venezuela**

This slightly elongated, hopper-shaped octahedron has deep ribs within the cavities. It weighs 9.88g. The specimen is pictured in *Mineralogical Record*, Volume 18, Number 1, 1987, page 90.

**#3738 3.4×2.6×1.4cm Plate #7-57**

**Kilometer 88 District, west of Santa Elena, La Gran Sabana, Venezuela**

This cluster of lightly abraded, complex tetrahexahedral crystals has remnant quartz and dark clay in crevices. It weighs 48.75g.

**#3737 1.8×1.25×0.9cm**

**Kilometer 88 District, west of Santa Elena, La Gran Sabana, Venezuela**

This slightly water-worn, hopper-shaped, elongated and flattened trapezohedron shows small modifying octahedral faces, probably twinned. It weighs 10.88g.

**#3735 3.7×1.3×1.0cm**

**Kilometer 88 District, west of Santa Elena, La Gran Sabana, Venezuela**

This elongated and twinned multiple crystal has individual flattened octahedra on two ribs. A slightly water-worn specimen, it weighs 21.40g.

**#3742 1.3×1.2×0.5cm Plate #7-57**

**Kilometer 88 District, west of Santa Elena, La Gran Sabana, Venezuela**

A crystal showing only the trapezohedron and cube is twinned and flattened with raised edges along the trace of twin plane. This rare complex crystal weighs 3.51g and is pictured in *Mineralogical Record*, Volume 18, Number 1, 1987, page 90.

**#3732, #3734, #3757** - not described.

Chapter 8

# Platinum and Other Native Metals

*by Gene L. LaBerge*

Platinum has been used for centuries in jewelry, where it was used as a substitute for gold or silver. In fact, it was originally named "platina" (from *plata*, the Spanish word for silver) because of its similarity to silver. It has also been used extensively in laboratory experiments because of its high melting point and resistance to chemicals. Currently, its major use is in catalytic converters in automobile exhaust systems and in the refining of petroleum. It is also widely used in "high tech" electronic equipment. Thus, this noble metal touches our lives in many ways.

Platinum is the most common member of a group of noble metals known as the platinum group elements. It was first discovered in placer deposits in Colombia, South America early in the 18th century. When first encountered in South America, it was thought to be "immature gold," so it was thrown back into the river to "develop." A small amount of platinum is still recovered from these deposits in Colombia. In the early 19th century platinum was discovered in deposits in the eastern Ural Mountains of Russia. For the next century, most of the world's supply of platinum came from these placer deposits in the Nizhni Tagil District of the Urals. These early sources were native platinum occurrences.

Native platinum was also recovered from placer deposits at Goodnews Bay on the Bering Sea coast of Alaska from the 1930s to 1980. This locality was the only area in the United States mined solely for platinum. Like platinum from the earlier Russian sources, the specimens from this site occurred in irregular nuggets.

Early in the 20th century it was discovered that significant amounts of platinum could be recovered from the copper-nickel sulfide ores at Sudbury, Ontario, the Bushveld igneous complex in the Republic of South Africa, and elsewhere. Platinum in these occurrences was mainly in the form of sperrylite, a platinum arsenide. Discovery of the large deposit at Noril'sk in northern Siberia again placed Russia as a major source of platinum metal.

*The recent discovery of spectacular crystals of sperrylite in the ores at Noril'sk, Siberia sparked an interest in collecting platinum minerals.*

Traditionally, platinum has not been a popular collector mineral because it occurs as irregular nuggets of native platinum or as small crystals of sperrylite in copper-nickel sulfide ores. However, the recent discovery of spectacular crystals of sperrylite in the ores at Noril'sk, Siberia sparked an interest in collecting platinum minerals. This interest was further enhanced by the introduction of exquisite native platinum crystals from Konder, in the Khabarovsk area in eastern Siberia. Because native platinum crystals were previously unknown, a major initial concern was to establish that the crystals were natural. The distribution of manufactured "crystals" of gold, attributed to placer deposits in Venezuela in the 1980s was the cause for this concern. However, an article in *Lapis* (October 1995) by Fehr, Hochleitner, and Weiss seems to establish that the crystals are indeed natural.

The appearance on the market of these well-crystallized forms of platinum minerals has also sparked an interest in the scientific community to explain how, and where, such crystals could be formed. Unlike gold, which forms in a wide variety of hydrothermal environments, platinum originates only in dark-colored igneous rocks rich in olivine and pyroxene minerals. Platinum is present in exceedingly low concentrations of a few parts per billion in these basaltic magmas that originate deep within the earth. The major problem is to concentrate the platinum sufficiently so meaningful deposits, and crystals, might be formed. The following mechanisms may be modified as more research is conducted, but serve as a general guide.

When sulfur is present in (or added to) the basaltic magma, platinum, along with copper, nickel, and iron may be concentrated as a sulfide-rich liquid that settles to the bottom of the intrusion and eventually crystallizes. Thus, a tiny amount of platinum in a huge volume of molten basalt may be greatly concentrated in the sulfide-rich liquid. If, as often happens, some arsenic is present in the sulfide-rich liquid, platinum may combine with it and crystallize as sperrylite. Well-formed sperrylite crystals may develop in fluid within the sulfide-rich zone. This condition evidently occurred at Noril'sk where well-formed sperrylite crystals are surrounded by the sulfide minerals pentlandite, chalcopyrite, and pyrrhotite.

Platinum may also be concentrated with chromite (iron chromium oxide) and magnetite (iron oxide), which may separate as a liquid from the basaltic magma. In this environment, platinum forms an alloy with iron, which may exist as molten platinum-iron liquid and fill irregular spaces between the common minerals in the rock. As the rocks cool, these areas solidify, and, when the enclosing rocks are weathered, the heavy, chemically resistant, platinum forms nuggets. Barlow specimen #5775 (Plate #8-1) strongly supports this concept, because crystals of chromite, green chrome pyroxene, and maybe altered olivine are embedded in this large, 852g, significant platinum nugget from Konder, Ajano-Maiskiy region, Khabarovsk Kray, Russia.

Native platinum nuggets and crystals characteristically contain up to 28% iron as part of their crystal structure. The platinum-iron alloys range from three parts platinum, one part iron to nearly equal portions of platinum and iron. Interestingly, the iron content of native platinum nuggets and crystals often renders them somewhat magnetic.

Whatever the mechanisms for concentrating the native platinum in the source rocks, platinum's resistance to chemical attack and high density ensure that it will remain as a heavy mineral in placer deposits when the enclosing rocks are broken down chemically and eroded.

*The Barlow Collection contains an outstanding suite of platinum minerals (27 specimens), including nuggets from most of the major sources, an incredible selection of native platinum crystals, one of the finest sperrylites known, and an excellent specimen of the rare platinum-bearing mineral potarite.*

*Plate #8-1* **PLATINUM**
**Upper left:** ***Specimen #T455*** **– Tura River, Perm, Ural Mountains, Russia.**
**Center right:** ***Specimen #5775*** **– Konder, near Neikan, Ajano-Maiskiy Region, Khabarovsk Kray, Russia.**
**Lower left:** ***Specimen #5737*** **– Fox Gulch, Goodnews Bay, Alaska. (MH)**

## PLATINUM (NATIVE ELEMENT)

**#5775 7.5×5.5×3.0cm**
**Plate #8-1 (center right)**

**852g**

**Konder, near Neikan, Ajano-Maiskiy Region, Khabarovsk Kray, Russia**

An outstanding, large, slightly rounded nugget with minor matrix. The surface of the nugget is pitted with subrounded cavities that contain minor black chromite, a green chrome-rich pyroxene, and a buff colored mineral that may be altered olivine. The surface of the specimen contains numerous areas that appear to be slightly rounded cubic crystalline areas of platinum. The luster varies from dull silvery white on the rounded areas to a brilliant metallic luster in the protected areas within the cavities where possible crystal faces of platinum are preserved. The grains of chromite, chrome pyroxene, and olivine included in the nugget suggest that liquid platinum was present within the host rock. The nugget is slightly magnetic due to the iron present in alloyed form.

**#T455 1.8×1.6×0.9cm Plate #8-1 (upper left)**

**14.3g**

**Tura River, Perm, Ural Mountains, Russia**

This platinum specimen from the classic Ural Mountains locality is a generally disc-shaped nugget with a pitted surface. The high areas are silver-white, while the depressions are coated with a fine black earthy material. The nugget shows little evidence of being water worn, but it is moderately magnetic, suggesting a high iron content.

**#5737 2.0×1.1×1.0cm Plate #8-1 (lower left)**

**13.0g**

**Fox Gulch, Goodnews Bay, Alaska**

This specimen is an elongate 13g nugget with an irregular, pitted surface. Unlike the Russian nuggets, it is not magnetic. The high points on the nugget are silver-white and slightly water worn. The depressions are partially filled with brown limonite-stained mud. This nugget is from the collection of the former Goodnews Bay Mining Company. It is the second largest nugget recovered in their mining operations at the now defunct Bering Sea coast locality. The largest nugget went to the Harvard University Collection shortly after the collection was acquired by longtime dealer David New.

**RARE LARGE ALASKAN NUGGET**

The rare platinum nugget from Goodnews Bay, Alaska comes from the only locality in North America where successful mining solely for platinum was done. This small community on the Bering Sea coast at Bethel, Alaska was the site of placer mining operations by Goodnews Bay Mining Company from 1934 to 1980. A large floating dredge recovered native platinum (and some gold) from buried stream channels near the mouth of the Salmon River. The platinum was presumably derived from a peridotite intrusion in the area called Red Mountain. During World War II the platinum was shipped to the Johnson-Matthey metal works in England for bomb fuses.

I first saw and photographed the platinum nuggets in 1962 while working as an exploration geologist in Alaska. When mining operations ceased in 1980, the larger platinum nuggets not retained by the mine owners were purchased by mineral dealer David New. The largest nugget acquired by New was sold to Harvard University. The nugget in the Barlow Collection, from David New, is one of the larger nuggets to come from this unique, now exhausted, locality.

*G.L.*

The recent find of platinum crystals from eastern Siberia is also well represented in the collection. Although the platinum nuggets from Russia are magnetic, the crystals are not magnetic.

***Plate #8-2*** **PLATINUM**
***Specimen #T455 (upper left)*** **– Tura River, Perm , Ural Mountains, Russia (not described in text).**
***Specimen #5737 (upper right)*** **– Fox Gulch, Goodnews Bay, Alaska (not described in text).**
***Specimen #5683 (lower left)*** **– Konder, near Neikan, Ajano-Maiskiy Region, Khabarovsk Kray, Russia.**
***Specimen #5682 (lower right)*** **– Konder, near Neikan, Ajano-Maiskiy Region, Khabarovsk Kray, Russia. (MH)**

## PLATINUM

**#5682 1.0×0.7×0.5cm**
**Plate #8-2 (lower right)**

**2.3g**

**Konder, near Nelkan, Ajano-Maiskiy Region, Khabarovsk Kray, Russia**

An outstanding cluster of five, sharp, dull-white, crystals that appear to be penetration twins. The largest crystal is a 0.6cm cube surrounded and intergrown with 0.3-0.4cm cubes. A superb specimen.

**#5683 0.8×0.6×0.7cm**
**Plate #8-2 (lower left)**

**2.3g**

**Konder, near Nelkan, Ajano-Maiskiy Region, Khabarovsk Kray, Russia**

A spectacular, sharp, 0.6cm, 2.4g, cubic crystal with a spinel twin. The platinum crystal is a bright metallic silver-gray color. One side shows the twinned cubic crystal very well. The other side shows a more complex intergrowth of the spinel twin. A sharp, interesting twinned crystal.

*Plate #8-3* **PLATINUM**
***Specimen #5772 (first row #1)*** **– Konder, near Neikan, Ajano-Maiskiy Region, Khabarovsk Kray, Russia.**
***Specimen #5773 (first row #2)*** **– Konder, near Neikan, Ajano-Maiskiy Region, Khabarovsk Kray, Russia.**
***Specimen #5771 (second row #1)*** **– Konder, near Neikan, Ajano-Maiskiy Region, Khabarovsk Kray, Russia. (MH)**

## PLATINUM

**#5773 9×8×7mm**
**Plate #8-3 (first row #2)**

**8.6g**

**Konder, near Nelkan, Ajano-Maiskiy Region, Khabarovsk Kray, Russia**

A superb, large cubic crystal of platinum with two 3×4mm cubic crystals included in the large crystal as penetration twins. This is the largest platinum crystal found to date.

**#5771 8×7×5mm**
**Plate #8-3 (second row #1)**

**4.7g**

**Konder, near Nelkan, Ajano-Maiskiy Region, Khabarovsk Kray, Russia**

A rare, large platinum crystal partially coated with gold. The large platinum crystal is intergrown with two smaller platinum cubes, one 5×3mm and the other 3×3mm.

**#5772 16×8×6mm Plate #8-3 (first row #1)**

**8.0g**

**Konder, near Nelkan, Ajano-Maiskiy Region, Khabarovsk Kray, Russia**

A slightly rounded, intergrown cluster of platinum crystals. Two large (7.0×5.0×5.0) crystals with six smaller crystals are intergrown in a complex grouping. An unusual, spectacular specimen.

Aside from the crystals pictured and described, the collection has 18 additional specimens.

Interestingly, results of a study by Tistl published in the journal *Economic Geology* in 1994 reported that cubic crystals of platinum-iron alloy to 1mm and crystal aggregates to 13mm are present in the Alto Condoto Complex in northwestern Colombia. Thus while the recent discovery of platinum crystals in the Konder area of Siberia has been of special interest to mineral collectors, cubic crystals of platinum also occur in the region of Colombia where platinum was first discovered several hundred years ago.

*Plate #8-4, Specimen #3077* **PLATINUM – Chatham Research Laboratory, San Francisco, California. (MH)**

## PLATINUM

**#3077 5.0×4.0cm Plate #8-4**

**Chatham Research Laboratory, San Francisco**

In 1983, a rare laboratory error at Chatham Research Laboratory resulted from an attempt to produce ruby and chrysoberyl crystals using platinum as a catalyst. Loss of control of temperature and pressure exploded the crucible into many pieces creating an extraordinary result.

This specimen was the largest and finest sample secured from that incident. It is pictured on page 135 of *Planet Earth Noble Metals Book*, 1984.

## POTARITE (PLATINUM MERCURY INTERMETALLIC COMPOUND)

**#T143 1.5×1.4×1.0cm Plate #8-5**

**14.7g**

**Limeira Farm, Pilar, Minas Gerais, Brazil**

This rare platinum-mercury specimen has a general tear-drop shape. The silver-white potarite forms globular (botryoidal) mounds to about 6mm. Some mounds are composed of smaller 1mm globular segments. The globular mounds are a lustrous silver-white, and the depressions between the mounds have a dull black or brown coating that serves to emphasize the mounds. This outstanding example of this rare mineral was found in the bed of a narrow river near the Limeira farm. It was purchased from Carlos Barbosa, well-known Brazilian mineralogist and dealer, in Tucson, 1974.

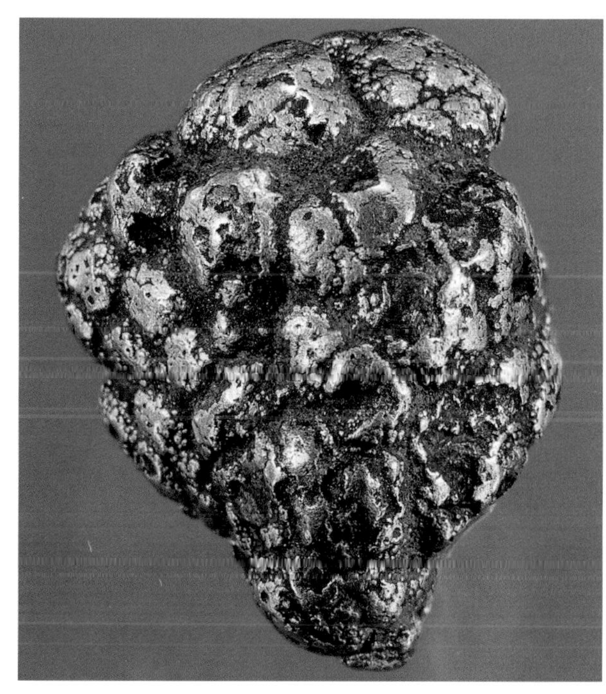

*Plate #8-5, Specimen #T143* **POTARITE – Limeira Farm, Pilar, Minas Gerais, Brazil. (MH)**

## SPERRYLITE (PLATINUM ARSENIDE)

**#5073 9.0×8.0×6.0cm Plate #8-6**

**October Mine, Noril'sk, Siberia, Russia**

This superb specimen consists of a 4.5cm group of parallel, mirror-bright, silver-white crystals on a brassy chalcopyrite matrix. The largest individual crystal is 1.5cm. *This is without question one of the most spectacular sperrylite specimens in existence.* Originally brought out of Russia by dealers Star Van Scriver and Evjen Pliaskov, it was acquired by collector William Pinch in 1992. A few months later, Pinch sold the piece to Barlow with an X-ray that revealed the presence of additional crystals buried within the matrix. A thorough cleaning and trimming by dealer Cal Graeber exposed the underlying crystals. The sperrylite crystals are sharp cubes modified by the octahedron. They rest in a stacked group atop a complicated mixture of paolovite, talnakhite, taimyrite, and mooihoekite.

*Plate #8-6, Specimen #5073* **SPERRYLITE – October Mine, Noril'sk, Siberia, Russia. (VP)**

## OTHER NATIVE METALS

Crystals larger than microscopic size of native antimony, arsenic, bismuth, and tellurium, as well as native lead, occur very rarely in nature. Native antimony, arsenic, bismuth, and tellurium generally form in hydrothermal gold and silver deposits. Native lead crystals form occasionally in alkalic igneous rocks related to nepheline syenites. Crystals of several of these rare native elements are represented in the Barlow Collection.

The metals constituting these rare native elements, as well as their more common neighbors, have very diverse uses in our everyday lives. For example, antimony is used in making "type-metal," for plates in storage batteries, in toothpaste tubes, and as a component in pewter, to name a few uses. Bismuth is used in common products like Pepto-Bismol, in cosmetics, in many glazes, and in electrical fuses and sprinkler systems for fire control because of its very low melting point. Tellurium is used to make tough, hard rubber for hoses and cables, as an additive in glass to produce blue and brown colors, and to make lead harder. Lead, of course, has a wide variety of uses in our industrial society.

***Plate #8-7, Specimen #5001*** **ANTIMONY – Ohi Railroad Quarry, Seina, Finland. (MH)**

***Plate #8-8, Specimen #5439*** **BISMUTH – 585 Meter Level, Shaft 38, Opal Vein, Schlema, Saxony, Germany. (MH)**

### ANTIMONY (NATIVE ELEMENT)

**#5001 4.0×3.0×0.8cm Plate #8-7**

**Ohi Railroad Quarry, Seina, Finland**

A superb, lustrous silver-white crystal plate. Prominently striated, steplike crystal faces are formed by stacking of the trigonal crystals. Some faces are strongly striated in a pattern reminiscent of rhombohedral cleavage. This is an outstanding miniature specimen.

### BISMUTH (NATIVE ELEMENT)

**#5439 4.0×3.5×2.0cm Plate #8-8**

**585 Meter Level, Shaft 38, Opal Vein, Schlema, Saxony, Germany**

Although bismuth is a well-known mineral in the German silver and cobalt-nickel mines, having been found in rich to pure masses many centimeters across, crystals are exceptionally rare and usually only a few millimeters across. These rhombohedral, but pseudo-isometric, crystals are skeletal. This specimen closely resembles a flattened octahedron but is actually a tabular rhombohedron composed of numerous minor rounded faces. Low power (10×) microscopic examination of the bismuth crystal faces reveals a marvelous topography. Rounded, etched faces comprising the triangular rhombohedral form are beaded to corrugated and resemble a simplified artist's rendition of bark on an elm tree. The rhombohedral faces that are elongated show skeletal clefts, a few ridges parallel to their length, as well as a few curved ridges that follow the adjacent face intersections. The elongated rhombohedral faces are also beaded in topography and somewhat resemble closely spaced Mayan temple glyphs. The main silvery crystal is enormous (1.8×1.6×0.9cm) and rests on a vuggy bismuth matrix (4.0×3.5×2.0cm). Some of the adjacent vugs are home to "ram's horn" and pyramidal acanthite, cubic skutterudite, and wedge-shaped crystals (safflorite?) coated by minor erythrite. This is an important specimen from the 1958 find.

***Plate #8-9, Specimen #4721*** **LEAD – Langban, Sweden. (MH)**

### LEAD (NATIVE ELEMENT)

**#4721 4.0×3.2×2.5cm Plate #8-9**

**Langban, Sweden**

An exceptionally fine, somewhat distorted crystal with prominent striations is perched on a base of dodecahedral lead crystals. The main crystal is 2.3cm long, with a dull gray color. The smaller dodecahedral crystals are about 1.0cm across. The crystal faces are partially coated with lustrous black hausmannite crystals about 1mm across, along with several brownish red pyrochroite crystals from 5 to 12mm across.

***Plate #8-10, Specimen #5078*** **TELLURIUM – Tashkent, Uzbekistan. (MH)**

### TELLURIUM (NATIVE ELEMENT)

**#5078 8.0×4.7×5.0cm Plate #8-10**

**Tashkent, Uzbekistan**

An exceptionally large crystal of this rare semimetal. The lustrous silver-white crystal is 5.0×2.7×2.5cm with prominent rhombohedral cleavage. The matrix is milky quartz with disseminated crystals of the gold and silver telluride minerals calaverite, sylvanite, petzite, and kostovite.

CHAPTER 9

# Silver

*by Terry C. Wallace, Jr. and Robert W. Jones*

Native silver often forms the cornerstone of important mineral collections, and the Barlow Collection is no exception. It was Barlow's fascination with silver from the Upper Peninsula of Michigan that got him interested in building a serious silver collection. The proximity of Appleton to one of the world's classic copper and silver localities made it only natural that native copper and silver and silver-bearing minerals became one of Barlow's collecting focuses. Between 1973 and 1975 John acquired 17 native silver specimens, and the chase was on to build one of the world's finest collections. This chase extended to include silver- (and gold-) bearing species described in Chapter 10.

The appeal of silver to collectors goes far beyond the beauty of the specimens. Perhaps more than the pursuit of any other mineral, the quest for silver has driven great changes in society. Silver financed empires and major wars. It drove colonialism and gave rise to international banking. Potosi, the great Bolivian silver camp developed under Spanish control, was the largest city in the *world* for 30 years — and it is situated at 13,000 feet elevation. Silver specimens are markers of many great human endeavors; they represent an artistic bridge between man and the mineral kingdom. Only in this context can one evaluate a silver specimen.

*Perhaps more than the pursuit of any other mineral, the quest for silver has driven great changes in society. Silver financed empires and major wars. It drove colonialism and gave rise to international banking.*

For a collector, the place of origin and time of collection of a specimen are often more important than its appearance. A humble wire of silver from St. Joachimstal in western Bohemia, Czech Republic invokes memories of the father of mineralogy, Georgius Agricola, who was the town physician. Or a small silver from Leadville may invoke images of Colorado's raucous past.

The F. John Barlow silver collection is true to the traditions of a great collection. It is extremely rich in localities with more than 40 different sites represented and also rich in duplication. The collection contains many outstanding display specimens, but its real strength is in its size: more that 125 silvers. Many of these specimens show subtle variation in crystal habit or mineral assemblages. There are 11 silvers from Batopilas, Mexico alone. Barlow's outstanding native silver collection naturally led to his silver-bearing minerals collection. Taken together, these two collections make up the finest private silver species collection in North America.

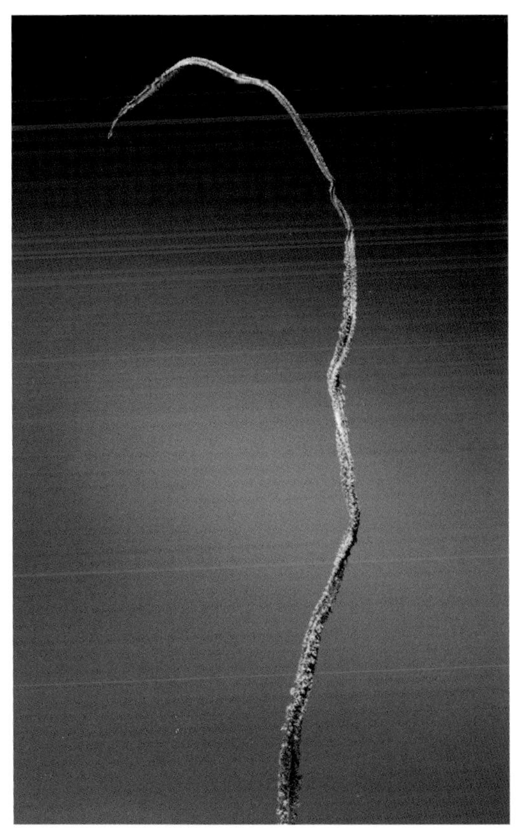

*Plate #9-1, Specimen #5377* **SILVER – Alura Mine, New South Wales, Australia. (MH)**

## AUSTRALIA

**#5377 19.0×0.4cm Plate #9-1**

**Alura Mine, New South Wales, Australia**

This 12.5cm long wire has modified cubic crystals sprouting continuously along all sides from bottom to top.

*Plate #9-2, Specimen #1624* **SILVER – Colquechaca, Bolivia. (MH)**

## BOLIVIA

**#1624 8.0×7.0cm Plate #9-2**

**Colquechaca, Bolivia**

This is a loosely tangled mass of fine silver wires forming a nest-like specimen. There is no matrix. The specimen is simply held together by the tangles. Colquechaca is about 100 km northeast of the famous Potosi area. Although the mines of Colquechaca are of minor economic importance in comparison to Potosi, the Gallofa Mine produced an extraordinary suite of silver minerals at the turn of the century. Some of the finest acanthites (see specimen #1135, Chapter 10, page 241) and pyrargyrites in museums around the world are from Colquechaca.

## CANADA

**#1151 5.0×3.0cm Plate #9-3**

**South Victoria Island, Hope Bay, Northwest Territories, Canada**

Sharp euhedral cubes of silver are not at all common as the metal tends to form branching wires and herringbone twins. This specimen, however, is made up of diverging stacks of small cubes that range in size from 2 to 5mm. Most are very well exposed and sharp-edged, but with a rather dull luster due to a coating of unknown composition. The interior of the specimen is a gray-white matrix to which the crystals and stacks are solidly attached. This is a superb specimen of cubic silver from an uncommon source.

*Plate #9-3, Specimen #1151* **SILVER – South Victoria Island, Hope Bay, Northwest Territories, Canada. (MH)**

**#1780 8.0×7.0cm**

**#3 Shaft, Castle Trethewey Mine, Cobalt, Ontario, Canada**

This is a crude crystallized mass of silver with minor amounts of safflorite as matrix. In addition, coatings of acanthite blacken much of the silver. The silver is generally short (to 1.0cm), sturdy, tapering wires. A few crudely formed cubes and some very minor calcite are also present. The specimen was obtained from Jack Everett, the geologist who discovered the deposit.

**#3351 13.0×5.5×9.0cm**

**Buffalo Mine, Timiskaming District, Ontario, Canada**

This is a classic example of Cobalt ore vein material. The face of the calcite vein, which is about 2.5cm thick, has been sliced and polished, revealing the typical long needlelike silver wires geometrically patterned in the calcite. Each slender wire has a halo of brilliant safflorite around it. The entire effect is cuneiform or graphic in appearance. The back side of the silver-bearing calcite vein is solidly attached to a matrix of typical gray-black basaltic rock into which the ore veins have been introduced. Some of these ore veins ran over 10,000oz. of silver to the ton. The specimen comes via the collection of M.H. Froberg from the Royal Ontario Museum.

***Plate #9-4, Specimen #4845*** **SILVER – Echo Bay Mine, Port Radium, Northwest Territories, Canada. (MH)**

**#4845 4.0×1.0cm Plate #9-4**

**Echo Bay Mine, Port Radium, Northwest Territories, Canada**

This is a textbook example of "ram's horn" silver. It is actually a group of wires, some 1-2mm thick, bonded together to form a 1.0cm wire that stands in a compound curve like a bent and crooked finger, nearly curling back on itself. Scattered over the face of the horn are perfect milky-white calcite crystals to 1mm in size. The entire horn is also heavily coated with crystallized microscopic argentite, which gives the wire a sparkly appearance when rotated under a light source.

***Plate #9-5, Specimen #4955*** **SILVER – Cobalt, Ontario, Canada. (MH)**

**#4955 7.0×3.0×2.0cm Plate #9-5**

**Cobalt, Ontario, Canada**

An unusual aesthetic wire from Cobalt. Its evolution was obviously not easy. The main wire is made of multiple parallel growth wires.

***Plate #9-6, Specimen #4957*** **SILVER – Cobalt, Ontario, Canada. (MH)**

### #4957 4.0×4.0×2.0cm Plate #9-6

**Cobalt, Ontario, Canada**

Much wire silver from Cobalt was completely enclosed in calcite when mined. The calcite was easily removed, sometimes revealing aesthetically arranged delicate wires of the native metal. This is clearly the case with this choice miniature. The base of the piece is an acid-etched white calcite from which hairlike to 1mm wires of bright silver emerge. The longest wire is about 3.0cm. This is a delightfully aesthetic specimen, an uncommon survivor of this silver camp, since the bulk of the metal went to the crusher without ever being treated to reveal the hidden beauty of the silver.

***Plate #9-7, Specimen #5141*** **SILVER – Highland Bell Mine, Beaverdell, British Columbia, Canada. (MH)**

### #5141 5.5×5.0cm Plate #9-7

**Highland Bell Mine, Beaverdell, British Columbia, Canada**

This specimen consists of a flattened mass of acanthite crystals, probably from a narrow vein filling. The acanthite is held together by bright curling wires of silver that gracefully protrude along the edges of the piece. All enclosing calcite has been etched away, revealing the complex silver network.

### #5367 12.5×5.5cm

**Christopher Mine, Cobalt, Ontario, Canada**

This piece is a classic example of a silver vein in sheet form still attached, in part, to the dark matrix. The silver is coated with massive argentite. No apparent crystallization of either mineral is evident. The silver sheet is freestanding to 7.0cm at the apex of the block of matrix. The specimen was obtained by collector Harry Quaas from the mine superintendent in 1967. The specimen became part of the Evan Jones Collection in 1991 and was acquired by Barlow in 1994.

## CHILE

### #5064 4.5×1.0cm

**Inca de Oro, Copiapo, Chile**

This loose 5.0cm wire of silver shows a typical twisting form with striations along much of it. The ends, however, crystallized crudely with one end showing an unusual, flattened, "spear-point" form.

## COBALT, CANADA

Prior to 1903 the area around Lake Timiskaming, on the edge of the Canadian Precambrian Shield north of Toronto, was just a farming region. The nearest important town was North Bay, a farming community in need of transportation to the outside world. The government agreed to build a railroad to service the area.

J.H. McKinley and Ernest Darragh were contracted to supply wooden ties for the roadbed. On August 7, 1903, at mile post 103 of the railroad, they found several veins of soft metallic ore. Samples sent to Montreal proved to be silver ore assaying at $4,000.00 per ton, an amazing sum when you consider that silver was selling for only pennies an ounce. Needless to say, the two men staked claims and proceeded to get rich!

The LaRose Mine was discovered in quite a different way. A local blacksmith, Fred LaRose, is reported to have thrown his blacksmith's hammer at an inquisitive fox. To his good fortune, he missed, and the errant hammer chipped a nearby outcrop, exposing a silver vein.

Coincidentally, the Nipissing veins were found about that same time. Before the O'Brien Mine property was staked in 1904, one of those mines eventually produced a half million dollars in silver. Each of these mine sites developed into a major producer, and the rush was on. The ground was so rich in places that one one-meter wide vein, exposed during roadwork in the new town of Cobalt, was nicknamed the "Silver Sidewalk." One experienced miner, William Trethewey, had been in Cobalt only two days when he found one rich vein, and then another later in the week!

One problem with the Cobalt silver ores was that they contained considerable arsenic, which presented problems with known smelting processes. Today it might well have created an environmental stalemate. A treatment process using cyanide for the ore was finally developed, and the area boomed. Huge masses of silver were encountered while mining veins that could assay a thousand ounces of silver per ton and more.

Mineralogically, the silver was generally sheet-like in seams, or enclosed in calcite as lovely arborescent wire groups. Etched free, these wire groups are Cobalt's most attractive specimens. Along with native silver, small amounts of the silver sulfosalts proustite and pyrargyrite were found, but few specimens of these minerals survived.

Arsenic, cobalt, and bismuth minerals were also abundant. Large masses of crystallized arsenic, bismuth, and their sulfide and oxide minerals were once common. The bismuth crystals from Cobalt are particularly noteworthy.

Today, the Cobalt camp is closed, the obtainable veins exhausted; but in its day, it was one of the great rushes of Canada and served to open up the Canadian Shield to mining.

***Plate #9-8, Specimen #4764*** **SILVER – Pribràm, Bohemia, Czech Republic. (MH)**

## CZECH REPUBLIC

**#4764 5.5×3.0cm Plate #9-8**

**Pribràm, Bohemia, Czech Republic**

The base of this specimen is a 5.0×3.0×1.5cm mass made up largely of white, well-crystallized calcite in rhombohedral form. Embedded at its base in calcite is a deeply striated 1.0cm thick wire of silver that starts flat on the calcite and curls up in hook form so the upper two thirds are free-standing. Two tiny white calcite crystals are attached to the final curl of the silver. This is a lovely classic from a classic source.

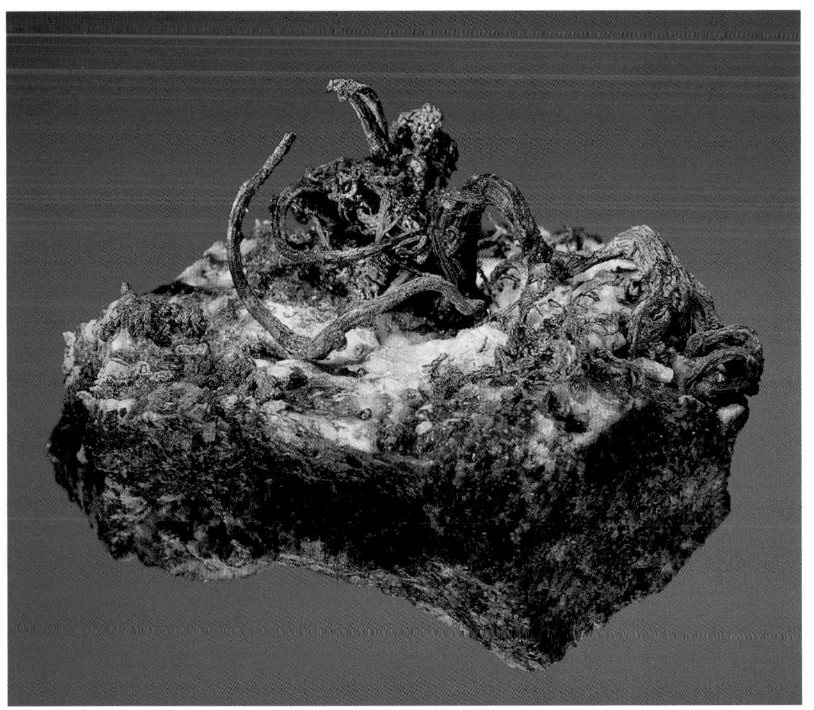

Plate #9-9, Specimen #4776 **SILVER – Pribràm, Bohemia, Czech Republic. (MH)**

**#4776 9.0×7.0×6.0cm Plate #9-9**

**Pribràm, Bohemia, Czech Republic**

A solid mass of matrix wall rock forms the pedestal for a superb tangle of silver wires that have been exposed by etching away the enclosing calcite vein. Remains of the acid-polished calcite are evident. The silver wires are tarnished bronze in color. They curve up from the rock matrix in all directions, varying in length from a few millimeters to some 8.0cm long. Much silver remains embedded in the calcite vein within the wall rock.

*Plate #9-10, Specimen #4909* **SILVER, pseudomorph after dyscrasite – Shaft 21, Pribràm, Bohemia, Czech Republic. (MH)**

**#4909 7.0×8.0×4.0cm Plate #9-10**

**Shaft 21, Pribràm, Bohemia, Czech Republic**

This specimen is a truly exceptional pseudomorph of native silver after dyscrasite. The replacement is complete, yet the original, long, slender lathlike form of the dyscrasite has been preserved. The crystals, glowing with a silver patina, interlock like a pile of jackstraws. The longest crystal across the face of the specimen exceeds 6.0cm. This is one of the best such pseudomorphs to come from an extraordinary cache brought to market in 1987.

## Germany

The Barlow Collection contains native silver from classic and historically important German mines of the Erzgebirge and Harz Mountains. The rich mining lore associated with these deposits, which have been mined for over 1,000 years, makes these specimens "classics," eagerly sought by collectors worldwide.

It would be difficult to overestimate the importance of the discovery of silver and related minerals in the Erzgebirge and Harz Mountain regions of Germany. These deposits have had a profound impact on the development of Germany itself and on the emergence of the sciences of mining, engineering, and mineralogy.

The earliest discoveries here occurred before the end of the first millennium A.D., during the Dark Ages. It can be said that the discovery of this wealth of precious metal contributed to the end of that era.

When mining commenced, knowledge of mining methods was largely based on early Roman techniques. This knowledge was handed down from father to son. Because knowledge was wealth, it was guarded jealously.

All this changed when Georg Bauer, a physician in the region, published a book in Latin called *De Re Metallica* under his Latinized name, Georgius Agricola. In it, he documented for the first time the traditional and closely guarded secrets of mining and ore treatment. His work was to have a signal impact on the exploration for, and mining of, ores for centuries thereafter. Even as late as the 19th century, Bauer's book was at the elbow of many successful mining engineers. Later translated into English by former President Herbert Hoover (who was also a world-renowned mining engineer) and his wife, Lou, the book is still interesting to read.

A second major impact of silver mining in the Erzgebirge was the emergence of science as an important aspect of mining. Freiberg was the home of the first mining academy, founded in 1765. It was here that Abraham Gottlob Werner (1750–1780) developed the foundations for modern mineralogy.

German miners were one of the most highly prized resources in the pursuit of mineral wealth. When mineral deposits were discovered elsewhere, it was German miners and their advanced techniques that were imported as skilled labor. Saxon miners helped develop the Kongsberg Mine in Norway. In the tin and copper mining region of Cornwall, Queen Elizabeth I brought German miners into the southwestern corner of England to train local fishermen and farmers in the ways of mining. Later, Cousin Jacks from Cornwall became imported skilled labor in many New World mining districts, including Bisbee and elsewhere in Arizona, Colorado, Michigan's Upper Peninsula, and the silver mines of Mexico and South America.

Of passing note is the fact that this region in Europe even gave us the name for our most used currency, the dollar. Silver mined at Joachimsthal, now Jachymov, Czechoslovakia, was minted into coins called *thalers*. You can see what an easy transition it was to go from *thaler* to *dollar*.

So, when you admire a lovely silver specimen from the Harz or study a complex species from the Erzgebirge, remember how pervasive and important to world development were the deposits here. Pause for a moment and wonder of the early miners.

## GERMANY

**#4777 5.0×2.3cm**

**White Elk Mine, Freiberg, Saxony, Germany**

This piece is a flattened mass of silver the upper surface of which is completely covered with the finest of hairlike silver threads to 1.0cm. The threads are so closely packed that individual wires are difficult to see. The underside of the specimen forming the base matrix is a solid plate of silver coated with microscopic acanthite.

**#5061 6.0×5.0×3.0cm**

**Refer to Chapter 3, Plate #3-81**

**Himmelsfürst Mine, Freiberg, Saxony, Germany**

This extraordinary silver specimen is further described in Chapter 3, page 89. Its thick curling wire, coated with acanthite, forms a complete loop on itself several times. The wire is deeply striated with bright microscopic acanthite crystals still visible in protected areas. The Himmelsfürst Mine produced Freiberg's finest silvers, of which this is one. The greatest production here occurred between 1760 and 1860.

*Plate #9-11, Specimen #5440* **SILVER – Level 120m, Tunnel 921, Pohla, Saxony, Germany. (MH)**

**#5440 8.0×4.5×4.0cm Plate #9-11**

**Level 120m, Tunnel 921, Pohla, Saxony, Germany**

This superb specimen of fernlike and "herringbone" silver crystals is a solid mass of black native arsenic well covered with the bright silver that stands in stark contrast to it. The crystal growths lie flat on the faces of the matrix and are more or less uniform in size, each about 1.5–2.0cm in length. The crystals are extremely delicate, barely attached in some places. They cover well over 50% of the black matrix and are well distributed, rendering an artful geometric design to the piece. Collected by Barlow in 1991.

*Plate #9-12, Specimen #1883* **SILVER – St. Andreasberg, Harz Mountains, Germany. (MH)**

**#1883 4.5×3.0cm Plate #9-12**

**St. Andreasberg, Harz Mountains, Germany**

This miniature consists of a twisted mass of striated silver wires, without matrix, coated in part by acanthite, particularly along the base. The freestanding wires are bright in areas. The longest wire group exceeds 3.0cm. The Harz silver deposits were discovered in 968 A.D., and the great mines of St. Andreasberg began operations in the 15th century. A superb miniature from a classic source. The piece was obtained by Barlow from the Lindstrom Collection.

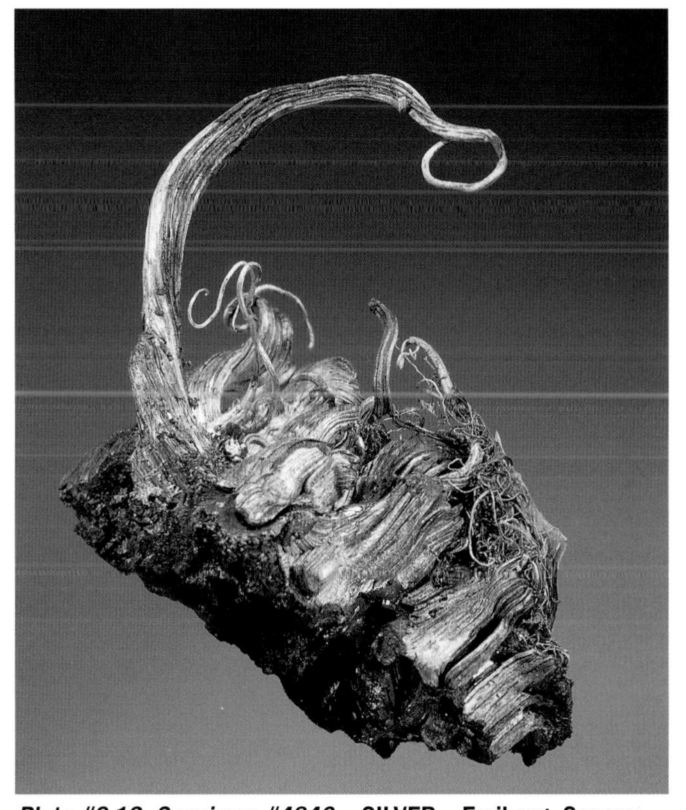

*Plate #9-13, Specimen #4846* **SILVER – Freiberg, Saxony, Germany. (MH)**

**#4846 3.5×2.0cm Plate #9-13**

**Freiberg, Saxony, Germany**

On a small base of acanthite, bright curving silver wires in near-parallel arrangement form a nearly solid mass. From that, a few wires stand free, the major wire exceeding 4.0cm as it curls above the mass. Silver was discovered at Freiberg in 1163 A.D.

***Plate #9-14, Specimen #4910*** **SILVER – Schneeberg, Saxony, Germany. (MH)**

**#4910 7.0×4.0×4.0cm Plate #9-14**

**Schneeberg, Saxony, Germany**

This is a very unusual specimen in which silver is replacing a galena crystal. The original galena core remains and may be seen on the reverse side of the specimen. Bright masses of silver are attached at random to the replaced core galena. Schneeberg is in Saxon Erzgebirge and was discovered in the mid 15th century. It is noted for producing Europe's finest proustites.

***Plate #9-15, Specimen #5222*** **SILVER – Himmelsfürst Mine, Freiberg, Saxony, Germany. (MH)**

**#5222 5.0×2.5cm Plate #9-15**

**Himmelsfürst Mine, Freiberg, Saxony, Germany**

This tangled mass of silver wires with a Medusa-like form is heavily encrusted with 1mm white calcite crystals, giving the specimen a sugary appearance. The thin silver wires, black with an acanthite coating, curl in all directions; the longest wire exceeds 4.0cm.

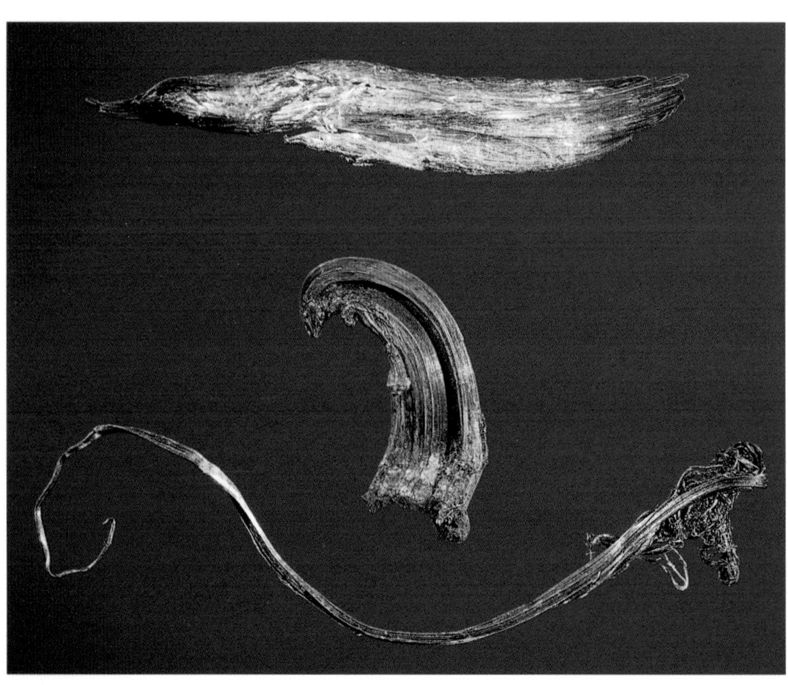

***Plate #9-16***
***(Upper) Specimen #5704*** **SILVER – Monte Narba Mine, Sarrabus, Sardinia, Italy.**
***(Middle) Specimen #5063*** **SILVER – Tuviois Mine, Sarrabus, Sardinia, Italy.**
***(Lower) Specimen #5415*** **SILVER – Tuviois Mine, Sarrabus, Sardinia, Italy. (MH)**

## ITALY

**#5704 6.5×1.0cm Plate #9-16 (upper)**

**Monte Narba Mine, Sarrabus, Sardinia, Italy**

This aesthetic wire in the shape of a feather forms an elongated spray. It was obtained from the Bally Collection. This classic European silver locality is well described in de Michele's *Guida Mineralogia d'Italia*.

**#5063 3.0×2.0cm Plate #9-16 (middle)**

**Tuviois Mine, Sarrabus, Sardinia, Italy**

This heavy "ram's horn" wire is an aesthetic specimen.

**#5415 11.0×1.0cm Plate #9-16 (lower)**

**Tuviois Mine, Sarrabus, Sardinia, Italy**

This piece is a fine wire compared with the "feather" #5704.

## Mexico

The Barlow Collection is rich in Mexican silver. This is fitting considering that since the conquest of Mexico in 1521 by Cortez, this country has produced one third of the total silver mined in history. The Barlow index of Mexican localities is impressive: Taxco, Zacatecas, Guanajuato, and Batopilas.

The Batopilas mining district in the southwestern corner of the state of Chihuahua is especially well represented. Although the vein deposits of Batopilas were discovered in 1632, significant quantities of herringbone silver from the Nevada Valenciana Mine there have reached collectors in the last 20 years. Much of the Nevada Valenciana material was obtained by John Whitmire and Gene Schlepp — and the better pieces found their way to the Barlow Collection.

By Spanish standards, Batopilas was a latecomer. Huge quantities of native silver, much of it in delightfully delicate herringbone crystal groups etched from calcite veins, have been produced. The deposit has also produced superb proustite, acanthite, and other sulfosalts. But, in spite of these other species, the native silver groups from here are what come to mind when someone mentions Batopilas. Most notable are the elongate wire crystals, almost always twinned, found in groups to a dozen centimeters and more. This twinning, which so often results in a herringbone pattern, is characteristic of Batopilas silvers. Few specimens of this type are now seen from any other Mexican source.

The first deposit was found by prospectors who stumbled on the water-polished surface of a silver-bearing ledge in the Batopilas River. That the silver veins were especially rich is attested by one historical comment wherein the miners reported that the ore could not be blasted as the vein contained too much metal and had to be hand-chiseled out. The mining district also bears the distinction of having been raided by Pancho Villa.

Today, little mining activity is taking place at Batopilas, although the ore body in the Valenciana is not yet exhausted. If the price of silver climbs to U.S. $6.00/oz., extensive exploration will resume in the district. The potential for more silver specimens would then be excellent.

***Plate #9-17, Specimen #1152*** **SILVER – New Nevada Valenciana Mine, Batopilas District, Chihuahua, Mexico. (MH)**

**#1152 8.5×5.5cm Plate #9-17**

**New Nevada Valenciana Mine, Batopilas District, Chihuahua, Mexico**

Brought out of Batopilas in 1975 by dealers Si and Ann Frazier, this specimen is one of a large quantity obtained at that time. The silver is classic "herringbone," typical of Batopilas, though these twinned wires are sturdier than most. The 6.0cm silver stands vertically from a block of white calcite that was partially etched away to reveal the metal. Black acanthite encrusts some of the silver. Pictured in *Mineralogical Record*, Volume 7, Number 1, 1976, page 59.

**#1156 18.0×12.0cm Plate #9-18**

**New Nevada Valenciana Mine, Batopilas District, Chihuahua, Mexico**

This is an enormous plate of interlocking and intergrown silver crystals. There is virtually no matrix. The crystals and wires are extremely small, mainly in the 1-2mm range. They are bright and sharp, often showing nearly perfect crystal form of "spear-point" habit with small pieces of gray rock and black matrix. This is one of the largest plates recovered from the New Nevada Valenciana Mine.

***Plate #9-18, Specimen #1156*** **SILVER – New Nevada Valenciana Mine, Batopilas District, Chihuahua, Mexico. (MH)**

***Plate #9-19, Specimen #1857*** **SILVER – Batopilas District, Chihuahua, Mexico. (MH)**

***Plate #9-20, Specimen #2342*** **SILVER – New Nevada Valenciana Mine, Batopilas District, Chihuahua, Mexico. (MH)**

**#1857 5.0×5.0cm Plate #9-19**

**Batopilas District, Chihuahua, Mexico**

Jutting from a 2.0×2.0cm piece of white calcite matrix are four thick silver wires, each approximately 4.0cm long. Three of them have 2-3mm tapered silver crystals growing at an angle from the wires.

**#2342 3.0×3.0cm Plate #9-20**

**New Nevada Valenciana Mine, Batopilas District, Chihuahua, Mexico**

This small, flat, specimen is an excellent example of spinel twinned silver crystals. It is near perfect in arrangement with crystals to 1.0cm twinning along other crystals, the longest of which is 3.3cm. Pictured in *Mineralogical Record*, Volume 12, Number 3, 1981, page 179.

***Plate #9-21, Specimen #2615*** **SILVER – Batopilas District, Chihuahua, Mexico. (MH)**

**#2615 5.0×4.5cm Plate #9-21**

**Batopilas District, Chihuahua, Mexico**

Etched from the white calcite that still forms the core of the piece, silver crystals in long tapering "spear-point" form compose some of the specimen. The longest wire is 3.5cm, while the bulk of the silver is in small clusters of 1-2mm arborescent crystals.

***Plate #9-22, Specimen #4954*** **SILVER – Batopilas District, Chihuahua, Mexico. (MH)**

**#4954 6.5×4.0cm Plate #9-22**

**Batopilas District, Chihuahua, Mexico**

A small classic plate of intergrown wires and twinned "herringbone" silver is completely etched from the enclosing calcite.

**#2692 8.0×6.0cm Plate #9-23**

**Zacatecas, Mexico**

Zacatecas has produced a considerable amount of leaf silver intimately mixed with massive covellite. Here, the silver leaf protrudes from the covellite about 3.5cm. Minor pyrite is also in the covellite. The combination of silver and covellite is unique to Zacatecas.

***Plate #9-23, Specimen #2692***
**SILVER – Zacatecas, Mexico. (MH)**

## MOROCCO

**#4800 12.5×5.0cm**

**Imiter, near Tizniti, Anti-Atlas Mountains, Morocco**

This is actually a specimen of silver-mercury amalgam wherein the two elements occur in nearly equal parts. It is a bright, undulated, sheet of metal of enormous size with only very minor matrix evident. The amalgam was originally labeled kongsbergite, a now discredited name.

## Kongsberg, Norway

The silver mines of Kongsberg, Norway were discovered in 1623 by two young shepherds. In the 320 plus years after discovery, the Kongsberg mines produced more than 1350 metric tons of the white metal and the world's finest examples of wire silver. The specimen silver from Kongsberg is legendary commonly as heavy wire coils, twisted and bent into surrealistic forms. If you had to pick a single mineral specimen to define a collection, most curators would chose a Kongsberg silver. The Barlow Collection has 14 Kongsberg pieces, four of which are truly outstanding.

The silver mines of Kongsberg are in a band of strongly metamorphosed Precambrian gneisses, granites, and quartzites. Within this sequence are lenses of sulfide-rich rock known as "Fahlbands." Where veins of quartz and calcite derived from the gneisses and granites cut the Fahlbands, silver was deposited. Originally, the silver was probably deposited as sulfides and sulfosalts along with sphalerite, chalcopyrite, pyrite, fluorite, and calcite. Weathering of the silver-bearing sulfides resulted in concentrating the silver in its native state. A remarkable amount of weathering must have occurred because native silver is so common at Kongsberg, accounting for 95% of the ore.

The silver of Kongsberg is divided into three types: massive, wire, and crystalline. Wire silver is the most common. Wires "grow" from a sulfide/sulfosalt substrate and vary in width from a fraction of a centimeter ("moss silver") to several centimeters ("baroque silver"). The Barlow Collection's first major silver was a Kongsberg wire purchased at the Tucson Gem and Mineral Show in 1974.

Much rarer than wire silver are well-formed silver crystals. The crystallized silver is occasionally extraordinary and is considered the world's finest. Sharp cubes with dimensions in excess of 2.0cm are known. Barlow specimen #1157 (Chapter 3, Plate #3-83) contains a magnificent stack of cubes 2.5cm on a side. The Kongsberg silver crystals are often twinned and show octahedral modifications.

Kongsberg is also famous for "platy" silver. These plates are laminar crystals flattened on {111} (octahedron) and showing triangular striations due to oscillatory development of {100} (cube) and {111} (octahedron).

**Plate #9-24, Specimen #769** **SILVER – Kongsberg, Norway. (MH)**

*This was the first important silver added to the Barlow Collection.*

### NORWAY

**#769 14.0×7.5cm Plate #9-24**

**Kongsberg, Norway**

This was the first important silver added to the collection. A small amount of matrix holds thick curving horns of silver along with the typical fine wires to 5.0cm long. A few of the horns terminate as cubes of silver rather than as the more typical tapered point. Within the tangle of wires, several white rhombohedral calcites are caught. Elsewhere, rhombic casts of silver reveal where other calcite had been. This is a superb example of Kongsberg silver and is typical of the better specimens in private hands. It was secured in Tucson in 1974 with the help of Peter Bancroft.

*Plate #9-25, Specimen #2448* **SILVER – Kongsberg, Norway. (MH)**

**#2448 4.0×3.0cm Plate #9-25**

**Kongsberg, Norway**

This is a mass of silver showing minor crystallization. Its primary feature is the myriad silver casts after calcite over much of the specimen. One small, 1.5cm, hornlike projection terminates in an uncommon tetrahexahedron form.

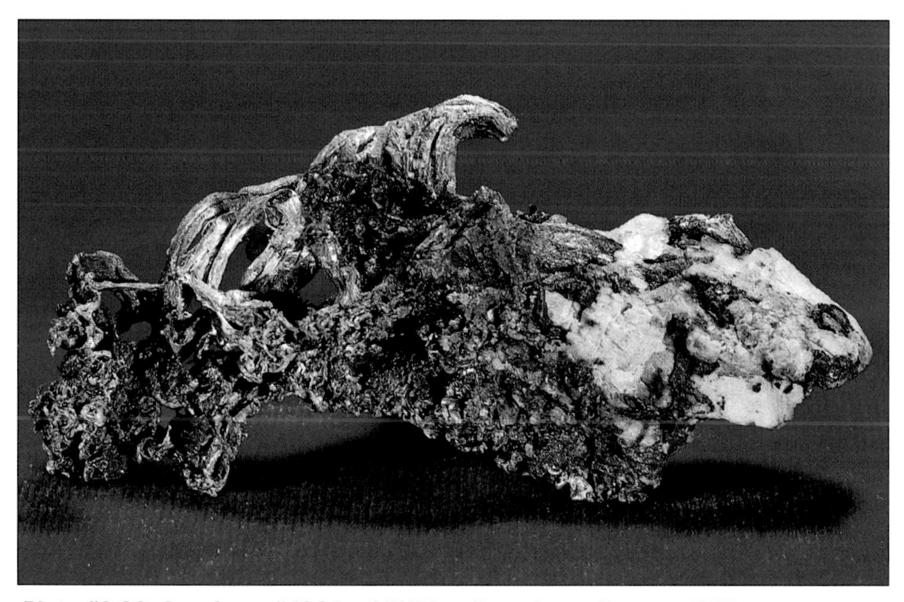

*Plate #9-26, Specimen #4924* **SILVER – Kongsberg, Norway. (MH)**

**#4924 7.0×3.2cm Plate #9-26**

**Kongsberg, Norway**

Projecting horizontally from a massive calcite matrix are several thick "ram's horns" of silver that curve and tangle into an aesthetic series of wires.

**#1157 4.0×2.5×1.5cm**

**Refer to Chapter 3, Plate #3-83**

**Kongsberg, Norway**

Fine, sharp cubes of silver from this locality are highly prized. This miniature is a choice example of Kongsberg's best cubes, with the entire specimen made up of crystals. The largest crystal in the group is 1.3cm across.

**#3348 6.0×4.0cm**

**Kongsberg, Norway**

This specimen is a series of slender, arborescent crystals to over 4.0cm. It is encrusted with minor white calcite. The silver crystals are generally cubic when observed with a lens. The arborescent form of silver is unusual for Kongsberg. The old label with the specimen indicates it was formerly in the collection of Duke Stephen.

**#4735 4.0×4.0cm**

**Kongen Mine, Kongsberg, Norway**

The matrix of this specimen is made up of calcite and purple fluorite, a classic association for the locality. The two features on the specimen are a crude flat horn of silver and a modified black acanthite cube.

**#4932 7.5×6.0cm Plate #9-27**

**Kongsberg, Norway**

The matrix of this specimen marks it as uniquely Kongsberg. It is massive to crystallized black sphalerite, very typical of the "Fahlbands" type occurrence. Within and protruding from the sphalerite are gnarled silver wire masses with two prominent slender horns, 4.0 and 2.5cm.

***Plate #9-27, Specimen #4932*** **SILVER – Kongsberg, Norway. (MH)**

***Plate #9-28, Specimen #5338*** **SILVER – Kongsberg, Norway. (MH)**

**#5338 6.0×3.5cm Plate #9-28**

**Kongsberg, Norway**

This specimen is dominated on the one hand by a large, 2.5×2.0cm, distorted cube of silver. On the other hand is an elongated, rather distorted crystal showing the dodecahedral form. Central to the piece is a tapering horn with a thick, flattened crystal twin. The piece was obtained by Barlow from the M.R. Institute of Hamburg, Germany.

**#5350 3.0×2.5cm Plate #9-29**

**Kongsberg, Norway**

This miniature is composed of sharp silver cubes mostly under 1mm with a few exceeding that size. Along with the cubes are several small plates of spinel twinned cubes. Though small, this is an aesthetic silver with most interesting crystal forms.

***Plate #9-29, Specimen #5350*** **SILVER – Kongsberg, Norway. (MH)**

## Peru

Although European localities like Kongsberg and Freiberg are extremely famous among collectors, the actual silver production from these mines is tiny compared to the fabulous mines of the New World. In 1532 Francisco Pizarro captured the Inca Atahualpa and demanded a ransom — three rooms full of gold and silver were paid. This led to a wave of plunder and, eventually, exploration for mines in the mid 1500s. Until 1776 Bolivia was part of Peru, and the most famous silver mine in the world was at Potosi. The total production at Potosi was 30,000 tons of silver.

During the colonial time, silver production in Peru was dominantly from four districts: Hualgayoc, Cerro de Pasco, Castrovirreyna, and Cailloma. Hualgayoc, was discovered in 1771, certainly ranks as one of the world's major silver mines, with more than 1,500 tons of silver produced. Alexander von Humbolt, the great German naturalist, visited the district on his tour of "New Spain" in the early 1800s and commented on seeing numerous specimens of native silver.

Cerro de Pasco is about 170 km northeast of Lima; the most important silver specimen producing mining area in the district is Uchucchacua. These mines produce about 3 million ounces of silver per year. Many of the fine pyrargyrites that came from Peru in the last decade are from this area. East of Lima is the Huancavelica district, which has also recently produced some fine silver sulfosalt specimens.

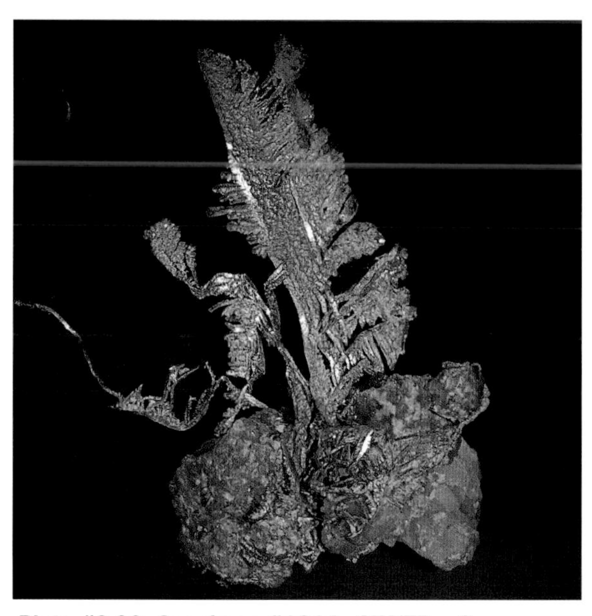

***Plate #9-30, Specimen #1614*** **SILVER – San Cristobal, Peru, (MH)**

## PERU

**#1614 6.0×2.0cm Plate #9-30**

**San Cristobal, Peru**

Projecting from a small, rounded mass of quartz is a delicate "herringbone" of silver showing the typical twinning of this form. This locality is currently an active producer of silver mineral specimens and is considered a potentially classic locality.

***Plate #9-31, Specimen #4953*** **SILVER – San Janero Mine, Huancavelica, Peru. (MH)**

**#4953 9.0×4.0cm Plate #9-31**

**San Janero Mine, Huancavelica, Peru**

Multiple layering of repeating sheets and "herringbone" crystal plates stack up to make this specimen. The layers are loosely attached and there is virtually no matrix. Acanthite coats the bulk of the delicate wires and crystals that make up this piece. It should be noted that the San Janero Mine is well known for extraordinary pyrargyrite crystals.

*The San Janero Mine is well known for extraordinary pyrargyrite crystals.*

## KAZAKHSTAN

**#4717 4.0×2.5cm Plate #9-32**

**Dzhezkazgan, Kazakhstan**

From a flat plate of matrix topped by minor crystallized quartz protrudes a tangled mass of delicate to thick wires of generally indeterminate length. Thc thickcst wire is at least 6.0cm long.

*Plate #9-32, Specimen #4717* **SILVER – Dzhezkazgan, Kazakhstan. (MH)**

**#4960 4.0×3.0cm**
**Plate #9-33 (left)**

**Dzhezkazgan, Kazakhstan**

Lacking matrix, this specimen has a thick mass of tangled wires for a base from which slender curving wires of silver rise. The entire piece is coated with sooty black acanthite.

**#3235 2.5×1.0cm**
**Plate #9-33 (center)**

**Dzhezkazgan, Kazakhstan**

This is a bundle of thick, curving silver wires without matrix. They are sooty black due to a sulfide coating. This was one of the first silver specimens to reach America when specimens began to emerge from Russia in the mid 1980s.

*Plate #9-33, Specimen #4960 Left; #3235 Center; #5137 Right* **SILVER – Dzhezkazgan, Kazakhstan. (MH)**

**#5137 5.0×3.0cm Plate #9-33 (right)**

**Dzhezkazgan, Kazakhstan**

Looking like a tangled skein of wool or thread, this matrix-free specimen of silver wires has a very graceful and artful design to it. The wires are coated with black acanthite for the most part.

## SPAIN

**#4933 6.0×5.0cm**

**Almena, Spain**

This specimen is largely a solid mass of black sphalerite showing some crystal faces and cleavage planes. Small amounts of silver wires in patches throughout the sphalerite make this an appealing specimen from a relatively rare source.

## Aspen, Colorado

The silver mines of the Aspen area were among the most famous American silver producing localities long before the area became a ski resort. In the early 1870s the Hayden survey visited the area and mapped a large section of dolomite. When the *Hayden Atlas* was published, some Leadville prospectors noted that the dolomite was the same rock that hosted the Leadville ores. A group of prospectors traveled from Leadville to Aspen in 1879 and located the Smuggler claims. The following year the Mollie Gibson, Emma, Della, and Aspen Mines were staked. These mines were extraordinarily rich in silver masses. In 1894, a single mass in excess of 2,200 pounds was found at the Smuggler Mines. The district, which was originally known as the Roaring Fork, produced a large number of aesthetic silver specimens — principally sculpted wires.

Although the specimen heritage from Aspen is important, the district is probably better known for bitter legal battles. The legal machinations began almost immediately with charges of claim jumping and stealing ore. In 1886 Judge Moses Hallet ruled that all the mines on Aspen Mountain were essentially one orebody, and he established the concept of "apexing." This ruling established the concept that owning the top of a system of veins and lodes implies ownership of the entire orebody, irrespective of property lines. The ultimate consequence of the apex law was wholesale claim consolidation and lawyers obtaining huge fees.

***Plate #9-34, Specimen #1159*** **SILVER – Aspen, Pitkin County, Colorado. (MH)**

**#1159 6.0×2.0cm Plate #9-34**

**Aspen, Pitkin County, Colorado**

"Herringbone" habit is the dominant feature of this specimen, for it is made up of a myriad of such twinned crystals. There is no matrix save for several nondescript dabs of calcite at the base. The longest crystals are about 1.5cm, but the majority are under 1.0cm.

## Creede, Colorado

In the early 1870s miners rushed to the rich San Juan Mountains in southern Colorado. The jumping-off point for most prospectors was the present-day town of Del Norte. The miners traveled west from there along the Rio Grande River and passed by the present site of Creede. Of course, prospectors did what comes naturally on their journeys, and silver was discovered. J. McKenzie located the first claim in 1876 (the Alpha Mine).

Although several other claims were staked soon after, the district did not really develop until 1889, when the Holy Moses claim was discovered by N.C. Creede, a notorious Del Norte gambler. Creede sold his claim to Dave Moffat, who was president of the Rio Grande Railroad. Moffat built a spur of the railroad to Creede, and the site became one of the last Wild West boomtowns. Between 1890 and the silver panic of 1893, Creede was a town "with no night."

The principal mines in the Creede area are the Amethyst, Bachelor, Commodore, Last Chance, and Bulldog. By far, the bulk of specimen material comes from the Bulldog Mine. This is due to the relative youth of the mine. In 1964 a shaft was driven into Bulldog Mountain at 10,000 feet, and the last great American silver mine was discovered (at least for specimens). The silver from the Bulldog is generally wires or dendritic masses (reminiscent of Batopilas). The wires are often tangled nests filling vugs and voids in quartz or barite. Creede specimens are easily damaged, owing to the delicate nature of the wires. A remarkable number of frauds have been constructed by gluing a silver wire into vugs of attractive quartz. Sometimes these specimens are labeled "repaired," because the silver wires are commonly only loose in the vug. Who is to know?

*Plate #9-35, Specimen #4959*
**SILVER – Bulldog Mine, Creede, Mineral County, Colorado. (MH)**

**#4959 4.0×2.0cm Plate #9-35**

**Bulldog Mine, Creede, Mineral County, Colorado**

This is a typical mass of crystalline silver crowned by wires to 2.0cm long, which twist, curl, and turn in the usual fashion for silver. The wires are striated and quite bright.

**#1168 3.5×3.0cm**

**Creede, Mineral County, Colorado**

This small mass of solid silver is crowned with a tangle of small silver wires from which protrudes one 2.0cm curling wire.

**#5225 6.5×3.5cm** **Refer to Chapter 3, Plate #3-80**

**Bulldog Mine, Creede, Mineral County, Colorado**

This is a truly exceptional specimen, one of the best in the collection. It consists of very bright, delicate wires, some almost hairlike, freestanding from a quartz matrix. Minor amounts of crystalline polybasite and acanthite are evident. The milky-white quartz is a grouping of spherules made up of bright crystal points, almost drusy in appearance. The length of the wires is difficult to judge as the longer ones curl and loop back on themselves repeatedly. As a cluster, they stand some 4.0cm. The specimen is reported to have been mined at Creede in 1972 and, if it was, is the best ever from that site. Regardless of locality, however, the arrangement of the bright silvers on sparkling quartz makes this an exceptionally aesthetic specimen.

**#5062 6.0×0.25cm**

**Leadville, Lake County, Colorado**

Leadville is probably the most notorious of the Colorado silver camps, thanks to H.A.W. Tabor, Baby Doe Tabor, Molly Brown, and their ilk. It was certainly one of the richest of the camps. Yet, wire silvers from this locality are decidedly uncommon.

The collection contains only one Leadville silver, #5062, a single curving wire, topped by an unusual crescent of slender wires. For this historic camp, it is a fine specimen.

Of all the silver camps in Colorado, Leadville probably produced more legends and lore per shaft than did any other district. Characters abound in the history of the place. H.A.W. (Haw) Tabor, Chicken Bill Lovell, Unsinkable Molly Brown, Baby Doe (Tabor), and more dot the history of the camp.

Tabor was one of the luckiest cusses in Leadville. He grubstaked himself into possession of the Little Pittsburgh and other rich deposits. Even when he did something really dumb, he came out a winner. For example, Chicken Bill, who got his name when he was snowed in with a load of chickens and ate up the load to survive, decided to work a little salting deal on Tabor, who was already getting wealthy. Bill got hold of some rich ore (legend says it was from Tabor's Little Pittsburgh claims) and salted a worthless hole in the ground. He promptly sold the hole to Tabor, who never even bothered to check it out. He named it the Chryolite and installed some miners. They promptly hit a rich vein, much to the chagrin of Chicken Bill!

Baby Doe, Tabor's wife, lived and died by the Little Pittsburgh. Harking to her husband's dying advice to never sell the mine, she lived virtually penniless and destitute on the property until her death.

## Michigan

The Keweenaw Peninsula, the upper reaches of Michigan, is known as "Copper Country." Indeed, more than 13 billion pounds of the red metal have been produced from the Keweenaw mines in their long history; these same mines produced some of the finest crystallized silver specimens known. The classic Michigan silver is a series of long "feathers" of warm, white crystals growing on a copper matrix. Specimen #1930 is a fine example of this habit. (Refer to Chapter 3, Plate #3-84.) In fact, it is one of the finest Michigan silvers known.

The Keweenaw mineralization is hosted in a thick stack of Precambrian volcanic and sedimentary rocks. The volcanic rocks, which include basalts, amygdaloids, and ash flows, are interbedded with conglomerates. The copper and silver were probably leached out of the volcanic rocks by ascending saline fluids that were produced by the progressive burial of the bottom of the lowest basalts. This "burial," caused by continuous extrusions and sedimentary deposits, leads to low-grade metamorphism that drives out interstitial fluids, which, in turn, migrate toward the surface. Apparently, the source rocks have a very low sulfur content, so the hydrothermal fluids concentrated and deposited the silver and copper in the native state rather than as the more common sulfides. This was unusual and led to the remarkable and very distinctive crystallization of "Michigan" silver.

The silver from the Keweenaw shows a range of crystalline habits, all typical of low-temperature deposition. The most common form for silver crystals is an elongated, twinned tetrahexahedron. This type of crystal growth often produces spectacular branching silver masses with a very coarse herringbone texture. Cubes are a less common form of Michigan silver. Dodecahedra are rare, and octahedra are observed only as modifications of other crystal forms. The silver is almost always associated with copper and is almost exclusively deposited "on" copper. Occasionally, the copper and silver are intergrown and are known as "half-breeds."

More than 20 Keweenaw mines produced excellent specimen silver. The best specimens come from the Kearsarge Lode, an amygdaloid body tapped by the Kearsarge and Wolverine mines. Other famous producers of fine specimens are the Cliff, Copper Falls, Quincy, and White Pine Mines. The Barlow Collection has 40 silvers from Michigan from at least six different mines. The specimens range in size from thumbnail to cabinet and show at least five different crystal habits.

**#1154 8.0×7.5×5.5cm**
**Refer to Chapter 3, Plate #3-82**

**Kearsarge Lode, Houghton County, Michigan**

This aesthetic crystal group forms a "bridge" on a matrix of copper and wall rock. The crystals are elongated and distorted tetrahexahedron twins, with the largest crystal reaching 3.5cm in length. The entire "bridge" measures 10.0cm across. This is the first truly fine Michigan silver in the collection, and the tale of its acquisition is typical of the lengths John Barlow will go to obtain classic specimens.

John was alerted to the existence of the silver in 1974 by dealer Don Olson, who had spotted it in a rock shop in Magdelena, New Mexico. John immediately flew his own plane into the Magdelena airport (a gross misnomer for the dirt strip a few miles south of the small town), hitched a ride into town, and bargained with the owner of the rock shop, Tony Otero, to acquire the piece after much haggling.

**#1161 4.5×3.0cm**

**Houghton County, Michigan**

This is an intergrown mass of both copper, which shows no crystal forms, and crystallized silver. The entire mass is crudely formed yet is an excellent example of what has become known as a Michigan "half-breed." Literally thousands of these masses were encountered during mining, but the vast majority of them went through the ore crusher before they could be retrieved. Since they simply flattened in the crusher, most half-breeds seen have no distinguishable form other than a flattened, intermixed lump of the two metals.

**#1162 7.0×6.5cm**

**Keweenaw County, Michigan**

This is a solid mass of bright silver crystals with roughly equidimensional (0.5-0.8cm) modified cubes. Some of the crystal faces show triangular striations that are typically associated with twinning. The base of the specimen is a nondescript mass of epidote rock. A small residual mass of calcite in the core of the specimen shows it may have been heavily encrusted with that mineral, later removed by acid etching. This could account for the apparent minor casts in the silver, which look like calcite cleavage planes.

**#1167 6.0×4.0cm**

**Keweenaw County, Michigan**

This is a specimen of massive and crudely formed silver crystals that are arranged in an open "winged" stance. Nested in that spread is a small "Christmas tree" of twinned silver crystals. Distinguishable crystals on the specimen exhibit a distorted tetrahedral habit.

**#1930 9.0×8.6×4.8cm**
**Refer to Chapter 3, Plate #3-84**

**Quincy Mine, Houghton County, Michigan**

While specimen #1154 ranks high on the Barlow list of fine silvers, it is exceeded by this specimen, the most aesthetic specimen in the silver suite. Proof of this lies in the fact it was also selected for inclusion in Chapter 3 here and has been pictured in several publications including the cover of *Rocks and Minerals*, Volume 52, Number 10, 1977, and on page 31 of Peter Bancroft's book *Gem and Crystal Treasures*, 1984. But you can judge for yourself. The piece is a series of brilliant winglike chains of slender curving crystals, many of which taper to a point. Their general appearance is that of a feather. These "feathers," of which there are about 10 on the specimen, fly off at all angles from a blackened copper matrix. The arrangement of these crystal groups, the brilliance of the metal, and the generous size of some crystals (the largest is 3.0cm long) make this one of the most attractive silvers in private hands.

Marc Wilson, expert on Michigan minerals and author of Chapter 11 ("Copper") in this book, writes, "This world-class specimen of sharp, elongated, twinned tetrahexahedral silver crystals on copper matrix originates from Houghton County, Michigan. The arborescent crystal grouping is of a type first described by Dana in 1886 and later coined as 'Type 2' by Wilson and Dyl (1992). Measuring 9.0×8.6×4.8cm, and exquisitely aesthetic, this is one of the finest examples of crystalline silver from a district known for top-quality crystal silver specimens and is possibly second only to 'the Buffalo' of the A.E. Seaman Mineralogical Museum in Houghton, Michigan." The specimen was purchased from Dr. Richard Kelly at the Tucson Show in 1979.

*Plate #9-36, Specimen #2112* **SILVER – Keweenaw County, Michigan. (MH)**

**#2112 10.0×3.0cm Plate #9-36**

**Keweenaw County, Michigan**

Centered on a 5.0cm long spine-like silver crystal are the typical Type 2 "feather" crystals showing tetrahexahedron twinning habit. One end of the piece terminates with a single distorted crystal, while the central portion has a solid barlike elongated crystal from which project the twinned crystals. This is an aesthetic specimen. It was acquired through a trade with Don Olson in 1980.

**#2293 5.0×3.5cm**

**Keweenaw County, Michigan**

This specimen is an open network of chainlike crystals, an especially aesthetic arrangement.The chain-like habit is due to small crystals of silver of either the cube or tetrahexahedron habit repeating again and again as contact or penetration twins.

**#4674 8.0×6.0cm Plate #9-37**

**Wolverine Mine, Houghton County, Michigan**

The copper forming the base of this specimen is massive to crystalline and provides a platform for a marvelous group of bright, superb, sharp, silver crystals. Some of the crystals are cubes, some tetrahexahedra. All are sharply formed and attractive. The silver cluster is 4.0×3.0×2.5cm, with the larger crystals exceeding 2.0cm. These are very pleasing silvers, among the best in the collection. The specimen was obtained from W.F. Ferrier, Ottawa, Canada.

*Plate #9-37, Specimen #4674* **SILVER – Wolverine Mine, Houghton County, Michigan. (MH)**

**#5416 8.0×6.0×5.5cm**

**Kearsarge Lode, Houghton County, Michigan**

The matrix is massive copper showing very minor crystal faces. On the copper is a spectacular group of bright silver crystals showing modified cubic and dodecahedral forms. The entire crystallized mass of silver is 4.5×3.5cm, with the larger crystals to 2.0cm. Below and to the left of the main silver group is a crude mass of silver that formed in contact with a since removed rough rock surface that precluded crystal face development.

**#2804 4.5×3.0cm**

**Central Mine, Houghton County, Michigan**

Sitting atop a nondescript mass of epidote-coated copper are three bright feathers of twinned silver. The crystals are small; the overall size of the feathers is about 1.0×1.5cm.

***Plate #9-38, Specimen #4837*** **SILVER – Houghton County, Michigan. (MH)**

***Plate #9-39, Specimen #5627*** **SILVER – White Pine Mine, Ontonagon County, Michigan. (MH)**

**#4837 6.5×5.0cm Plate #9-38**

**Houghton County, Michigan**

Only traces of copper fringe this open network of branching silver crystals that are typically elongated and twinned. The elongations follow the plane of the octahedron. The entire bright mass is typically treelike, with individual crystals exceeding 2.0cm.

**#4798 4.0×0.5cm**

**Wolverine Mine, Houghton County, Michigan**

This is a simple, flattened stack of four modified silver cubes and octahedra strung like beads on a copper wire "string." This habit is an unusual occurrence for the locality.

**#4884 7.0×5.0cm**

**Gratiot Mine, Houghton County, Michigan**

A single, large twinned silver crystal perches on a copper matrix. The twin is "V" shaped, joining two distorted tetrahexahedra.

**#5627 6.5×2.5cm Plate #9-39**

**White Pine Mine, Ontonagon County, Michigan**

This is a most delicate arrangement of small, mostly cubic silver crystals. These are typical of silver from the White Pine Mine. They show the normal twinning resulting in an arborescent growth pattern. This specimen is quite aesthetic, with a bright, "tree" crystal pattern and pink calcite.

**#4892 10.0×5.0cm**

**White Pine Mine, Ontonagon County, Michigan**

This is an exceptional specimen from a recently closed mine. It is an open network of arborescent silver crystals, lightly pitted and rounded, but bright and aesthetic. Small fragments of calcite cling to the silver in spots. The pitting gives the specimen a frosted appearance.

**#5056 4.0×3.0cm**

**Adventure Mine, Ontonagon County, Michigan**

On a matrix of greenish epidote rock sit two rounded crystal clusters of ball-like habit. The larger is 1.5cm in diameter, the other 1.3cm. What appears to be triangular etching on the spheres is actually due to twinning. Minor copper provides some contrast for the white metal.

**#5059 6.0×6.0cm Plate #9-40**

**Adventure Mine, Ontonagon County, Michigan**

This specimen is composed of two nearly equal masses of silver and copper. Both the metals are crudely crystallized, having a "woolly" appearance due to fine wires and crystals. The matrix is a pedestal of quartz-impregnated epidote rock.

***Plate #9-40, Specimen #5059*** **SILVER – Adventure Mine, Ontonagon County, Michigan. (MH)**

**#5223 6.0×4.0cm Plate #9-41**

**Quincy Mine, Houghton County, Michigan**

From a large, 3.0cm, crude copper crystal rises an exceptional series of crystal twins, tetrahexahedral in habit with an overall "V" form. On the main silver group are small second-generation arborescent growths of the metal.

**#5422 13.5×7.0cm**

**Kearsarge Lode, Houghton County, Michigan**

This lovely copper specimen is best described in another chapter (Chapter 11, page 286), as it is largely made up of superb, sharp copper crystals of varying habit. Yet, the piece is included here for it is endowed with a scattering of sharp bright silver crystals to 4mm along the upper edge of the specimen. The silvers are nearly perfect cubes and dodecahedra.

***Plate #9-41, Specimen #5223*** **SILVER – Quincy Mine, Houghton County, Michigan. (MH)**

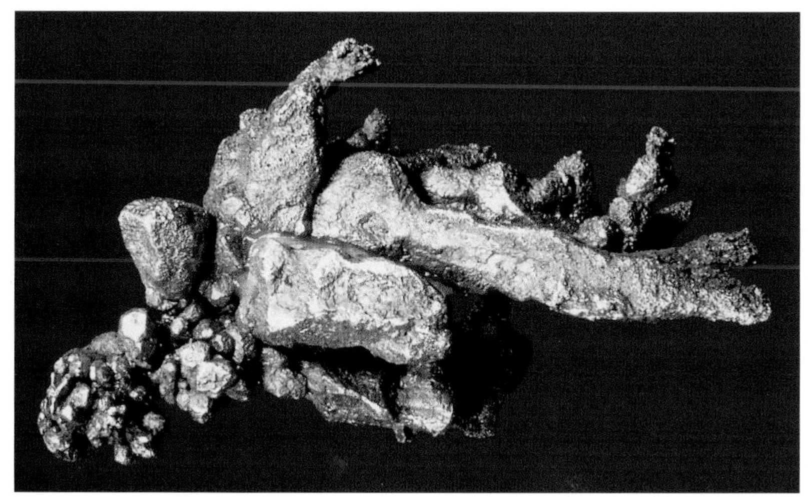

***Plate #9-42, Specimen #5386*** **SILVER – Elkhorn Mine, Jefferson County, Montana. (MH)**

## MONTANA

**#5386 2.0×1.0cm Plate #9-42**

**Elkhorn Mine, Jefferson County, Montana**

This fine thumbnail, which is a sharply crystallized "thorn," comes from a famous American silver district. The Elkhorn Mine, located about 30 km south of Helena, was discovered during the great gold rush to Alder Gulch in the 1860s. The mine produced more than 10 million ounces of silver and 4 million pounds of lead. This specimen was obtained from McGill University, Toronto, Canada.

## SILVER-BEARING MINERALS KNOWN BY THE AUTHOR AS ACCEPTED SPECIES 1995

With the grateful help of Dr. Joseph A. Mandarino,
retired curator of mineralogy of the Royal Ontario Museum

|  |  |
|---|---|
| Acanthite | S Ag |
| Aguilarite | S Ag Se |
| Allargentum | Ag Sb |
| Andorite | S Ag Sb Pb |
| Antimonpearceite | S Cu As Ag Sb |
| Aramayoite | S Ag Sb Bi |
| Arcubisite | S Cu Ag Bi |
| Argentojarosite | H O S Fe Ag |
| Argentopentlandite | S Fe Ni Ag |
| Argentopyrite | S Fe Ag |
| Argentotennantite | S Fe Cu Zn As Ag Sb |
| Argyrodite | S Ge Ag |
| Arsenpolybasite | S Cu As Ag Sb |
| Aurorite | H O Ca Mn Ag |
| Balkanite | Cu Ag Hg S |
| Baumhauerite 2A | S As Ag Sb Pb |
| Benjaminite | S Cu Ag Pb Bi |
| Benleonardite | S As Ag Sb Te |
| Berryite | S Cu Ag Pb Bi |
| Bideauxite | H O F Cl Ag Pb |
| Billingsleyite | S As Ag Sb |
| Bohdanowiczite | Se Ag Bi |
| Boleite | H O Cl Cu Ag Pb |
| Borodaevite | S Fe Ag Sb Pb Bi |
| Bromargyrite | Ag Br |
| Cameronite | Cu Ag Te |
| Canfieldite | S Ag Sn |
| Capgaronnite | S Cl Br Ag Hg I |
| Cervelleite | S Ag Te |
| Chlorargyrite | Cl Ag |
| Criddleite | S Ag Sb Au Tl |
| Crookesite | Cu Se Ag Tl |
| Cupropavonite | S Cu Ag Pb Bi |
| Danielsite | S Cu Ag Hg |
| Dervillite | S As Ag |
| Diaphorite | S Ag Sb Pb |
| Dyscrasite | Ag Sb |
| Empressite | Ag Te |
| Eskimoite | S Ag Pb Bi |
| Eucairite | Cu Se Ag |
| Eugenite | Ag Hg |
| Fischesserite | Se Ag Au |
| Fizelyite | S Ag Sb Pb |
| Freibergite | S Fe Cu As Ag Sb |
| Freieslebenite | S Ag Sb Pb |
| Furutobeite | S Cu Ag Pb |
| Geffroyite | S Fe Cu Se Ag |
| Giessenite | S Cu Ag Sb Pb Bi |
| Giraudite | S Cu Zn As Se Ag Sb |
| Gustavite | S Ag Pb Bi |
| Hatchite | S As Ag Tl Pb |
| Henryite | Cu Ag Te |
| Hessite | Ag Te |
| Heyrovskyite | S Ag Pb Bi |
| Hocartite | S Fe Ag Sn |
| Imiterite | S Ag Hg |
| Incaite | S Fe Ag Sn Sb Pb |
| Iodargyrite | Ag I |
| Jalpaite | S Cu Ag |
| Krennerite | Ag Te Au |
| Kutinaite | As Cu Ag |
| Laffittite | S As Ag Hg |
| Larosite | S Cu Ag Pb Bi |
| Lengenbachite | S Cu As Ag Pb |
| Luanheite | Ag Hg |
| Makovickyite | S Cu Ag Pb Bi |
| Marrite | S As Ag Pb |
| Matildite | S Ag Bi |
| Mckinstryite | S Cu Ag |
| Miargyrite | S Ag Sb |
| Miersite | Cu Ag I |
| Moschellandsbergite | Ag Hg |
| Mummeite | S Cu Ag Pb Bi |
| Muthmannite | Ag Te Au |
| Naumannite | Se Ag |
| Neyite | S Cu Ag Pb Bi |
| Novakite | Cu As Ag |
| Ourayite | S Ag Pb Bi |
| Owyheeite | S Ag Sb Pb Bi |
| Paderaite | S Cu Ag Pb Bi |
| Paraschachnerite | Ag Hg |
| Pavonite | S Cu Ag Pb Bi |
| Pearceite | S As Ag |
| Penzhinite | S Cu Se Ag Au |
| Perroudite | S Cl Br Ag I Hg |
| Petrovskaite | S Se Ag Au |
| Petzite | Ag Te Au |
| Pirquitasite | S Zn Ag Sn |
| Polybasite | S Cu Ag Sb |
| Proustite | S As Ag |
| Pyrargyrite | S Ag Sb |
| Pyrostilpnite | S Ag Sb |
| Ramdohrite | S Ag Sb Pb |
| Rayite | S Ag Sb Tl Pb |
| Roshchinite | S Cu Ag Sb Hg Pb |
| Samsonite | S Mn Ag Sb |
| Schachnerite | Ag Hg |
| Schirmerite | S Ag Pb Bi |
| Selenostephanite | S Se Ag Sb |
| Silver 2H | Ag |
| Silver 3C | Ag |
| Silver 4H | Ag |
| Smithite | S As Ag |
| Sopcheite | Ag Pd Te |
| Stephanite | S Ag Sb |
| Sternbergite | S Fe Ag |
| Sterryite | S As Ag Sb Pb |
| Stetefeldtite | H O Ag Sb |
| Stromeyerite | S Cu Ag |
| Stutzite | Ag Te |
| Sylvanite | Ag Te Au |
| Telargpalite | Ag Te Pd |
| Tocornalite | Ag I Hg |
| Toyohaite | S Fe Ag Sn |
| Treasurite | S Ag Pb Bi |
| Trechmannite | S As Ag |
| Tsnigriite | S Se Ag Sb Te |
| Uchucchacuaite | S Mn Ag Sb Pb |
| Uytenbogaardtite | S Ag Au |
| Vikingite | S Ag Pb Bi |
| Volynskite | Ag Te Bi |
| Wallisite | S Cu As Ag Tl Pb |
| Weishanite | Au Ag Hg |
| Xanthoconite | S As Ag |
| Zoubekite | S Ag Sb Pb |

All the above 125 species are represented in the Barlow Collection except the following:

Arcubisite
Argentotennantite
Bideauxite
Borodaevite
Criddleite
Dervillite
Larosite
Paderaite
Roshchinite
Selenostephanite
Tsnigriite
Zoubekite

References:


Fleischer, M., and J. Mandarino (1995) *Glossary of Mineral Species*, Mineralogical Record, Tucson, Arizona, 280 pp.
Nickel, E.H. and M.C. Nichols (1991) *Mineral Reference Manual*, Von Nostrand Reinhold, New York, 250 pp.
Wallace, T. (1994) Silver and Silver-Bearing Minerals, *Rocks & Minerals*, Volume 69(2): 16–38.

CHAPTER 10

# Gold- and Silver-Bearing Minerals

*by Bill Smith*

This group of minerals stands in a special relation to the John Barlow Collection. In general, Barlow's collection has broad representation of all mineral groups. Regardless of how they are categorized, no element, country, chemical class, crystal system, or paragenesis is ignored. Opportunity and personal taste are demonstrated by concentrations in certain areas (Mexico, Africa), but, in general, John has cast his net widely. In two areas, John has gone beyond general collecting to an exhaustive representation of everything in that area. One such domain is the tourmaline group, and the other is the precious metal group. The tourmaline group is discussed in Chapter 4; this section will discuss the second area, the compound minerals of gold and silver.

The earliest native silver acquisition still in the collection was obtained in 1973, and two more silvers were obtained in 1974. In that year, Barlow also added 14 silver-bearing minerals (acanthites, allargentum, pyrargyrites, smithite, stephanite, etc.), and in 1975 he added at least 21 more native silvers (some may have since been deaccessioned), as well as eight silver compound specimens. Then came 1976-77, the Boleo years, when John added dozens of superb specimens from the newly reopened Amelia Mine at Santa Rosalia, Baja California Sur, Mexico. By now John had unconsciously prepared for a new challenge. It came in a conversation that must have taken place near the end of 1977, with Dr. Pete Dunn, of the U.S. National Museum (Smithsonian). John remarked that it might be interesting to collect all the silver minerals, whereupon Dr. Dunn replied to the effect that such a quest would almost certainly fail. Anyone familiar with John's character and earlier life will recognize this statement as an irresistible challenge to him, and he at once set out to prove Dr. Dunn wrong. In 1978 he added 30 more silver-bearing minerals (and ten silvers), and in 1979, 20 more silver minerals (and seven silvers). By the end of the seventies, he was fairly launched on his effort to obtain all the silver species.

*John remarked that it might be interesting to collect all the silver minerals, whereupon Dr. Dunn replied to the effect that such a quest would almost certainly fail. Anyone familiar with John's character and earlier life will recognize this statement as an irresistible challenge to him, and he at once set out to prove Dr. Dunn wrong.*

But Barlow's desire to acquire all these rare species had to remain within the overall canon of the collection: each specimen must be the best possible for the species. A chip of canfieldite here and a smear of xanthoconite there would not do; better no representation than inferior representation. Of course, some species have been found only as grains or laths wholly included in other minerals; if this is the best there is, it must do. In cases when nature has been more generous, however, the results may be seen in the collection.

*Of the approximately 125 silver species and varieties known to science, 12 are still missing (the "Curse of Dr. Dunn"?) from the collection.*

Comprehensiveness is a difficult goal; comprehensive perfection is even more difficult. Of the approximately 125 silver species and varieties known to science, 12 are still missing (the "Curse of Dr. Dunn"?) from the collection. On John's "want list" are such improbables as bideauxite, sopcheite, and zoubekite. The remaining 113 species are represented by 314 cataloged specimens as of this writing.

It is interesting to track the growth of the silver suite through the years. After the 1978-79 burst, acquisition slowed for several years, though there has been no year devoid of acquisitions. The next big influx started in 1982 (13 specimens), and 1983 (12), with peak years 1984 (20), 1985 (36), and 1986 (22). The 103 specimens of these five years represent a major acquisition of fine and unusual Guanajuato material, and the start of a program of correspondence and visits in an attempt to obtain rare species. This is when Barlow's great Guanajuato pyrargyrites and polybasites were obtained, as well as such rarities as fischesserite, stutzite, eskimoite, volynskite, and vikingite.

Since the mid 1980s, John has continued to add selected fine pieces, especially from relatively new localities like Uchucchacua, Peru, while still pursuing the missing species. In 1992, he added 10 more rare species including benleonardite, capgaronnite, cervelleite, and eugenite. Science is not discovering new silver species as fast as John is collecting them. For a complete list of silver-bearing species in the collection, including those still being sought, see page 236.

If one is attempting a comprehensive collection of silver minerals, it seems a natural extension to also include the minerals of gold. John's program for encompassing the gold minerals ran in parallel with his silver program. Though there are far fewer gold species in the collection (21 at this writing, not including the native metal), many are so rare that the growth of his gold mineral suite has consumed nearly as much time and effort as the building of his much larger silver collection. At this time, John still lacks three gold species. Many gold minerals also contain silver (e.g. sylvanite, petzite, electrum), and such species have been counted among the silver minerals. (Two of the missing gold species are also on the silver want-list.) Thus to the previously counted 250+ silver (or silver-gold) specimens we can add 15 more, representing the gold minerals with no, or minor, silver. These 265 plus specimens in the collection thus cover all but 15 of the known gold and silver species.

What follows is a description of some of the notable specimens in the Barlow subcollection of gold- and silver-bearing minerals. The selection is subjective and arbitrary. Anyone inspecting the complete precious metal suite will surely find some very pleasant surprises that are not described here.

The dimensions given for the following specimens are the two longer ones.

*John's program for encompassing the gold minerals ran in parallel with his silver program. Though there are far fewer gold species in the collection (21 cataloged at this writing, not including the native metal), many are so rare that the growth of his gold mineral suite has consumed nearly as much time and effort as the building of his much larger silver collection. At this time, John still lacks three gold species.*

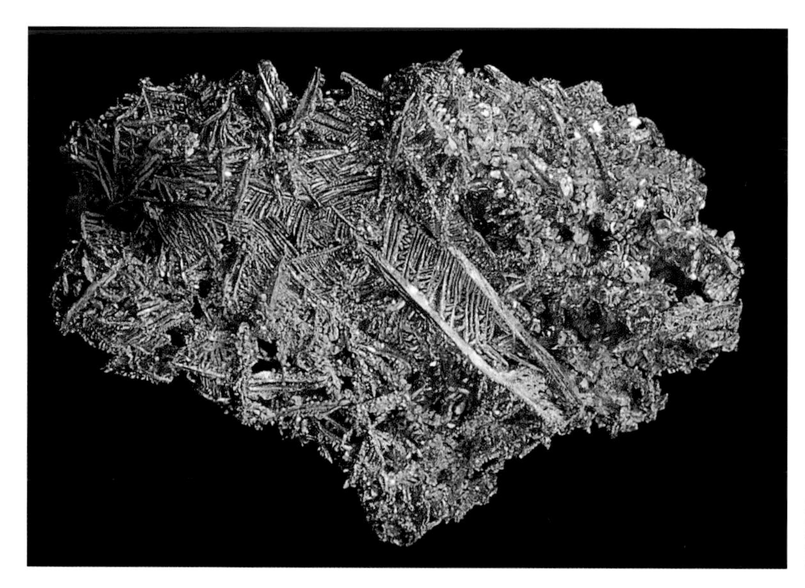

## ELECTRUM
### (NATURAL GOLD-SILVER ALLOY)
**#1394 3.1×2.2cm Plate #10-1**

**Ophir District, Placer County, California**

This piece from the Ophir District is all electrum in small wires (extended spinel twins) with skeletal octahedral terminations. The color is between pure silver and pure gold, and the luster is very bright.

***Plate #10-1, Specimen #1394***
**ELECTRUM – Ophir District, Placer County, California. (MH)**

## Colorful Cripple Creek

Cripple Creek was the scene of the last great gold rush in the continental United States and for good reason. The ores here are decidedly different from anything encountered in California or elsewhere in the West.

Instead of native gold in quartz veins, the Cripple Creek ores were mainly "gold tellurides," sylvanite and calaverite. And instead of being in quartz they had invaded the fractured phonolite plug of a huge caldera along with fluorite and other unusual associates.

Named for a creek in which a calf had fallen and broken a leg, Cripple Creek was discovered by Bob Womack, who like John Marshall of Sutter's Mill, died virtually penniless. Womack was a cowboy and sometimes prospector. He was stubbornly convinced that the hills behind Pike's Peak held a bonanza in gold. Almost to no avail, he prospected, displaying what he found in store windows in Colorado Springs which was by then a vital city. Finally, he hit the El Paso Lode, and the rush was on.

Noteworthy names that emerged from this area include Jack Dempsey, who was born in nearby Manassa, and George de la Bouglise, noted collector of crystallized gold, who was a milling expert here. Much of his collection constitutes the heart of the great Harvard gold collection, a specimen from which now resides in the Barlow Collection.

Another famous name was Spenser Penrose, who rose to fame in Colorado railroading. Yet another was Winfield Scott Stratton, whose Independence Mine was one of the bonanzas. Then there was the Cresson Mine and its rich gold-lined cavern.

One story about Stratton that is particularly intriguing occurred early in his mining career. A carpenter by trade, he developed the Independence Mine, a marginal operation at best. Stratton gradually lost confidence in the property and finally decided to lease it out. The lease was signed and the day before it was to go into effect, Stratton was in the mine cleaning up. Going into a small tunnel he had thought to work but never had, he stumbled onto a vein so rich he immediately realized that the lessee would be instantly rich if he found it. Carefully covering it up, Stratton left the mine and crossed his fingers.

Six months later the lessee had still not found that vein, or anything else noteworthy. The night before the lease was up, Stratton was sitting with the lessee in front of the fireplace of a local hotel. The lessee, disgusted that he hadn't struck it rich, pulled out the lease and in effect said, "I've had enough. Here's the lease." Stratton was so shook up at his good fortune that he could not reach out for the paper for fear his shaking hand would give him away. He simply waved a hand toward the blazing fire and said, "Toss it in there." Stratton extracted some nine million dollars in gold from the "worthless" Independence.

The Cresson Mine is a story unto itself. During underground operations a cavity was opened that was as big as a room and whose walls were fully lined with crystals of gold, calaverite, and sylvanite. The owners promptly put a steel door on the vug and proceeded to scrape a million and a half in gold off the walls. Some geode! *R.W.J.*

## CALAVERITE (GOLD TELLURIDE)

**#1392 13.1×6.8cm Plate #10-2**

**Cripple Creek, Teller County, Colorado**

This piece displays a classic Cripple Creek assemblage: the phonolite country rock; the veins of amethystine quartz, both massive and drusy; and the combination of native gold and calaverite crystals. Only fluorite is missing. Judging by color, the gold is high in silver content. There is a 1.0cm long tabular calaverite crystal bridging a small amethyst vug; the crystal is set with tiny crude gold crystals. There are also many tiny black telluride masses, some partly replacing calaverite crystals. This is an exceptionally attractive specimen obtained from dealer Ron Vance.

***Plate #10-2, Specimen #1392*** **CALAVERITE – Cripple Creek, Teller County, Colorado. (MH)**

**#5175 11.0×9.1cm**

**Cripple Creek, Teller County, Colorado**

This piece is remarkable for the perfection of its unusual calaverite crystals. Though they are only 1-3mm long, they are equant prismatic crystals, rather than the typical tabular flakelike crystals. The prisms are striated, and the terminations are often sharp, planar, and shiny. Well-terminated calaverite crystals are extremely rare. Unusually massive granular purple fluorite makes up the matrix.

## KRENNERITE (GOLD SILVER TELLURIDE)

**#1948 6.5×3.6cm**

**Emperor Mine, Viti Levu, Fiji Islands**

This specimen consists of carbonate rock with drusy quartz vugs containing much krennerite in shiny, silvery, crystal masses to 1.0cm across. The krennerite crystals are equant to tabular, with surfaces that appear to be so full of holes as to seem porous.

***Plate #10-3, Specimen #1693*** **NAGYAGITE – Nagyag (now Sacaramb), Romania. (MH)**

## NAGYAGITE (LEAD GOLD TELLURIDE SULFIDE)

**#1693 5.0×2.8cm Plate #10-3**

**Nagyag (now Sacaramb), Romania**

This specimen comprises very thin, tabular, flakelike crystals, to 1.2cm across, standing on their edges on and in pinkish carbonate masses, with blebs of clove-brown crystalline sphalerite. There is a druse of tiny sulfide crystals partially coating each nagyagite flake. This piece from the type locality was obtained from dealer Brad Van Scriver.

*Plate #10-4, Specimen #2228* **SYLVANITE – Cripple Creek, Teller County, Colorado. (MH)**

*Plate #10-5, Specimen #1677* **SYLVANITE – Nagyag (now Sacaramb), Romania. (MH)**

## SYLVANITE (SILVER GOLD TELLURIDE)

**#2228 10.0×8.0cm Plate #10-4**

**Cripple Creek, Teller County, Colorado**

This sylvanite is a mesh of brilliant tabular-to-thin acicular crystals, grouped as rather sinuous arborescent partial coverings of dark brown drusy quartz crystals. The piece shows great contrast. It is from the collection of Lazard Cahn.

**#1673 8.8×5.5cm**

**Cripple Creek, Teller County, Colorado**

A pale granular drusy quartz vein has been split to expose a long vug lined with brilliant silvery crystals of sylvanite. They have many forms, from equant to elongated flat tabular. The largest crystal is 5mm long. The specimen was obtained from the Lazard Cahn Collection.

**#1677 6.5×4.9cm Plate #10-5**

**Nagyag (now Sacaramb), Romania**

The base of this piece is siliceous vein matter topped by a pinkish carbonate surface. On this surface are numerous points decorated with radiating delicate silvery sylvanite needles with tellurium. The longer needles are 9mm. Formerly in the Lazard Cahn Collection, the identity of the piece was verified by John White of the Smithsonian.

# Silver Species

## ACANTHITE (SILVER SULFIDE)

**#1135 4.1×3.2cm Plate #10-6**

**Colquechaca, Bolivia**

This is a very well defined group of acanthite in lustrous gray crystals, mostly simple octahedra to 1.8cm on edge. The piece was formerly in the Peter Bancroft Collection.

*Plate #10-6, Specimen #1135* **ACANTHITE – Colquechaca, Bolivia. (MH)**

## ACANTHITE

**#988 11.0×8.0×8.5cm**

**Refer to Chapter 3, Plate #3-1**

**Segen Gottes Mine, Rossvein, Gersdorf, Saxony, Germany**

The specimen is an old classic described in Chapter 3, Plate 3-1.

**#4719 4.1×4.0cm Plate #10-7**

**Freiberg, Saxony, Germany**

An unusual group of rounded crystals with curved wires of acanthite to 2.0cm long. The wires resemble native silver, but they have acanthite luster and terminations. This specimen was acquired from dealer Cal Graeber.

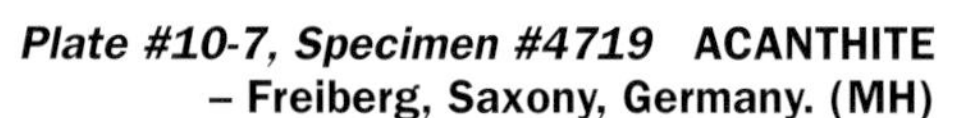

***Plate #10-7, Specimen #4719*** **ACANTHITE – Freiberg, Saxony, Germany. (MH)**

***Plate #10-8, Specimen #2111*** **ACANTHITE – Freiberg, Saxony, Germany. (MH)**

**#2111 3.2×3.2cm Plate #10-8**

**Freiberg, Saxony, Germany**

Two large, 2.2cm, shiny, black, penetration spinel twins form this piece. The upper twin is in high relief, with almost all faces of the twin exhibited. The piece came from the Northland College Collection, Ashland, Wisconsin.

**#1684 6.9×5.3cm**

**Freiberg, Saxony, Germany**

This piece consists of fat, equant, moderately shiny crystals cementing brecciated quartz and rock. The biggest acanthite is 2.2cm across. The piece was formerly in the Lazard Cahn Collection.

**#2887 15.2×10.8cm Plate #10-9**

**Guanajuato, Gto., Mexico**

A comb-structure quartz vein opens into a large vug lined with 5mm quartz points. On these are perched a 6.5×3.7cm group of shiny, dark, complex crystals, to 1.2cm long, of acanthite resting on clean, bright drusy quartz covering a pale matrix. This matrix piece has splendid contrast.

**#2307 3.5×1.5cm**

**Guanajuato, Gto., Mexico**

This brilliant piece consists of fully developed equant cubic crystals (no skeletal crystals here!) to 7mm on the edge. Acquired from dealer Harvey Gordon.

***Plate #10-9, Specimen #2887*** **ACANTHITE – Guanajuato, Gto., Mexico. (MH)**

## Guanajuato, Mexico

Guanajuato is one of the great silver camps in Mexico. Local Indians worked the surface outcrops before the arrival of the Spaniards in 1548. News of an odd discovery of silver brought about the region's development. The discovery is reported to have occurred when travelers built a campfire against a rich silver vein without realizing it. The heat of the fire caused the metal to liquefy, leading to the discovery of productive veins.

The San Juan de Rayas vein, still producing, and the source of a number of Barlow specimens, was started in 1558, the same year that the extremely rich Veta Madre vein was opened. In places, this latter vein exceeded 150 feet in width. The ore veins intrude a foliated slate and several other rock types. The veins are deeply oxidized, but the deposits are rich in sulfides as well.

For the collector, the silver species are of particular note, but this deposit has also produced delicate amethyst groups, fluorite, adularia, silver-bearing galena, a variety of calcite crystal forms, apophyllite, and even some zeolites. Fortunately for collectors, fine acanthites and rarer polybasite and stephanite crystals are still being salvaged from the mining operation. *R.W.J.*

***Plate #10-10, Specimen #3557* ACANTHITE – Guanajuato, Gto., Mexico. (MH)**

## ACANTHITE

**#3557 3.9×2.7cm Plate #10-10**

**Guanajuato, Gto., Mexico**

This is a group of partly complete, partly skeletal, brilliant crystals, with no matrix. The crystals have cube edges to 1.2cm.

**#3562 2.8×1.9cm**

**Guanajuato, Gto., Mexico**

This is a group of shiny lead-gray skeletal cubic crystals, with cube edges to 1.9cm.

**#3563 3.5×1.9cm**

**Guanajuato, Gto., Mexico**

This piece is a group of brightly reflective semi-skeletal sharp cubes to 1.8cm on edge.

**#2337 6.2×5.8cm**

**Guanajuato, Gto., Mexico**

Pseudomorphous after pyrargyrite, this specimen bears a close resemblance to pyrargyrite #2881 (see below). It consists of sharp, tapering, black 1.0cm crystals in very high relief.

## New Aguilarite Find at Guanajuato, Mexico 1995

*Peter Megaw and Chris Tredwell*

In the fall of 1994, approximately 20 specimens of suspected aguilarite were obtained in a mixed lot of sulfides and sulfosalts from the Penafiel Mine, Guanajuato, Mexico. The specimens are composed of skeletal dendritic pseudo-octahedral crystals to 4mm across with a surface coating of chalcopyrite. Specimen size ranges from micros (16) to thumbnails (4). Initial identification was made based on morphology and similarity to historic material from the district. However, optimism was muted because recent work on several supposed aguilarites from Guanajuato has shown them to be either erroneously identified or completely pseudomorphed to acanthite. The recent material was heavily enough coated with chalcopyrite to suggest similar pseudomorphing. A few specimens were released at the Tucson Show, with the understanding that analysis was underway at the University of Arizona Mineral Museum. After Tucson, Dr. Terry Wallace informed us that significant selenium was detected in the samples, enough to indicate that aguilarite is definitely present under the chalcopyrite coating.

This is an exciting find, as it is the fourth known locality for aguilarite in the Guanajuato District and it comes from a new area. The San Carlos Mine, which is the type locality for aguilarite, lies along the Veta Madre in the central part of the district and exploited the upper part of the Rayas Oreshoot. Its workings were long ago incorporated into the Rayas Mine, as were the workings of the Flores de Maria Mine, another early aguilarite source (Inst. Geol. de Mex. Bol. #40, 1923). The original aguilarites thus come from the uppermost portions of the orebody, which produces the abundant acanthite pseudomorphs after cubic argentite and many elegant silver sulfosalts. The new material comes from the "El Amparo" portions of Industrias Peñoles' Penafiel Mine, which exploits a separate oreshoot farther to the southwest along the Veta Madre. This has historically been a relatively gold-poor portion of the district, but the mine personnel report that the gold grades in the area where the new aguilarites were found are very high. Unfortunately, no gold minerals have been identified with the new aguilarites yet. Previous aguilarite discoveries in the district have come from two separate areas: the Nino Perdido Mine (long since defunct) in the Nayal vein group portion of the Peregrina–El Cubo–San Rafael vein system, which lies in the northeastern part of the district; and on the western fringe of the district in the La Luz zone (Panczner, 1987).

No additional aguilarites have been forthcoming, so hopes are dim for additional specimens in the near future.

**Plate #10-11, Specimen #5679 AGUILARITE – El Amparo portion of the Penafiel Mine, Guanajuato, Gto., Mexico. (MH)**

## AGUILARITE (SILVER SELENIUM SULFIDE)

**#5679 1.8×1.3cm Plate #10-11**

**El Amparo portion of the Penafiel Mine, Guanajuato, Mexico**

The dark gray aguilarite crystals are typically hopper shaped or even skeletal. The edges outlining the diamond shaped faces are subparallel, forming an almost reticulated array (typical for aguilarite from the type locality). Individual crystals run to 3mm or 4mm across. A druse of dark golden chalcopyrite is sprinkled across the aguilarite crystal giving a pleasant color contrast. There is a tiny pyrite group at the point of attachment. This piece was obtained from Chris Tredwell.

**Plate #10-12, Specimen #5770 AGUILARITE – Sirena Mine, Guanajuato, Mexico. (MH)**

**#5770 2.5×2.5×2.0cm Plate #10-12**

**Sirena Mine, Guanajuato, Gto., Mexico**

This is one of the best specimens ever found of this exceedingly rare silver mineral. It is a roughly "L"-shaped aggregate of sparkly gray metallic herringbone-dendritic, elongate, pseudododecahedral crystals to 8mm in length. The crystals are sharp and elongate along the fourfold axes rather than along the threefold axis like most Guanajuato aguilarites. Doubly terminated white scalenohedral calcites to 5mm are locally perched on the aguilarite. Unlike many Guanajuato aguilarites, this specimen lacks a coating of chalcopyrite. This is the best specimen from a small lot mined in April 1995 and brought to Barlow's attention by Peter Megaw and Chris Tredwell. The miner who collected the specimen was fired for "high-grading" two weeks after he removed this specimen. This specimen was analyzed by laser-coupled mass spectrometry by Dr. Terry Wallace at the University of Arizona and found to contain 90% of the silver expected for aguilarite. There is a compensating excess of sulfur.

## ALLARGENTUM (SILVER ANTIMONIDE)

**#481 8.8×6.4cm**

**Glen Lake Mine, Cobalt, Ontario, Canada**

This specimen consists of a slab of bronzish silver with silvery allargentum with some black tarnish. The piece was obtained from dealer Sharon Cisneros.

## AMALGAM (MERCURIAN SILVER)

**#1679 2.4×2.3cm**

**Landsberg, near Ober Moschel, Pfalz, Germany**

This is a nearly pure mass of 180.81g from Landsberg. It is covered with a dark brown coating, which is heavily streaked with bright silvery amalgam. The specimen came from the Lazard Cahn Collection.

**#3179 2.0×1.5cm**

**Taxco, Mexico**

This is a 168g shiny silvery nugget, 2.0cm across, with a microcrystalline surface. Taxco is a noted silver mining district.

## ANDORITE (LEAD SILVER ANTIMONY SULFIDE)

**#1929 3.7×2.9cm**

**San José Mine, Oruro, Bolivia**

One doubly terminated shiny lead-gray crystal, as long as the specimen is wide (2.9cm) rests on a matrix of tiny brownish crusty quartz crystals, with tiny pyritohedra. This piece was acquired from dealer Terry Szenics.

***Plate #10-13, Specimen #4739*** **ANDORITE – San José Mine, Oruro, Bolivia. (MH)**

**#4739 6.3×5.9cm Plate #10-13**

**San José Mine, Oruro, Bolivia**

Large, fat, tabular, striated crystals to 3.0×2.7cm are on a dark brown mesh of tiny sparkly stained quartz crystals, with a few stibnite crystals (to 1.5cm) on the back.

***Plate #10-14, Specimen #1917*** **ANDORITE – San José Mine, Oruro, Bolivia. (MH)**

**#1917 3.5×3.0cm Plate #10-14**

**San José Mine, Oruro, Bolivia**

This piece is 3.5cm across and is the termination of one exceptionally large crystal. The crystal is composed of tiny rodlike elements parallel to the *c*-axis, producing a glistening luster on the prism faces, with a dull, light-absorbing, luster on the termination. The piece was obtained from the William Pinch Collection.

## The San José Mine

The San José Mine is several kilometers west of Oruro, in the foothills of the eastern branch of the Andes known as the Cordillera Real. This range is also the home of other famous mining centers such as Potosi and Huanuni. Oruro mines were originally exploited for silver, but the end of the 19th century their tin content became important.

The mine works steeply dipping mesothermal (medium-temperature hydrothermal) veins that once averaged 70cm in width. The vein filling consisted of quartz, pyrite, stannite, tetrahedrite, zinkenite, and the "feather ores," jamesonite and boulangerite. Galena and sphalerite are not common, and chalcopyrite is rare. Values appear to diminish with depth.

Although secondary enrichment produced some bonanza ore that was exploited in earlier mining, San José is most famous for its amazing variety of primary sulfides and sulfosalts. The world's finest andorites came from this mine, as well as outstanding zinkenites and bournonites. Good teallite and wurtzite specimens come from the San José, as well as exceptional cubic and octahedral pyrite crystals. Other sulfur-bearing species include arsenopyrite, franckeite, plagionite, semseyite, cylindrite, chalcostibite, and the other "feather ore," meneghinite. The thinly scattered silver minerals produce the silver values of the ore. They include miargyrite, stephanite, freieslebenite, pyrargyrite, and rarely, silver. Beautiful wavellite specimens are also found at the San José Mine.

***Plate #10-15, Specimen #2117*** **ANDORITE – Mibladen, Khenifra District, Morocco. (MH)**

### ANDORITE

**#2117 4.1×2.7cm Plate #10-15**

**Mibladen, Khenifra District, Morocco**

These crystals, from an unusual source, are fat and compound, and much less tabular than the typical San José Mine (Oruro, Bolivia) crystals. They are a shiny dark gray, the largest measuring 1.9×1.0×0.7cm. They are on a small matrix of tiny glassy quartz crystals. Acquired from dealer Victor Yount.

### ARAMAYOITE (SILVER ANTIMONY BISMUTH SULFIDE)

**#2296 7.0×5.5×5.0cm**

**San Genero Mine, Huancavelica, Peru**

This piece is mostly a crystalline mass of miargyrite, some in tiny flakelike crystals, and dark gray crystalline aramayoite. Near a 2.0×2.0cm quartz mass are several 1.0cm groups of subparallel platy aramayoite crystals with the diagnostic silvery sheen. This specimen was obtained from dealer Sharon Cisneros.

## ARGENTOPYRITE (SILVER IRON SULFIDE)

**#5326 3.8×2.9cm Plate #10-16**

**Freiberg, Saxony, Germany**

This specimen is from 750 meters deep in the Brahma vein, Shaft 366, Aue-Alberoda. There is essentially no matrix with this mass of small, 2mm, short prismatic crystals. The crystals are shiny, with a bronzish iridescence, and are formed of thin, steplike growths. There are a few tiny dolomite crystals. This piece was mined in 1962.

*Plate #10-16, Specimen #5326* **ARGENTOPYRITE – Freiberg, Saxony, Germany. (MH)**

**#5697 7.2×4.7×3.5cm**

**Junghäuer Zechengang Mine, Jachymov, Czech Republic**

This is a rich specimen with patches of crystals up to 2mm as a crust. It was discovered by the mineralogist Walther Sortorius van Waltershausen in 1866 and was formerly in the collection of the University of Göttingen.

*Plate #10-17, Specimen #1690* **ARGENTOPYRITE – St. Andreasberg, Harz Mountains, Germany. (OM)**

**#1690 2.0×0.5cm Plate #10-17**

**St. Andreasberg, Harz Mountains, Germany**

There are many choice sharp microcrystals to 2mm resting in a small vug. This specimen was obtained from dealer Richard Kelly in an odd way. Kelly, along with dealer Gene Schlepp, obtained the Lazard Cahn Collection. It contained a number of rare silver species, many obtained by John, as you will note in reviewing this chapter.

John had flown to Rochester to view the Cahn Collection at Kelly's home. The collection was in a dimly lit cellar. Among the treasures was this small, unlabeled specimen. Because John had recently been studying argentopyrites at the British Museum, he recognized the mineral. Kelly, however, disagreed. So a deal was struck. John would buy the unknown specimen for $150.00. If it proved to be argentopyrite, he would obtain the piece for nothing. The piece was sent to the Smithsonian and John was proven correct, so he obtained the delightful and rare species for free.

> *If it proved to be argentopyrite, Barlow would obtain the piece for nothing. The piece was sent to the Smithsonian and John was proven correct, so he obtained the delightful and rare species for free.*

## ARGYRODITE

**(SILVER GERMANIUM SULFIDE)**

**#1607 5.2×3.9cm Plate #10-18**

**Colquechaca, Bolivia**

Germanium minerals are very rare; this one has germanium combined with silver. Sharp shiny gray crystals cover the entire specimen; one dodecahedron is 2.0cm across. The piece was formerly in the Dave Wilber Collection.

**#5700 3.7×3.4×1.7cm**

**Himmelsfürst Mine, Freiberg, Saxony, Germany**

This specimen, with a crust of 1mm crystals, is part of the type material at the Bergakademie, Freiberg, where C. Winkler discovered the new element germanium in 1885.

***Plate #10-18, Specimen #1607*** **ARGYRODITE – Colquechaca, Bolivia. (MH)**

***Plate #10-19, Specimen #5783*** **ARGYRODITE – Oruro, Bolivia. (MH)**

**#5783 4.5×3.4×2.4cm Plate #10-19**

**Oruro, Bolivia (according to the old label)**

Argyrodite is the only silver-bearing mineral containing germanium as a major constituent, and it is very rare. This superb specimen consists of a 2.4×2.5cm dodecahedron of argyrodite with a few other small argyrodite crystals surrounding it. The main crystal is somewhat crude, but for the most part is complete. *This remarkable specimen is the largest I have ever seen for the species (Wm. Pinch).* The old label (a Foote label) has his number #163 for the species on it and is probably from about 1909 to 1920. The species appears in his 1909 catalog. The price of $40.00 is also on the back of the specimen, and the price indicates that this was considered to be a major specimen even back then. The locality listed on the label, Oruro, Bolivia, is probably incorrect. It is most likely from Colquechaca, Dept. of Potosi, Bolivia.

*Argyrodite is the only silver-bearing mineral containing germanium as a major constituent, and it is very rare. This remarkable specimen is the largest I have ever seen for the species.*

## The Boleo District, Baja California Sur, Mexico

The Boleo District was discovered in 1868 by a Mexican rancher, Jose Villavicencio. While out riding in the Arroyo del Purgatorio, he noticed small rounded nodules of blue and green material that proved upon analysis to be the copper-rich minerals azurite and malachite. Since the minerals occurred in small rounded masses that the Mexicans called *boleos*, the district was so named.

Legend has it that Villavicencio sold his discovery for 16 pesos, about U.S. $1.28 at the time.

Be that as it may, the district developed into a prosperous copper mining district operated by the French Compagnie de Boléo, called Compania del Boleo, S.A. in Mexico. It was during their tenure of operation that the three rare minerals boleite, pseudoboleite, and cumengite were found and analyzed, making this the type locality for them. Refer to Plate #10-22.

The deposit is a recently uplifted group of Pliocene sediments seen in mesas dissected by arroyos. Mineralization is of the replacement type, resulting from ascending hydrothermal solutions.

The ore occurs in layers interspersed with conglomerate and sandy and clayey tuff. The age of the ores is thought to be about 8 to 15 million years.

The deposit has yielded about 60 different species but of these, only boleite, cumengite, and pseudoboleite are important to collectors. Only boleite is silver bearing. The species occur in the volcanic tuff above bed #3 near the Cumenge shaft of the Amelia Mine. (See Chapter 13.) *R.W.J.*

## BOLEITE (LEAD COPPER SILVER CHLORIDE HYDROXIDE HYDRATE)

The three rare copper minerals, boleite, pseudoboleite and cumengite, from the Amelia mine, Boleo District, Santa Rosalia, Baja California Sur, Mexico are internationally known. Barlow has visited the district and self-collected some of the minerals described here.

***Plate #10-20, Specimen #767*** **BOLEITE – Amelia Mine, Santa Rosalia, Baja California Sur, Mexico. (MH)**

### #767 4.5×3.3cm Plate #10-20

**Amelia Mine, Santa Rosalia, Baja California Sur, Mexico**

This piece consists of 5mm cubes scattered on a hard matrix intermixed with atacamite and other copper minerals in an unusually attractive combination. This piece from the Dave Wilber Collection was mined prior to the reopening of the mine by Bill Larsen and Ed Swoboda.

***Plate #10-21, Specimen #1433*** **BOLEITE – Amelia Mine, Santa Rosalia, Baja California Sur, Mexico. (MH)**

### #1433 7.7×5.2cm Plate #10-21

**Amelia Mine, Santa Rosalia, Baja California Sur, Mexico**

It is unusual for a crystallized specimen to be notable for its matrix, but such is the case here. These 10 fine cubes (to 8mm on edge) are scattered on a cream-gray hard rock (not the usual tan clay) with concentric layers of atacamite and other copper minerals, and quartz. This is a very fine specimen. from the Bill Larson Collection.

*Plate #10-22* **BOLEITE – Amelia Mine, Santa Rosalia, Baja California Sur, Mexico.**
***Top Row: Specimen #T243, #1608; Bottom Row: #T245, #T374, #T372, #T392.*** **(MH)**

## BOLEITE

**#T243 2.3×1.5cm Plate #10-22 (top row, left)**

**Amelia Mine, Santa Rosalia, Baja California Sur, Mexico**

This piece has two dark blue cubes on small matrix. The larger cube is a bit more than 1.0cm on the edge. Bill Larson, the dealer who mined here, regards this specimen as the best boleite thumbnail from the site.

**#1608 3.5×2.5cm Plate #10-22 (top row, right)**

**Amelia Mine, Santa Rosalia, Baja California Sur, Mexico**

This piece has a 1.0cm cube at one end of the matrix and a slightly smaller cube at the other end.

**#T245 1.2×2.0cm Plate #10-22 (bottom row, 1st)**

**Amelia Mine, Santa Rosalia, Baja California Sur, Mexico**

This specimen, 1.2cm across, has two intergrown blue crystals, one an octahedron modified by the cube and dodecahedron, the other a cube modified by the octahedron.

**#T374 0.7×0.7cm Plate #10-22 (bottom row, 2nd)**

**Amelia Mine, Santa Rosalia, Baja California Sur, Mexico**

This is a perfect 7mm cubo-octahedron on a white matrix.

**#T372 0.9×0.9cm Plate #10-22 (bottom row, 3rd)**

**Amelia Mine, Santa Rosalia, Baja California Sur, Mexico**

This is a 9mm perfect deep blue cube.

**#T392 1.6×1.6cm Plate #10-22 (bottom row, 4th)**

**Amelia Mine, Santa Rosalia, Baja California Sur, Mexico**

This is a complete dark blue compound group of crystals 1.6cm across. This specimen is not typical of the area; some of the faces may be pseudoboleite.

*Plate #10-23, Specimen #1915* **CANFIELDITE – Aullagas, Colquechaca, Bolivia. (UK)**

*Plate #10-24, Specimen #1928* **CHLORARGYRITE – Broken Hill, New South Wales, Australia. (MH)**

## CANFIELDITE (SILVER TIN SULFIDE)

**#1915 2.3×1.5cm Plate #10-23**

**Aullagas, Colquechaca, Bolivia**

This specimen is all canfieldite in black, moderately shiny crystals with a bluish luster. The crystals range to 4mm and are pseudocubic, in dodecahedra, with octahedral modifications. This is a fine specimen of an unusual silver-tin sulfosalt. It was formerly in the William Pinch Collection.

## CHLORARGYRITE (SILVER CHLORIDE)

**#1928 6.4×5.2cm Plate #10-24**

**Broken Hill, New South Wales, Australia**

Brownish crystals on a tan matrix are up to 1.3cm across. Large for the species, the crystals are distorted; some are tabular and almost flakelike, and several resemble spinel twins. The specimen has an A.E. Foote label and was obtained from the Northland College Collection, Ashland, Wisconsin.

**#1738 5.3×4.3cm**

**Lake Valley, Lake County, New Mexico**

This is pure horn silver from Lake Valley. The original surface is dull dark gray, with many tiny waxy blebs and tiny rounded crystals. The surface exposed by a miner's chisel mark is a waxy translucent pale gray. The piece was acquired from the Lazard Cahn Collection.

*Plate #10-25, Specimen #1246* **CUMENGITE – Amelia Mine, Santa Rosalia, Baja California Sur, Mexico. (MH)**

## CUMENGITE (LEAD COPPER CHLORIDE HYDROXIDE)

**#1246 3.3×2.5cm Plate #10-25**

**Amelia Mine, Santa Rosalia, Baja California Sur, Mexico**

Cumengite is not a silver mineral, but because of the fine combination of cumengite, pseudoboleite, and boleite (a silver mineral) on the specimen, it is included here. This specimen has a 1.8cm combination crystal, with cumengite pyramids on most of the boleite's cubic faces. One face, however, has a millimeter-thick pseudoboleite platform on it, with numerous tiny cumengites starting to grow up from the edges of the platform.

*This specimen is an example of the three classic minerals from this deposit; also see boleite and cumengite, Chapter 13.*

## DIAPHORITE (LEAD SILVER ANTIMONY SULFIDE)

**#1682 10.5×5.2cm Plate #10-26**

**Mina Banco de la Vascongada, Heindelaencia, Spain**

These shiny, black, striated, twinned, tabular crystals, to 1.0cm long, partly cover drusy quartz coating mica schist. Many crystals are in high relief and high contrast, making this one of the world's best diaphorite specimens. This piece was acquired from the Lazard Cahn Collection and was verified by X-ray at the University of Wisconsin and at the Smithsonian.

***Plate #10-26, Specimen #1682*** **DIAPHORITE – Mina Banco de la Vascongada, Heindelaencia, Spain. (MH)**

***Plate #10-27, Specimen #4774*** **DYSCRASITE – Pribram, Czech Republic. (MH)**

## DYSCRASITE (SILVER ANTIMONIDE)

**#4774 8.0×7.3cm Plate #10-27**

**Pribram, Czech Republic**

This is a modern classic from Pribram, Czechoslovakia, composed of a complex mass of bright, steely, long prismatic or thick acicular, crystals, with individuals to 3.0cm long. The specimen front has an interesting fish bone pattern, with long crystals growing en echelon from a spine of smaller crystals. There is only a little matrix present.

**#1686 3.1×2.0cm**

**St. Andreasberg, Harz Mountains, Germany**

On a quartz-sulfide matrix is a 3mm black equant twin crystal, with a strong re-entrant angle and chevron striations. The piece was obtained from the Lazard Cahn Collection and was verified by X-ray at the University of Wisconsin, June, 1978.

***Plate #10-28, Specimen #1672*** **DYSCRASITE – St. Andreasberg, Harz Mountains, Germany. (MH)**

**#1672 4.0×3.5cm Plate #10-28**

**St. Andreasberg, Harz Mountains, Germany**

This piece, acquired from the Lazard Cahn Collection and verified by X-ray at the Smithsonian, consists of fat dull black tabular crystals, embedded in white calcite on a pale gray matrix, with excellent contrast. The largest dyscrasite crystal is a compound one, 2.2×1.4cm.

## DYSCRASITE

**#5166 7.5×7.0cm**

**Grube Charlotta, St. Andreasberg, Harz Mountains, Germany**

The base of the specimen consists of large (around 2.0cm) modified cubic acanthite crystals. On the acanthites are several black dyscrasite crystals to 2.5cm long. These have a silky, slightly bronze luster and are prismatic barrel-shaped crystals, but never complete. There is generally a several millimeter gap between the dyscrasites and most of the acanthites, as if to allow for fluid flow, and there is a sprinkling of small (to 6mm) pyrargyrite crystals on the earlier minerals. There is a little quartz at one end, and some bright silver on the back of this piece, which was formerly in the Eric Asselborn Collection.

***Plate #10-29, Specimen #2294*** **DYSCRASITE – Catorcé, San Luis Potosi, Mexico. (MH)**

**#2294 4.9×2.7cm Plate #10-29**

**Catorcé, San Luis Potosi, Mexico**

This pale matrix with some small barite crystals has black equant pseudohexagonal tablets of dyscrasite to 1.7cm across, in high relief. This fine miniature comes from an unusual locality, Catorcé, San Luis Potosi, which is best known for fine stibiconite crystals. The identity was verified by Dr. Pete Dunn of the Smithsonian.

## FIZÉLYITE (LEAD SILVER ANTIMONY SULFIDE)

**#2050 6.2×4.4cm**

**Felsobanya, Romania**

This is an excellent example of "black on black," a combination not generally valued in mineral trophy competition. These Felsobanya fizélyites are 3mm equant crystals with poor terminations. They are associated with black semseyite laths and small glistening black sphalerite crystals. The piece was obtained from collector Abe Rosenzweig.

## FREIESLEBENITE (SILVER LEAD ANTIMONY SULFIDE)

**#1683 4.0×5.0cm Plate #10-30**

**Hiendelaencina, Spain**

The specimen consists of millimeter-sized steel-gray crystals of freieslebenite covering most of the surface. It is closely associated with pale brown siderite crystals to 5mm and massive argentite, all on a massive siderite matrix. A small (3mm) cluster of bright ruby-red pyrargyrite crystals completes the suite.

***Plate #10-30, Specimen #1683*** **FREIESLEBENITE – Hiendelaencina, Spain. (JS)**

## HATCHITE
**(LEAD THALLIUM SILVER ARSENIC SULFIDE)**
**#5688 4.6×4.5×2.5cm**

**Lengenbach Quarry, Binnental, Switzerland**

This specimen of an extremely rare mineral consists of beautiful, tiny crystals with contrasting colors. The label is handwritten by Professor Tony Stalder, former curator of the Museum of Natural History in Bern, Switzerland.

## HESSITE (SILVER TELLURIDE)
**#1090 8.5×4.0×5.0cm**
**Refer to Chapter 3, Plate #3-52**

**Botés Mine, near Zlatna, Transylvania, Romania**

The top of this piece is covered with clean, milky-to-transparent, 1.0cm quartz crystals. On the quartz crystals is a contrasting 7.0cm group of sharp hessite crystals. The hessite crystals display three habits: long prismatic (one crystal is almost 2.2cm) with steep terminations; equant-to-slightly-flattened-and-rounded dodecahedron-like crystals; and flat tabular-to-prismatic partly skeletal crystals. All crystals are well terminated, and the group, from Botes, has superior exposure and contrast. This is one of the world's great hessites. This piece was obtained from the British Museum via dealer Charlie Key.

## IMITERITE
**(SILVER MERCURY SULFIDE)**
**#4799 2.6×2.4cm**

**Imiter, Morocco**

A relatively new silver mercury sulfide from the namesake Imiter Mine is seen here in two small vugs in crystalline pale pinkish dolomite, with scattered brilliant silvery imiterite crystals to 2mm. The crystals look like tiny brilliant stibnites. Obtained from dealer John Patrick.

***Plate #10-31, Specimen #1905*** **IODARGYRITE – Proprietary Mine, Broken Hill, New South Wales, Australia. (MH)**

## IODARGYRITE (SILVER IODIDE)
**#1905 7.2×6.8cm Plate #10-31**

**Proprietary Mine, Broken Hill, New South Wales, Australia**

Pale greenish, equant, slightly rounded, iodargyrite crystals to 3mm across, reside in a 5.0cm vug in light-to-dark brown goethite. Also on this specimen is an attractive small group of prismatic azurite crystals to 1.6cm long, and a 1.0cm patch of crystalline selenite. From dealer Sharon Cisneros.

***Plate #10-32, Specimen #5336*** **JALPAITE – Level 407, San Juan de Rayas Mine, Guanajuato, Gto., Mexico. (MH)**

## JALPAITE (SILVER COPPER SULFIDE)
**#5336 5.7×4.7×4.5cm Plate #10-32**

**Level 407, San Juan de Rayas Mine, Guanajuato, Gto., Mexico**

This piece, found on level 407, San Juan de Rayas Mine, consists almost wholly of rare jalpaite in shiny lead-gray, thick tabular crystals to 2.5×2.1cm. These crystals look like flattened octahedra, with surface growth figures. There is also a small amount of gypsum, var. selenite. This is one of the world's finest jalpaite specimens. The piece was mined in the early 1990s and obtained by Barlow from dealer Dave Bunk.

*Plate #10-33, Specimen #1613* **MIARGYRITE – Pasto Bueno, Huancavelica, Peru. (JS)**

## MIARGYRITE (SILVER ANTIMONY SULFIDE)

**#1613 7.0×4.7cm Plate #10-33**

**Pasto Bueno, Huancavelica, Peru**

Five plump ping-pong-ball-like mounds are composed of bright black tiny flakelike crystals of miargyrite without matrix. The specimen was acquired from Vicente Quispe.

**#5696 3.5×6.2×2.4cm**

**Bräunsdorf, Saxony, Germany**

Two perfect 2mm crystals lie in a vug along a quartz seam. The piece was found in 1824 at the type locality mine Neue Hoffnung, Bräunsdorf, Germany. It was formerly in the Mineralogical Museum of Hamburg and still has its original labels.

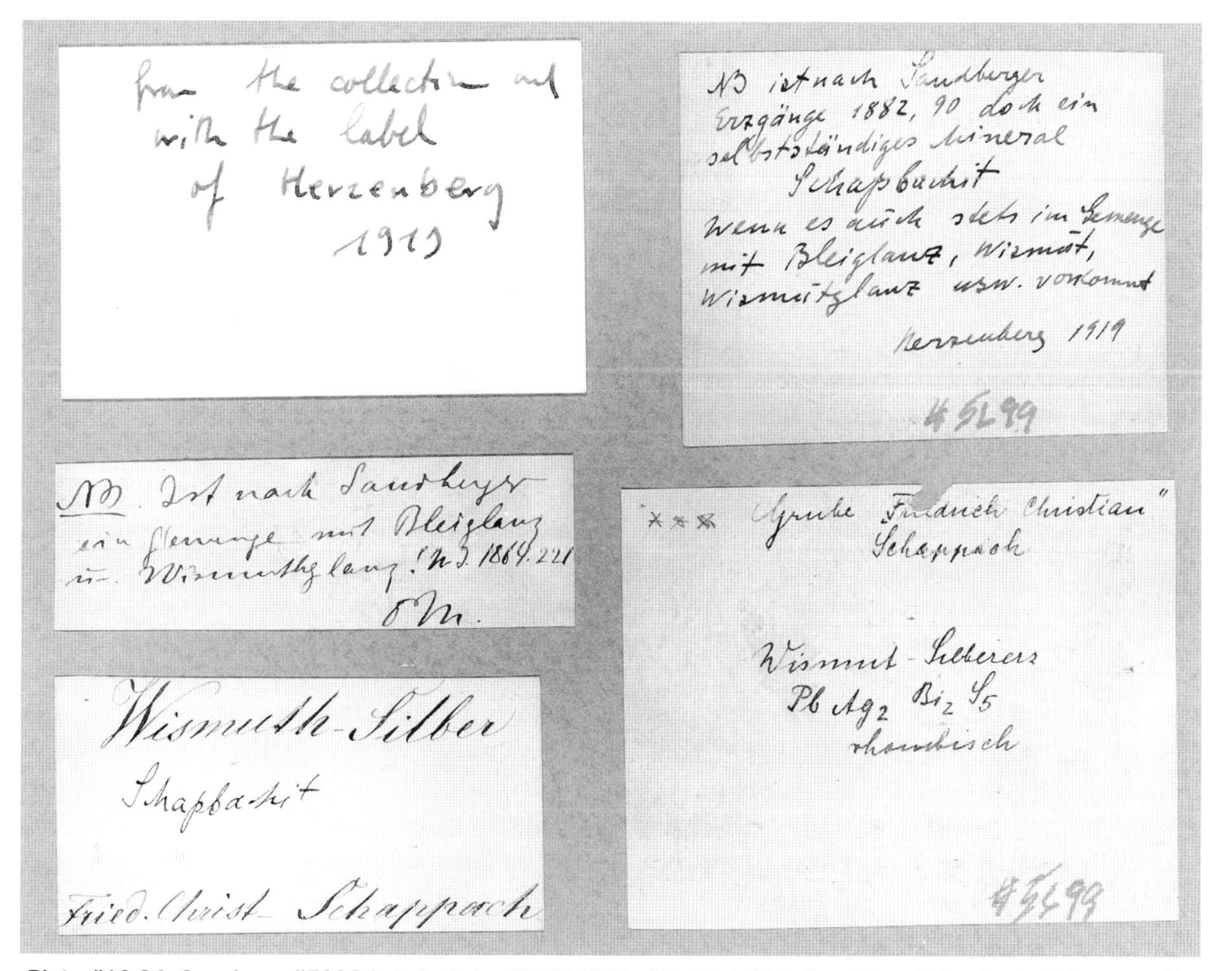


*Plate #10-34, Specimen #5699 Labels Only* **MATILDITE – Friedrich Christian Mine, Schapbach, Schwarzwald, Germany. (MH)**

## MATILDITE (SILVER BISMUTH SULFIDE)

**#5699 7.0×7.0cm Plate #10-34 Labels Only**

**Friedrich Christian Mine, Schapbach, Schwarzwald, Germany**

This large specimen is nearly pure matildite. Found in 1794, it is one of the best existing specimens from Schapbach. Plate #10-34 shows the labels that have been kept with this superb specimen.

## MIERSITE (SILVER COPPER IODIDE)

**#3709 4.1×2.3cm Plate #10-35**

**Proprietary Mine, Broken Hill, New South Wales, Australia**

Heavy greenish-to-grayish-black ore has a vug with brilliant canary-yellow millimeter-size crystals of rare miersite on a deep-red-to-black manganese oxide surface. The crystals are mostly compound octahedra and are in vivid contrast to the matrix. The piece was obtained from the South Australian Museum, Adelaide, via dealer Forrest Cureton.

***Plate #10-35, Specimen #3709*** **MIERSITE – Proprietary Mine, Broken Hill, New South Wales, Australia. (MH)**

**#2870 4.0×2.8cm**

**Proprietary Mine, Broken Hill, New South Wales, Australia**

The miersite is found on porous dark gossan with many 1-2mm equant tan miersite crystals. There is also a partial thin crust of pale greenish nantokite. The specimen was acquired from dealer Forrest Cureton.

### Broken Hill, New South Wales, Australia

Broken Hill is one of the great deposits of the world. When it was found, it was a hill that looked like it had a broken back, hence the name. Huge black outcrops proved to be silver bearing, and as mining continued, the hill also proved to be species rich. A great variety of minerals were found, some of them rare or uncommon silver species, as now seen in the Barlow Collection.

The property was first pegged by Charles Rasp in 1883. Rasp thought he had a tin claim, but assays proved to bear no tin and only marginal silver. Yet Rasp and his partners persisted and, a year later, penetrated a rich silver-bearing zone. Later the men realized that the nearby gossan outcrop was loaded with silver halides and lead carbonates. Their holdings become the Broken Hill Proprietary Company.

The deposit eventually yielded many new mineral species: marshite, a copper iodide; raspite, a lead tungstate; miersite, a silver copper iodide; costibite, a cobalt antimony sulfide; willyamite, a cobalt nickel antimony sulfide; paradocrasite, an antimony arsenide; and tocornalite, silver mercury iodide.

The origin of the orebody is still open to debate, but the leading theory suggests that the complex ores resulted from lead and zinc materials being deposited on the ocean floor from hot springs. This theory raises the suspicion that the deposit is related to an ancient "black smoker," a highly mineralized undersea hot spring surging to the surface of the crust along a major oceanic fault zone. In the case of this deposit, subsequent heavy metamorphic action and other forces moved and altered the original ores, resulting in an assemblage of over 300 species, varieties, and mixtures being formed. How fortunate for the species collector! *R.W.J.*

*Plate #10-36, Specimen #1735* **MOSCHELLANDSBERGITE – Moschellandsberg, Germany. (MH)**

## MOSCHELLANDSBERGITE (SILVER MERCURY INTERMETALLIC COMPOUND)

**#1735 4.9×3.8cm Plate #10-36**

**Moschellandsberg, Germany**

From the type locality, Moschellandsberg, this specimen contains a small vug displaying 7mm well-formed cubo-octahedra of this rare intermetallic compound ($Ag_2Hg_3$). It has a bronze tarnish and shows several minute parallel cleavage lines. The matrix for the vug is a typical mercury ore: dull red, yellow, and brown ocherous rock. The piece was formerly in the Archduke Stephan (of Austria-Hungary) Collection.

There is an interesting tale in the discovery of this species. Landsberg, on the Moschell River, has exploited mercury deposits at least since the 11th century, and these very symmetric isometric crystals have been known to mineralogy since the science began. Even though Germany has been hearth and home to the science of mineralogy, no one recognized these crystals as a unique mineral; they were always lumped with amalgam (which is indeed common in this district). Amalgam is a natural alloy of silver and mercury, and the elemental proportions vary through a wide range. In the 1930s, two Americans, H. Berman and G.A. Harcourt, became suspicious of these crystalline "amalgam" specimens. Sure enough, the crystals turned out to have a specific formula ($Ag_2Hg_3$) and a specific structure appropriate to an intermetallic compound. The two Americans had found a brand-new silver mineral from Germany, and they published its description in the *American Mineralogist,* Volume 23, 1938, page 761.

*Plate #10-37, Specimen #5743* **(top),** *Specimen #5742* **(bottom) NAUMANNITE – Eskoborner Stollen, Tilkerode, Harz, Germany. (MH)**

## NAUMANNITE (SILVER SELENIDE)

**#5743 5.0×5.0×3.0cm Plate #10-37 (top)**

**Eskoborner Stollen, Tilkerode, Harz, Germany**

Crystals of naumannite are practically nonexistant. This dolomite specimen, which came from the discoverer of naumannite by Bergrat Zinken (zinkenite was named after him), shows small rich crystals. He sold it to the famous dealer Krantz in 1824, the year of the discovery. Zinken's private collection went to a German museum from which Professor Georg Gebhard procured the piece for the Barlow Collection. All original labels remain with the specimen.

**#5742 5×5mm Plate #10-37 (bottom)**

**Eskoborner Stollen, Tilkerode, Harz, Germany**

This 5mm naumannite crystal is perhaps the only single crystal of this extremely rare mineral preserved in the world.

## PEARCEITE
**(SILVER COPPER ARSENIC SULFIDE)**
**#5276 7.9×4.0cm**
**Plate #10-38**

**Molly Gibson Mine, Pitkin County, Colorado**

The matrix of this specimen consists of galena, sphalerite, and silver minerals. The surface is formed by the flattened hexagonal pyramids of pearceite to 1.2cm across. The pearceites are liberally sprinkled with shiny silver wires, each several millimeters long. The piece came from dealer Cal Graeber.

***Plate #10-38, Specimen #5276*** **PEARCEITE – Molly Gibson Mine, Pitkin County, Colorado. (MH)**

***Plate #10-39, Specimen #5337*** **PEARCEITE – San Juan de Rayas Mine, Guanajuato, Gto., Mexico. (MH)**

*This is one of the largest and best pearceites in existence.*

**#5337 11.6×9.1cm Plate #10-39**

**San Juan de Rayas Mine, Guanajuato, Gto., Mexico**

This large, heavy piece from the San Juan de Rayas Mine consists almost solely of shiny black pearceite in thick tabular pseudohexagonal crystals, some more than 2.0cm across. There is a sprinkling of tiny (2mm) brassy chalcopyrite sphenoids, and both the chalcopyrite crystals and the pearceite crystals are partly coated by a druse of a dull black-gray mineral, perhaps primary acanthite. There are also several patches of drusy quartz. This is one of the largest and best pearceites in existence. It was mined in the early 1990s and acquired by Barlow from dealer Dave Bunk.

## POLYBASITE
**(SILVER COPPER ANTIMONY SULFIDE)**
**#3149 4.1×3.6cm**

**San Juan de Rayas Mine, Guanajuato, Gto., Mexico**

Thin tabular hexagonal polybasites to 1.5cm stand on the top of this specimen, with a heap of hexagonal pyrargyrite crystals to 8mm long on one side. There are also a few tiny quartz crystals, as well as a small amount of brown coating. The piece is well crystallized and shows an interesting association.

**#5180 3.6×3.6cm**

**San Juan de Rayas Mine, Guanajuato, Gto., Mexico**

This specimen is basically a single shiny compound hexagonal crystal, 3.6cm across. The crystal has step-like edges, and it is accompanied by a few small friends. Acquired by Barlow from dealer Ken Roberts.

*Plate #10-40, Specimen #1945*
**POLYBASITE – San Juan de Rayas Mine, Guanajuato, Gto., Mexico. (MH)**

## POLYBASITE

**#1945 5.3×3.8cm Plate #10-40**

**San Juan de Rayas Mine, Guanajuato, Gto., Mexico**

This piece is a vein selvage with the top coated by rounded, colorless fluorite cubes to 4mm. Almost covering the fluorites are paper-thin, sharp polybasite hexagons standing on edge; these are typically 1.1cm across. The polybasites are coated with tiny chalcopyrite crystals, giving them a bronze color. Here and there on the chalcopyrites are tiny equant galena-gray acanthite crystals.

**#2446 4.8×2.0cm Plate #10-41**

**Arizpe, Sonora, Mexico**

This is a stalk of shiny lead-gray skeletal acanthite crystals with a group of tabular hexagonal polybasite plates perched on one end. The specimen comes from one of Mexico's great polybasite localities. The polybasites are shiny black, and up to 1.4cm across. Obtained from dealer Sharon Cisneros.

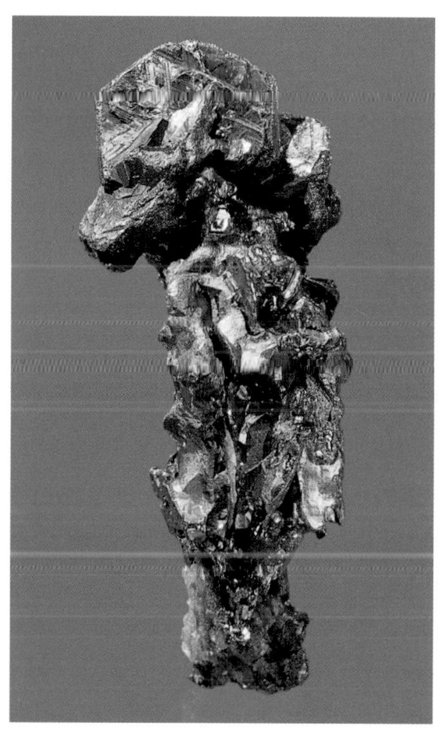

*Plate #10-41, Specimen #2446*
**POLYBASITE – Arizpe, Sonora, Mexico. (MH)**

*Plate #10-42, Specimen #T317* **PROUSTITE – Chañarcillo, Chile. (MH)**

## PROUSTITE (SILVER ARSENIC SULFIDE)

**#T317 1.8×1.5cm Plate #10-42**

**Chañarcillo, Chile**

A singly terminated bright red crystal, with a much smaller crystal attached. The major crystal has striated prism faces, with a low pyramidal termination. Acquired by Barlow from dealer Dr. Gary Hansen.

## PROUSTITE (SILVER ARSENIC SULFIDE)

After native gold and the gem minerals, the most desired and most valuable mineral species may be a group of beautiful blood-red proustite crystals. The pinnacle of proustite crystallization was attained at Chañarcillo, Chile; there the crystals of this uncommon silver arsenic sulfide reached a size, perfection, and color duplicated nowhere else.

The now defunct Chañarcillo mines are located about 50km south of Copiapo, Chile, in the heart of the bone-dry Atacama Desert. The first mine was located in 1832 by a muleteer, Juan Godoy, in pursuit of a guanaco. The usual rags-to-riches-to-rags pattern followed, and Godoy squandered his wealth.

This district was an example of secondary enrichment by which surface waters maintained pace with erosion, dissolving and redepositing the silver as the land surface eroded away. The primary ores from the original veins are of little importance, but Chañarcillo is a classic representation of what secondary enrichment can do. One mass of secondary silver weighed 6,000 pounds and had to be mined with chisels. One mass of chlorargyrite with native silver weighed 45,092 pounds and was 75% pure silver. By 1850, Chañarcillo had 18 producing mines whose ore ran about $5,000.00 per ton. The yield in that year was $2.7 million, of which $2.1 million was net profit.

Native silver and numerous other silver minerals such as chlorargyrite, stephanite, acanthite, dyscrasite, polybasite, and miargyrite formed the bonanza ore, but by far the most mineralogically desirable species was proustite. L.P. Gratacap described the proustites thusly: "The proustite from Chile will be seen in long fluted crystals with projecting spikes of smaller crystals, like a Hercules club; often acute and obtuse rhombohedra of a deep cochineal transparent tint; also in deeply channeled prisms."

A demonstration of the intrinsic value of the Chañarcillo proustites can be had from the sale of a 7.0cm specimen to Perkins Sams in 1980 for $30,000.00. In the 1850s this specimen was sold to William Vaux for between $500 and $1,000, so these proustites were always expensive. It is to the everlasting glory of the Chilean miners that they realized the value, aesthetic and scientific as well as commercial, of these great crystals, and that so many were saved.

## PROUSTITE

**#1649 6.9×4.5×3.0cm**

**Refer to Chapter 3, Plate #3-73**

**Dolores Tercera Mine, Chañarcillo, Chile**

This piece consists entirely of brilliant deep crimson proustite crystals, all of perfect form, with no damage. There are three major prismatic crystals stacked in parallel, but climbing from front to rear. Myriad smaller crystals grow from the three major crystals at an angle close to 90°. Traded from the Harvard Collection by mineral dealer Gene Schlepp, and then to the Barlow Collection.

**#2520 3.3×2.3cm**

**Dolores Tercera Mine, Chañarcillo, Chile**

From the same mine, this is a group of six larger (to 1.3cm long) and a number of smaller crystals. All are lustrous deep red hexagonal prisms, with flat trigonal terminations.

**#1931 4.1×3.2cm.**

**Jachymov (formerly Joachimsthal), Czech Republic**

This piece is all proustite in two generations. The first consists of one fat and several smaller crystals; these have turned almost black (but not through human mishandling). The second generation has bright deep red crystals, including two 1.6cm diameter compound crystals, that apparently surrounded a mineral now missing (calcite?), plus one short, plump proustite 1.7cm across, with a rhombohedral termination. Acquired from dealer Herb Obodda.

**#2166 4.7×2.5cm**

**Freiberg, Saxony, Germany**

This classic German specimen consists entirely of proustite crystals. These fat crystals, with simple rhombic terminations, are deep red with intense red interior flashes. The largest crystal is 2.3cm long.

## PROUSTITE

**#1187 3.8×2.6cm Plate #10-43**

**Niederschlema, Freiberg, Saxony, Germany**

This is a group of proustite crystals without matrix and with no damage, and only a tiny point of attachment. The prismatic crystals are transparent, deep red, and lustrous and have rhombohedral terminations. They are arranged in an open grouping. This piece, for its size, cannot be improved. Acquired by Barlow from dealer Gene Schlepp.

***Plate #10-43, Specimen #1187*** **PROUSTITE – Niederschlema, Freiberg, Saxony, Germany. (MH)**

**#2066 3.0×2.3cm**

**Freiberg, Saxony, Germany**

Only proustite is present here, in transparent deep red glassy prisms with medium rhombic terminations. The longest crystal is 1.8cm.

**#2949 8.1×7.2cm Plate #10-44**

**Uchucchacua, Oyon Province, La Libertad, Peru**

The base of this piece has a drusy pinkish carbonate-quartz crust over concentric layers of variably mineralized ore. Planted atop the crust is a 2.2cm group of deep red shiny crystals; one termination is 1.0cm across. As a grace note, there are several frosted calcite crystals to 1.7cm long, with included sulfide specks. Obtained from dealer Don Belsher.

***Plate #10-44, Specimen #2949*** **PROUSTITE – Uchucchacua, Oyon Province, La Libertad, Peru. (MH)**

### Uchucchacua

Uchucchacua (Oo"-chu-cha'-ua, or in "American," "ootchy-kootchy") is a polymetallic mine at an altitude of 14,569 feet in the central Andes, about 180 miles from Lima. The deposit has been known at least since the early years of the Republic. It was extensively explored in the 1950s by new management and then equipped for modern production. The mine is Peruvian owned and operated, producing a little less than 100,000kg of silver each year (about 6% of total Peruvian production) plus significant lead and zinc tonnages, from about 800 tons of ore per day.

Since Uchucchacua ore contains significant quantities of both arsenic and antimony, both proustite and pyrargyrite can, and do, occur. Fine large pyrargyrite twin crystals have been found, as well as native silver in wires at least 10cm long. The very rare silver mineral uchucchacuaite was described in material from this mine, and a specimen of this species is also in John's collection.

## PYRARGYRITE (SILVER ANTIMONY SULFIDE)

**#1671 8.5×8.0cm**

**Refer to Chapter 3, Plate #3-74**

**Samson Mine, St. Andreasberg, Harz Mountains, Germany**

A surface of drusy white calcite covering very rich ore displays an unusually lustrous deep red crystal group to 3.0cm across. The pyrargyrites are very sharp and are terminated by two different rhombs. There is an iridescent small satellite group next to the larger one, and ruby slivers show in small vugs on the side of this specimen.

**#1413 4.6×4.0cm**

**St. Andreasberg, Harz Mountains, Germany**

A small quartz-rock base supports stacks of deep red prismatic crystals to 1.8×0.8cm. The crystals have an intense cupritelike sheen and terminations displaying as many as three different rhombs. This specimen has been repaired (carefully). It was obtained from dealer Mike Ridding.

**#3150 8.0×4.6cm**

**Durango, Mexico**

A drusy quartz surface underlies thin black polybasite plates to 1.2cm across. Piled on the polybasite are dozens of small (~3-4mm) barrel-shaped pyrargyrite crystals. The pyrargyrites are also dark but some have luster.

**#2883 3.6×2.1cm**

**Guanajuato, Gto., Mexico**

This is a single fat dark crystal sprinkled with tiny deep red lustrous pyrargyrites.

***Plate #10-46, Specimen #2881*** **PYRARGYRITE – Guanajuato, Gto., Mexico. (MH)**

***Plate #10-45, Specimen #2447*** **PYRARGYRITE – Guanajuato, Gto., Mexico. (MH)**

**#2447 5.3×3.7cm Plate #10-45**

**Guanajuato, Gto., Mexico**

This piece is a collection of brilliant, almost black, columnar crystals to 2.5cm long. The columns consist of 5mm growth stages, stacked one atop another, with each stage fluted, yielding an effect like that of palm tree trunks. There are also a few minor calcite crystals. Acquired from dealer Sharon Cisneros.

**#2882, #2884, #2888 4.4, 4.2, and 5.5cm**

**Guanajuato, Gto., Mexico**

Three superb prismatic pyrargyrites in compound crystal form are sprinkled heavily with secondary growth pyrargyrites. This secondary growth, plus their size, make these crystals above average for the species.

**#2886 3.3×2.5cm**

**Guanajuato, Gto., Mexico**

Two fat dark crystals with their c-axes at 30° have their terminations and part of their prisms heavily sprinkled with tiny lustrous pyrargyrites.

**#2881 5.8×5.7cm Plate #10-46**

**Guanajuato, Gto., Mexico**

This specimen consists of black, thin, tapering, fluted hexagonal prisms (probably very steep rhombs) up to 3.9cm long, sprinkled with younger tiny bright deep red pyrargyrites.

## PYROSTILPNITE (SILVER ANTIMONY SULFIDE)

**#1912 4.6×3.7cm**

**St. Andreasberg, Harz Mountains, Germany**

This rare dimorph of pyrargyrite is found here in a small vug in galena-rich ore; the vug is lined with drusy quartz on which are several pyrostilpnite crystals 2–3mm long. Dimorphs are two different crystal forms of the same chemical compound. The free-growing, thin, tabular, lathlike pyrostilpnite crystals are shiny, translucent, and cinnamon-red, with fine contrast and exposure. There is also a rare bornite-blue equant stromeyerite crystal on this specimen. The piece was obtained from the William Pinch Collection.

## SCHACHNERITE (SILVER AND MERCURY INTERMETALLIC COMPOUND)

**#5695 2.0×1.5×1.2cm**

One of three existing specimens, this piece consists of schachnerite crystals to 1mm with paraschachnerite and moschellandsbergite. A grayish-black crystal druse of somewhat bladed schachnerite crystals coats 1.0cm reddish-gray crystals of cinnabar, which are on a matrix of fine-grained black submetallic material formed by weathering of silver minerals. This specimen was found around 1780.

## SMITHITE (SILVER ARSENIC SULFIDE)

**#5692 3.0×2.5×2.5cm**

**Lengenbach Quarry, Binnental, Switzerland**

This specimen carries the label of Professor Nowacki, University of Bern. The lustrous vivid red-orange crystals to 1.5mm with prominent cleavage are on a matrix of finely crystalline white dolomite. Minor pyrite crystals are scattered in the matrix. With the specimen is the original capsule with crystal used for X-ray identification prepared by Professor Werner Nowacki.

***Plate #10-47, Specimen #1136*** **STEPHANITE – St. Andreasberg, Harz Mountains, Germany. (MH)**

## STEPHANITE (SILVER ANTIMONY SULFIDE)

**#1136 4.0×3.4cm Plate #10-47**

**St. Andreasberg, Harz Mountains, Germany**

This piece is entirely stephanite, in a subparallel grouping of bright black fat tabular crystals, with the longest crystal 2.6cm. Pictured on page 421 of *Gem and Crystal Treasures* by Peter Bancroft, the piece came to Barlow from Bancroft's collection.

***Plate #10-48, Specimen #1675*** **STEPHANITE – Cananea, Sonora, Mexico. (JS)**

**#1675 9.6×5.4cm Plate #10-48**

**Cananea, Sonora, Mexico**

A fat milky quartz crystal supports a large mound of bluish-black stephanite, composed of larger crystals bounded by tiny bright crystals. In and on the stephanite are 5mm long quartz needles, pinkish-tan carbonate crystals, and small pyrites. There are several half-centimeter carbonate rhombs, replaced by quartz casts. This specimen was obtained from the Lazard Cahn Collection.

***Plate #10-49, Specimen #5430*** **STEPHANITE – Rayas Mine, Guanajuato, Gto., Mexico. (MH)**

## STEPHANITE

**#5430 4.0×4.0×3.5cm Plate #10-49**

**Rayas Mine, Guanajuato, Gto., Mexico**

This is a massive, lustrous, fully terminated crystal of stephanite.

## STERNBERGITE (SILVER IRON SULFIDE)

**#1676 5.6×5.3cm Plate #10-50**

**Sampson Mine, St. Andreasberg, Harz Mountains, Germany**

Silver seldom combines with iron, but it does here, as shiny, very black prismatic crystals, usually deeply striated, and about 2mm across. They are sprinkled over very red pyrargyrite and dark gray galena crystals. Obtained from the Lazard Cahn Collection.

***Plate #10-50, Specimen #1676*** **STERNBERGITE – Sampson Mine, St. Andreasberg, Harz Mountains, Germany. (MH)**

***Plate #10-51, Specimen #1674*** **STROMEYERITE – Yellow Pine Mine, Boulder County, Colorado. (MH)**

## STROMEYERITE (SILVER COPPER SULFIDE)

**#1674 4.2×2.6cm Plate #10-51**

**Yellow Pine Mine, Boulder County, Colorado**

Stromeyerite, copper-silver sulfide, is not a truly rare mineral, but how often is it seen in crystals? This specimen, which has a 3.0×1.3cm patch of stromeyerite crystals, was collected from a portion of a rich stromeyerite-quartz-barite vein found in the Yellow Pine Mine. The crystals are hexagonal and deeply striated with usually pinacoidal terminations, and they look like tiny Ilfeld (Harz, Germany) manganites. They are all about 2mm long and have a bright blue tarnish. This is one of the best crystallized stromeyerites in captivity (and it has been verified by X-ray at the Smithsonian and the University of Wisconsin). Acquired from the Lazard Cahn Collection.

## WALLISITE
### (LEAD THALLIUM COPPER SILVER ARSENIC SULFIDE)

**#5694 6.3×3.5×4.7cm Plate #10-52**

**Lengenbach Quarry, Binnental, Switzerland**

This is interesting and rich specimen of the rare mineral wallisite. The bulk of the specimen is sugary white dolomite with realgar, pyrite, and a quartz crystal. The wallisite covers an area about 3.5×4.0mm within the realgar and consists of a group of dark gray crystals with a perfect cleavage and deep red internal reflections. Another area of realgar crystals has a small vug with excellent wallisite crystals coating much of the vug. There is a 6mm crystal of what is most likely hutchinsonite on the specimen. It is also associated with hatchite, another rare species with a chemical composition and structure similar to wallisite; the only difference is that wallisite contains copper as well as silver. The specimen comes from the famous Lengenbach Quarry, Binnental, Switzerland, and has the number L-22300 on it. It is probably from the Natural History Museum in Bern, or the museum in Basel, Switzerland.

***Plate #10-52, Specimen #5694*** **WALLISITE – Lengenbach Quarry, Binnental, Switzerland. (MH)**

## XANTHOCONITE (SILVER ARSENIC SULFIDE)

**#1685 4.5×4.3cm**

**St. Andreasberg, Harz Mountains, Germany**

This piece is a portion of a calcite-sulfide vein. There is a small vug of calcite crystals with a 3mm proustite crystal and, 1.5cm from that, a spray of thin-bladed orange-red xanthoconite (the dimorph of proustite) crystals, the largest being 3mm long. The contrast with the calcite makes this a very attractive specimen. This piece was formerly a Lazard Cahn specimen.

**#1913 5.7×4.5cm**

**Himmelsfürst Mine, Freiberg, Saxony, Germany**

This specimen is the epitome of low-temperature silver vein material, and it would repay a long study. The base of the specimen is formed by convex layers of microcrystalline arsenic and small crystals of calcite. On the upper surface are centimeter-size, curved, compound galena crystals (a few cleaved, to prove identity). Next comes a coating of tiny dark, brassy flakes of marcasite. On top of the marcasite are a number of red proustite crystals to 7mm long; these have sharp faces. Scattered all around the top of the specimen are orangish-red, thick and thin, tabular xanthoconite crystals, in singles and fanlike groups, to several millimeters in size. As in the Harz xanthoconite (#1685 above), both members of the dimorphic pair are found on the same specimen. Acquired from the Marion Godshaw Collection.

## REFERENCES

Bancroft, P. (1984) *Gem and Crystal Treasures*, Western Enterprises, Fallbrook, California, and Mineralogical Record, Tucson, Arizona, 488 pp.

Cook, R.B. (1979, July-August) Famous Mineral Localities: Chañarcillo, Chile." *Mineralogical Record*, 10:(4).

Gratacap, L.P. (1912) *A Popular Guide to Minerals*, Van Nostrand, New York, 330 pp.

CHAPTER 11

# Copper

*by Marc L. Wilson*

The Barlow Collection contains 96 cataloged copper specimens as well as numerous copper items of historical and archeological importance. John's interest in copper dates to the beginning of his collecting experiences. He fondly recalls vacations to the scenic Keweenaw Peninsula of Michigan, where he first became acquainted with the "red metal." Early trips to the A. E. Seaman Mineralogical Museum at Michigan Technological University in Houghton, Michigan and visits with then Curator Jean Petermann (Kemp) Zimmer stimulated his interest in the native metals and instilled a love of copper that led to an early expertise. One such trip was in the company of Dr. Peter Bancroft to collect photographs of the museum's specimens for illustration in Bancroft's book *Gem and Crystal Treasures*. (It was on this occasion that I first met John and became impressed with his determination to create "one of the finest privately owned mineral collections in the world.") Other trips established friendships with such famous Copper Country personages as Richard Whiteman and Donald Pearce, and resulted in the addition of many fine specimens to Barlow's collection. John's love of copper crystals, for which he maintains a "soft spot in the heart," led to copper's being among the first minerals to be established in the Barlow Collection. The copper suite is still vibrant, with several world-class specimens having been acquired only recently.

Copper crystallizes in the isometric system, typically in arborescent groupings of twinned and distorted crystals. More rarely, copper may occur as euhedral crystals in the form of the cube, octahedron, dodecahedron, or one of the various tetrahexahedra. Single crystals are often heavily modified and may exhibit numerous examples of the more common forms, as well as a host of rarer crystal faces.

Copper crystals are typically small, twinned, distorted, rounded, and pitted, making identification of individual crystal faces difficult or impossible. Identification is further complicated by the nature of the arborescent crystal groupings that most specimens exhibit. Sharp and well-developed copper crystals are the exception, and, if specimens from Michigan are excluded from consideration, it can be argued that top-quality specimens of copper are rarer than specimens of crystalline silver or gold. But, of course, any consideration of native copper must include Michigan specimens and the Barlow Collection is no exception, with over two dozen superb examples from localities throughout the Copper Country.

Native copper is a geological oddity and would be considered rare if it were not for the tremendous quantity mined from the unique deposits of Michigan's Keweenaw Peninsula. Here, native copper occurs with lesser amounts of native silver as open-space fillings in amygdaloidal basalts and conglomerates of the Portage Lake Volcanics. Somewhat similar occurrences, on a vastly smaller scale, are found in serpentinized greenstones in a number of localities worldwide and in low-grade metamorphosed basalts, such as those of the eastern traprock district of New Jersey, Connecticut, Massachusetts, and Pennsylvania.

More typically, native copper is found as a secondary mineral in the oxidized portions of ore deposits containing copper sulfides, including chalcopyrite, chalcocite, or bornite. Specimens occur with other secondary minerals such as cuprite and malachite, usually in a matrix of friable limonitic gossan or in fractures in crystalline intrusive rocks or limestone. Famous localities of this type are numerous and include porphyry copper deposits at Bisbee, Ajo, and Ray, Arizona and elsewhere in the western United States, Central and South America, and the great sulfide deposits of Broken Hill, New South Wales, Australia, the Tsumeb and Onganja Mines, Namibia, and the great Copper Belt of Zambia and Zaire. Most recently, superb specimens have become available from the states of the former Soviet Union.

In the Michigan traprock deposits, the native copper does not result from the weathering of primary sulfides; rather, it represents a primary mineralizing event in and of itself. It is easy to speculate that the abundance of native copper in the New World may have even retarded the early development of metallurgy here. The relative ease with which it could be obtained, either by mining or through extensive trade routes, which we know existed in North America long before the white man came (refer to ancient Indian copper hunting spears and tools, Plate #11-33) may have inhibited the development of smelting and subsequent metallurgical techniques for the recovery of copper from its ores. This may help to explain the relative lack of sophistication of mining and metallurgy among the Native Americans compared to other early cultures at comparable levels of cultural development.

In the Old World, once the production of metallic copper from its secondary ores like azurite, malachite, cuprite, and tenorite found in the weathered zones became common, quantities of the metal were available, and mining, smelting,

The copper minerals have long been an attractive collector suite. Azurite, malachite, cuprite, native copper, copper sulfides, copper sulfosalts, and many more have long impacted the collector scene, as most provide both colorful and beautifully crystallized examples of nature's handiwork. John's particular interest has been in the native element, as witnessed by this chapter. Barlow also has excellent representatives of the other copper minerals, including primary sulfides. Note his fine suite of chalcocites, particularly the excellent specimens recently mined at the Flambeau Mine, Wisconsin, and from the historically important Bristol, Connecticut mine. This mine yielded superb chalcocites and bornites rivaling those of Cornwall, England.

The importance of copper in the history of civilization cannot be overemphasized. It was one of two useful native metals available for early man. There was the occasional iron meteorite found, considered a gift from the gods, and there was gold in quantity found in stream beds. Native silver was rarely found, as it quickly alters during weathering. So, only gold and copper were found by early man in any useful quantities. Both are relatively soft and malleable. The former is too soft to be useful, even when cold hammered, save for ornamental and monetary purposes. Copper, on the other hand, tends to harden when cold-hammered, so it had limited use as a tool metal in early days.

The problem is that in much of the world native copper usually occurs as a secondary mineral and is not abundant, save in the upper reaches of weathered deposits. In the Near East, early man resolved the problem, albeit accidentally, when pottery makers, using high-temperature kilns that created a reducing atmosphere, realized that the blue and green coloring pigments that they applied to decorate their work yielded small beads of copper when left too long in the kiln. Thus came about the serendipitous birth of smelting and metallurgy. *R.W.J.*

and trading of copper throughout the Old World rose dramatically as the Chalcolithic Age emerged there. It was, however, short lived, for the Bronze Age emerged almost coincidentally.

Mineralogy had yet to wait thousands of years before becoming a science, for there was little understanding of minerals, mining, and metal ores in those early millennia. Copper ore mining was at its crudest, as was the separation of ores. Mineral recognition, likewise, was crude, being based on identification of physical properties, not an accurate way to distinguish between some copper-bearing minerals. Early metalsmiths knew only that certain colored ores yielded copper while others did not. Yet, they soon recognized that certain ores produced a "copper" that worked into harder tools and weapons. Without realizing it, they were confusing copper ores rich in arsenic with copper oxide ores. The latter gave copper; the former yielded a more useful arsenical bronze. Later, the early metalsmiths realized that still other ores produced an even better bronze alloy. These ores contained tin rather than arsenic.

So gradually, by trial and error, metallurgy evolved from crude accidental operations to highly refined and skilled processes, processes that were still being used when Georg Bauer (Agricola) wrote his monumental work *De Re Metallica* in the mid 1500s. The techniques described by Bauer were undoubtedly still in use when huge copper spikes were made to hold together Spanish galleons like the *Atochia* and the *Spring of Whitby*. The *Atochia's* beams were held by such copper spikes until a hurricane off the Florida coast sank it in 1743. One such spike now resides in the Barlow artifact collection, reflecting once again John's broad interest in things copper. (Refer to Plate #11-32.)

With the advent of Bauer's book, mining, metallurgy, and mineralogy entered a new era. The German silver mines became the key to the metal sciences' emergence from the Dark Ages. A mining academy later developed there. As new metal deposits were found, trained miners and engineers were in demand. Skilled German miners were highly regarded, and it was they who trained Cornish fishermen and farmers at the invitation of Queen Elizabeth I so that they could work the great copper and tin mines of Cornwall. The Cornish "Cousin Jacks," in turn, moved across the globe to Michigan, Arizona, Colorado, Montana, Utah, Australia, and South America, becoming a part of the romance and lore of the famous mining camps at Ajo, Bisbee, Morenci, Butte, Bingham Canyon, Broken Hill, Chiquicamata, and even Central Africa.

From their humble beginnings in the copper mines of the Near East and the silver mines of Eastern Europe, metallurgy, mining, and mineralogy have come of age.

## COPPER (NATIVE ELEMENT)

## AUSTRALIA

***Plate #11-1, Specimen #5629*** **COPPER – Sandy Flat Pipe, Redbank, Northwestern Territory, Australia. (MH)**

**#5629 5.5×5.2×4.0cm Plate #11-1**

**Sandy Flat Pipe, Redbank, Northwestern Territory, Australia**

This unique specimen consists of sharp, complexly crystallized and twinned copper crystals to 1mm in diameter in flattened, arborescent groupings to 1.2cm in length filling fractures in a brecciated pyro-bitumen matrix. Purchased in 1995, the unusual association and vivid contrast of bright copper crystals with an iridescent black pyro-bitumen matrix make this a most aesthetic and interesting addition to the collection.

**#T039 3.2×2.5×1.1cm**

**Broken Hill, New South Wales, Australia**

The rich, metamorphosed, oxidized ores of the Broken Hill silver-lead-zinc volcanogenic massive sulfide deposits, first developed in 1883, have produced some of the world's most striking mineral specimens. Although the deposit is generally not copper rich, the highly oxidized portions of the district have produced notable native copper specimens. Specimen #T039 is a small, aesthetic, dark brown miniature consisting of a jackstraw arborescent grouping of elongated, rough-textured branches of crystals repetitively twinned on the octahedral plane. Individual branches reach lengths of 3.2cm and widths of 1mm and culminate in cubic and tetrahexahedral terminations. While not as large as some, this is an excellent example of unusual twinning in copper crystals from a classic locality.

## CHILE

Since pre-Columbian times, copper has been known to exist in the rugged Andes Mountains of Chile and Peru. Extreme aridity and relatively low water tables have yielded well-developed oxidation and supergene enrichment zones as caps upon small high-grade vein and large low-grade disseminated sulfide deposits. Such arid, oxidized ore deposits, often above 15,000 feet in elevation, have produced some of the most interesting suites of minerals in the world. Native copper occurs in these oxidized and enriched zones above the great disseminated porphyry copper deposits, as well as in the high-grade veins with high copper values.

**#5319 12.0×8.1×7.2cm**

**Atacama, Chile**

Dark red-brown in color, this is a hackly mass of small, highly distorted and poorly formed copper crystals. Euhedral quartz crystals to 2.0cm occur partially embedded in the copper, and numerous hexagonal holes left by the removal of smaller quartz crystals are evident throughout the copper mass.

**#5323 12.3×10.5×6.4cm**

**Manto Cuba Mine, Inca de Oro, Chile**

This specimen consists of an intricate mass of bright, untarnished, sharp copper crystals of complex habit. Individual crystals reach 3mm. Euhedral quartz crystals to 2.1cm and a massive, black mineral (probably cuprite) occur as associated species. Of particular interest is an elongated cluster of small copper crystals to one side of the specimen. These form a "ram's horn" 4.8cm in length and 1.1cm at the base, which lends this specimen great aesthetic appeal.

## NAMIBIA

Namibia is best known for its splendid suites of colorful, often rare, secondary minerals from the multiple oxidation zones of the unique Tsumeb deposit. Nearly 200 miles to the south, however, lies a second deposit of great mineralogical interest. The Onganja Mine, near Seeis, has thrice made mineralogical news. First, came the production of malachite-encrusted cuprite crystals up to 5cm in diameter, and, later, came large, nonpseudomorphous malachite crystals of great beauty. From the same mine, thousands of exquisite groupings of euhedral copper crystals have been recovered. These often exhibit rare and complex crystal habits and form a necessary part of any copper suite. (See Chapter 12.)

***Plate #11-2, Specimen #2346*** **COPPER – Emke Mine, Onganja, Namibia. (MH)**

**#2346 5.7×5.1×2.3cm Plate #11-2**

**Emke Mine, Onganja, Namibia**

This superior specimen from the Emke Mine is an aesthetic combination of twinned and untwinned copper crystals of dark brown color. Individual crystals exhibit the form of the steep tetrahexahedron *l* {035} in a development that could easily be mistaken for the dodecahedron. Untwinned crystals are up to 1.1cm in diameter, while twins, of the platy, complex type famous from Ray, Arizona, reach lengths of 2.0cm. Obtained by Barlow from the Tom Gressman Collection.

*Plate #11-3, Specimen #3139* **COPPER – Seeis, Namibia. (MH)**

**#3139 4.2×3.0×1.8cm Plate #11-3**

**Seeis, Namibia**

This dusky-brown specimen is a small but superb cluster of sharp copper crystals from Seeis, Namibia. Individual crystals, some of which are twinned, exhibit the form of the cube and the tetrahexahedron in an aesthetic arborescent grouping. This is a fine example of copper crystals of a form typical to the locality.

### Russia

The Tur'inskie mines of the old Bogoslovsk district in Sverdlovsk Oblast, Russia have been a source of fine copper crystal specimens since the mid 1800s. Until recently, superb arborescent groupings of copper crystals from this classic Ural Mountains locality have rarely reached Western collections.

**#4951 8.5×3.0×2.7cm**
**Refer to Chapter 3, Plate #3-24**

**Bogoslovsk District, Northern Ural Mountains, Sverdlovsk Oblast, Russia**

The Ural Mountains produced this superb grouping of sharp, elongated cubic copper crystals with modifications by the dodecahedron, octahedron, and one or more tetrahexahedra, all in arborscent growth. The slightly tarnished copper color, large size, and exquisite crystal development make this an outstanding specimen from a classic Russian locality. Upon first seeing this specimen in a photograph sent by a nationally known mineral dealer, John immediately made arrangements for its purchase and still expresses a sense of gratitude that he was able to add it to his excellent copper suite.

## UNITED STATES— ARIZONA

Histories of the great Arizona copper mines are steeped in the types of legends characteristic of the Southwest. Rife with tales of gunfights and claim jumpers, stock swindles and frauds, gullibility and genius, science and spiritualism, good luck and bad, accounts of mining camps such as Bisbee, Ajo, and Ray symbolize a period of American history as colorful as it was brief. Mined only on the most primitive level by the pre-Columbian inhabitants of the region, the great deposits lay largely idle until the 1848 Treaty of Guadalupe Hidalgo and the 1853 Gadsden Purchase annexed them to the United States. Small-scale gophering of limited high-grade oxidized veins and supergene-enriched areas gave way to large-scale mining of vast reserves of low-grade, disseminated ore bodies following the successful operations of the Utah Copper Company at Bingham Canyon, Utah beginning in 1907. Superb specimens of crystallized native copper, cuprite, malachite, and a host of other attractive and interesting mineral species were produced when oxidized ores were encountered in the shafts and vast open pits of the many mines. Some of the finest of these are preserved in the collection.

### Ajo, Arizona

As reported by Ira Joralemon (1973), the New Cornelia Mine is the lineal descendant of the Cornelia Copper Company founded in 1900 by John R. Boddie and several friends. Named after Boddie's departed first wife, the Cornelia Copper Company, as well as the Rescue Copper Company and the Shotwell Tri-Mountain Copper Company, both of which the Cornelia Company later absorbed, was created as a result of a stock swindle of which Boddie and friends were the unwitting victims. Despite this dubious beginning, several episodes of being suckered by fraudulent ore processing schemes, and the unanimously negative opinions of the leading miners and engineers of the day, Boddie never lost faith in his deposit. He was rewarded beyond his wildest dreams when, in 1911, the Calumet &

Arizona Mining Company took an option on his property with full production beginning in 1917. In 1931, the deposit was taken over by Phelps Dodge Corporation and, with a total production of nearly six billion pounds of copper, ranks as one of the great copper districts of the country. Splendid specimens of crystalline copper, azurite, malachite, and cuprite, as well as an unusual suite of copper silicate minerals, have been produced from the New Cornelia Mine.

**#5305 9.4×9.0×2.1cm**

**New Cornelia Mine, Ajo, Pima County, Arizona**

Slender, twinned, copper crystals form a reticulated, arborescent mass with minor attached matrix on this specimen. Bright copper in color, individual crystals reach lengths of 2.7cm with widths of 1mm or less. The crystal grouping is arborescent and similar to that described by Wilson and Dyl (1992) as "Type 2" but with complex twinning on individual crystals resembling those of Ray, Arizona. While not as showy as some in the collection, this specimen is of great interest due to its unusual and well-developed crystal habit.

**#5306 11.8×7.1×3.5cm**

**New Cornelia Mine, Ajo, Pima County, Arizona**

This specimen consists of complex, blocky, distorted, and somewhat rounded crystals of copper with a tetrahexahedral form predominant. Twinned crystals of dark copper color reach 1.8cm in length and form a compact mass. On the back of the specimen, smaller, spiky crystals, to 1.0cm in length, exhibit twinning similar to that of #5305 above. Obtained by Barlow from the Carl Stentz Collection.

**#5307 9.5×8.7×2.7cm**

**New Cornelia Mine, Ajo, Pima County, Arizona**

The fissure-filling nature of flattened, arborescent copper crystal groups is clearly evident in this piece. One surface of the limonite-stained matrix hosts a fernlike, arborescent grouping of flattened and rounded copper crystals. Near the center of the specimen, a small piece of matrix overlies the copper. This form clearly demonstrates the nature and width of the original fracture in which the copper subsequently grew and renders this an important specimen not only for its aesthetics and locality, but for its instructional value as well.

## Bisbee, Arizona

Army scout John Dunn first discovered mineralization in what would become the Warren Mining District and Town of Bisbee in 1877. Two early mining endeavors by the Copper Queen Mining Company and Phelps, Dodge and Company on the Copper Queen and adjacent Atlantic claims, respectively, concentrated on the rich oxidized and supergene ores of the hosting limestones. In 1885, a rich orebody was discovered on the sideline of the claims, and the properties were merged to form the Copper Queen Consolidated Mining Company, eventually to become the great mining giant, Phelps Dodge Corporation.

In the early 1880s the Irish Mag claim, named after a notorious lady of fallen virtue, was located by a drunken Irish prospector by the name of Jim Daley. According to accounts preserved by Joralemon (1973), Daley found it prudent to flee Bisbee and head for Mexico following a fatal confrontation with a local constable over a right-of-way to his claim. After Daley drank himself to death in Mexico, Daley's Irish Mag claim ended up in the possession of a Tombstone saloonkeeper. The property was visited in 1898 by two Calumet & Hecla mine captains from the great copper district of Michigan's Keweenaw Peninsula. While examining the claim, "Cap'n Jim" Hoatson fell asleep and had a vision of a vast orebody at a depth of 900 feet beneath him. Based on his faith in the vision, he and his brother "Cap'n Tom" returned to Michigan to raise the capital to purchase and mine the property. Thus was born the Superior and Western Copper Company, later reorganized as the Calumet & Arizona Mining Company. Strangely enough, the richest ore deposit ever found in Bisbee was discovered exactly where "Cap'n Jim's" vision foretold, and only 50 feet deeper than expected.

The discovery of this orebody propelled Calumet & Arizona into contention with Copper Queen for the position of top mining company in the district, and a spirited competition for new claims developed between the two companies. Calumet & Arizona eventually hemmed in Copper Queen and forced a merger in 1931, at which time Phelps Dodge bought out Calumet & Arizona and assumed control of all its properties.

By the close of mining operations in 1975, nearly eight billion pounds of copper had been produced from the rich oxidized and, later, disseminated sulfide ores. Many of the finest copper mineral specimens known originated from this district, and several superb examples are preserved in the Barlow Collection.

*Plate #11-4, Specimen #1699* **COPPER – Calumet & Arizona Mining Company, Bisbee, Cochise County, Arizona. (MH)**

**#1699 4.1×4.1×2.2cm Plate #11-4**

**Calumet & Arizona Mining Company, Bisbee, Cochise County, Arizona**

An aesthetic, small cluster of copper crystals from one of the mines of the Calumet & Arizona Mining Company. It exhibits a partial black coating, possibly tenorite, which highlights the crystal structure. Individual crystals are of the form of the cube modified by the dodecahedron and minor octahedron with one or more tetrahexahedra in an unusual grouping similar to that described by Dana in 1886 and illustrated in his figure 46. This attractive specimen was added to the collection in April 1978 when John, in the company of dealer Gene Schlepp, was given an opportunity to select specimens from the Lazard Cahn Collection by dealer Dr. Richard Kelly of Rochester, New York.

**#2334 11.6×10.0×6.7cm Plate #11-5**

**Bisbee, Cochise County, Arizona**

This attractive specimen, also from Bisbee, consists of a mass of small, dendritic copper crystals partially altered to cuprite and malachite. One side of the arborescent mass has been polished to reveal the sequential alteration of copper to cuprite and cuprite to malachite that represents a classic assemblage from mineral deposits of this type. Formerly USNM (Smithsonian) #75539.

**#5348 6.7×4.1×2.5cm**

**Copper Queen Mine, Bisbee, Cochise County, Arizona**

Intergrown, elongated, twinned, spiky crystals of a dodecahedral or tetrahexahedral habit dominate this aesthetic copper specimen from the Copper Queen Mine. Individual crystals to 1.3cm in length, but only 1mm or less in width, form a compact, arborescent mass of delicate beauty. The upper portion of the specimen is only slightly tarnished and is a bright, reddish copper in color while the base, of somewhat coarser crystals, is a more typical dark copper. Obtained by Barlow from the Pohndorf Collection.

*Plate #11-5, Specimen #2334* **COPPER – Bisbee, Cochise County, Arizona. (MH)**

## Ray, Arizona

Hostile Apaches precluded serious investigations into the Mineral Creek mining district until the 1870s, despite the knowledge of the region's mineral potential as far back as the mid 1840s. The first serious mining commenced in 1880, and in 1883 the Ray Copper Company was formed.

A blanket of rich oxidized ore, much of it native copper, was discovered, and various small mining ventures flourished and faded. By 1905, a railroad was established nearby and, with shipping costs thus lowered, the prospect of mining the low-grade disseminated sulfide deposit discovered beneath the supergene-enriched cap became attractive.

In 1907 the Ray Consolidated Copper Company amalgamated most of the old mining properties in the district, and experts from Bingham, Utah came down to supervise the mining of the low-grade orebody. The first Ray Consolidated ore was shipped in 1911, and in 1933, after a series of mergers and reorganizations, the district came under the control of the Kennecott Copper Corporation. Kennecott began open-pitting the deposit in 1955 with total production through 1979 estimated at five billion pounds of copper (Jones and Wilson, 1983).

While fine specimens of copper and cuprite were preserved from the native copper mineralization of the deposit's oxidized upper portions, and fine specimens of chrysocolla and other copper silicates have been harvested from the unusual silicate orebody adjacent to the main deposit, the most striking copper crystal specimens were discovered in 1973 along an altered fault zone. The beauty of these elongated, complexly twinned copper crystals, some of which are twisted by what appear to be screw dislocations in the crystallographic structure, have brought more fame to Ray, Arizona than all the specimens previously collected there. The Barlow Collection contains three excellent examples from this famous find.

***Plate #11-6, Specimen #T271*** **COPPER – Pearl Handle Pit, Ray Mine, Pinal County, Arizona. (MH)**

**#T271 3.3×1.4×1.2cm Plate #11-6**

**Pearl Handle Pit, Ray Mine, Pinal County, Arizona**

This specimen is a moderately sized but aesthetic example of the best of the find. It is a reddish-brown, single, complex twin with excellent re-entrant development and sharp crystal faces. Pictured in *Mineralogical Record*, Volume 5,1974, page 234, Figure 3.

**#5308 10.0×5.8×2.6cm Plate #11-7**

**Pearl Handle Pit, Ray Mine, Pinal County, Arizona**

Larger but less well developed than #T271, this specimen exhibits the same classic complex twinning that has become associated with copper from Ray. The specimen is bright copper in color with its largest twinned crystal 5.6cm in length. Other crystals are much smaller and form an aesthetically pleasing, arborescent grouping.

***Plate #11-7, Specimen #5308***
**COPPER – Pearl Handle Pit, Ray Mine, Pinal County, Arizona. (MH)**

**#5805 7.6×4.5×2.9cm Plate #11-8**

**Ray Mine, Pinal County, Arizona**

This superb specimen consists of an arborescent grouping of complexly crystallized and twinned crystals of bright copper color. Reportedly collected in the 1950s, it is an exceptional example of aesthetic crystals from one of the Southwest's most important copper localities.

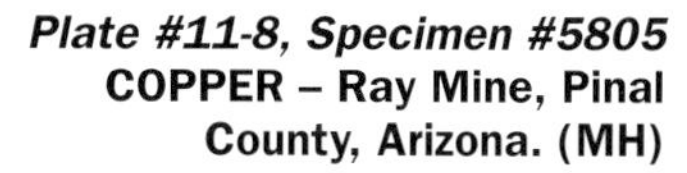

***Plate #11-8, Specimen #5805***
**COPPER – Ray Mine, Pinal County, Arizona. (MH)**

***Plate #11-9, Specimen #5816*** **COPPER – Ray Mine, Pinal County, Arizona. (MH)**

**#5816 5.4×5.4×1.9cm Plate #11-9**

**Ray Mine, Pinal County, Arizona**

This outstanding specimen is one of the most aesthetic examples of the complexly twinned copper crystals that, due to their perfection of development at this locality, have become a classic of the Ray Mine. Two flattened, twinned crystal groups intersect at nearly right angles to form a cross, which is delicately decorated with smaller crystal groupings. The lustrous, slightly tarnished copper-brown color combines with the unusual form to create the effect of a natural sculpture of great beauty and interest to both the mineralogist and lay collector. It is a masterpiece of crystalline copper by any standard.

## MICHIGAN
### Keweenaw, Houghton, Ontonagon Counties; Keweenaw Peninsula

The great native copper deposits of Michigan's Keweenaw Peninsula, the "Copper Country," were the object of the first mining rush and boom of the United States. Although known to the pre-Columbian inhabitants of the Great Lakes region for thousands of years, it was the famous report of Michigan State Geologist Douglass Houghton in 1841 that sparked the development of these unique deposits. By 1845, the district was in production with all of the color and wild character that would come to typify mining camps farther west 40 years later. Tales of gunfights, desperadoes, and notorious madams, mixed with stock swindles, fortunes made and lost, and even a town that turned to cannibalism during one particularly harsh winter, give an insight into conditions on what was then considered the nation's western frontier during the early years of the district.

The first half-century of mining in the Keweenaw was free of the violent labor disputes common elsewhere, but the district experienced one of the bitterest and most violent of strikes in 1913. A minor quarrel over the introduction of the one-man air drill to the district quickly grew to a major strike fueled by agitators in the Western Federation of Miners. As violence escalated, replete with shootings, rioting, and cold-blooded murder, the governor called out the state militia to help restore order and protect innocent citizens.

The strike appeared to be dying out when the most tragic event in the district's history fanned the flames of hatred to a feverish pitch. During a miners' Christmas party on the second floor of the Italian Hall in the Calumet suburb of Red Jacket on December 24, 1913, as a jolly Kris Kringle prepared to pass out presents to the expectant children, a false call of "fire" was heard. In the ensuing panic, 18 adults and 56 small children perished in the press of those behind them in the stairwell when the first-floor exit doors failed to open. The perpetrators were never identified, but the strike organizer, a bully-boy from outside of the district, was shot, dragged to a train bound for Chicago, and warned that he would be publicly hanged if he ever returned to Calumet (Murdoch, 1964).

The orebodies of the Keweenaw, which can be classified as fissure, amygdaloid, or conglomerate, depending upon the host rock and nature of the open-space fillings of the native copper, were opened by hundreds of shafts, pits, and trenches along the district's 100-mile length. Each of the dozen most important lodes produced enough copper to be considered, outside of Michigan, an important district in itself.

The fissure mines were the first to open; rich masses of native copper, sometimes weighing many tons, were found as vein fillings in the large fissures, or faults. The Phoenix, Cliff, and Central Mines were important producers of fissure ore and specimens in Keweenaw County in the northern portion of the district, and the Minesota Mine (so spelled because of a clerical error made as the incorporation papers were being drawn up by candlelight) was an important fissure mine in Ontonagon County in the southern portion. Between these two mines, in Houghton County, fissures were mined as they were encountered in other orebodies, the most important being the Mass fissure in the Ahmeek Mine.

Amygdaloid lodes, in which copper occurs as amygdule fillings and in the brecciated flow-tops of the hosting basalts, were the next to be discovered. By 1851 prospectors were exploring the lean amygdaloids, and in 1856 the Quincy Mining Company began production from the Pewabic amygdaloid lode. Other amygdaloid lodes, such as the Isle Royale, Ashbed, Osceola, Baltic, and Evergreen Series, were opened in quick succession. The greatest of the amygdaloid lodes, the Kearsarge, was discovered in 1874 and opened by the Wolverine Mine in 1882. Famous today for the numerous splendid specimens of crystallized copper and silver they produced, the Wolverine, North and South Kearsarge, Mohawk, Seneca, and Ahmeek Mines, and others on the Kearsarge lode, produced two billion pounds of copper and many of the finest copper and silver specimens in the world.

Conglomerate lodes, in which native copper replaced the fine-grained cement holding the various pebbles of the conglomerate together, were the third major deposit type to be opened on the Keweenaw. In 1861, so the story goes, a surveyor named E.J. Hulbert visited a poor farmer on the range. The farmer's best pig had become lost and, following its squeals for help, Hulbert located it trapped in an ancient pit. The pit was a prehistoric mine sunk in what would later be called the Calumet & Hecla Conglomerate: the richest lode in the district. Shafts of the Calumet, Hecla, Osceola, Centennial, and Tamarack mines produced over four billion pounds of native copper from the Calumet & Hecla Conglomerate.

In 1871, the Calumet and Hecla Mines merged to

form the Calumet & Hecla Mining Company. With the absorption of most of the other important mines in the northern portion of the district by 1923, the Calumet & Hecla Mining Company became the Calumet & Hecla Consolidated Copper Company, one of the three mining giants of the Copper Country. The others were the Copper Range Company, with extensive holdings mostly in the southern portion of the district, and the Quincy Mining Company, with important deposits on the Pewabic amygdaloid lode on the north shore of Portage Lake in the center of the district. The Quincy Mining Company operated some of the deepest shafts in the district with the No. 2 reaching a depth of two miles and being serviced by the largest single-drum hoist in the world.

Unlike other major copper deposits, Michigan native copper represents a primary mineralization rather than a product of oxidation and supergene enrichment. The native copper was locally graded by size into three classifications: (1) shot copper, so called because of its small size and resemblance to round shot, having formed as amygdule fillings in the hosting basalts; (2) barrel copper, which could be hand-sorted and shipped in barrels; and (3) mass copper, which consisted of solid chunks large enough to be shipped by flatcar directly to the smelters.

The largest piece of mass copper, reported variously at between 420 and 520 tons, was found in 1857 in the Minnesota Mine. Copper masses of such size posed a unique problem for the miners: how to get the copper to the surface. Too large and heavy to hoist up a shaft, and too ductile to be blasted apart by explosives, such masses were blasted loose with hundreds of pounds of powder emplaced in hollows excavated especially for the purpose, and then cut by hand. Miners spent months working by candlelight chiseling channels through the great masses until they were reduced to manageable sizes. Barrels of the pure copper fragments chiseled out were shipped to the smelters with other masses as high-grade ore. Such "chisel chips" may still be found on the dumps of the great fissure mines and represent unique mementos of Michigan's copper heyday.

The native copper of Michigan is unique in another respect as well. The copper as mined was typically pure; only silver, mined with the copper in vastly lesser amounts, was deliberately allowed to alloy with copper in the smelters. The resultant refined copper, known as "Lake Copper," was considered a premium product due to its combination of high strength and high conductivity. It set the standard against which other copper products were measured. Indeed, so highly regarded was Lake Copper that it was specifically requested for wiring in U.S. warships during the First World War.

A total of over 13 billion pounds of copper and 16 million ounces of silver have been produced from the Keweenaw Peninsula. As of this writing, the White Pine Mine, the only active mine left in the district, continues production of both ore and fabulous mineral specimens. Many of the finest specimens of copper and silver in the world originated from the mines of Michigan's Copper Country, and the Barlow Collection contains numerous splendid examples.

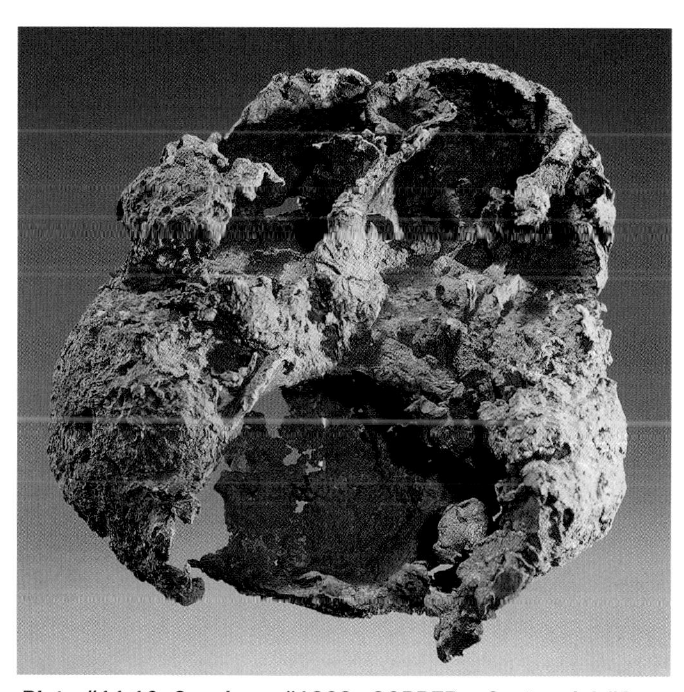

***Plate #11-10, Specimen #1368*** **COPPER – Centennial #6 Mine, Calumet & Hecla Conglomerate Lode, Calumet, Houghton County, Michigan. (MH)**

**#1368 9.0×8.5×3.5cm Plate #11-10**

**Centennial #6 Mine, Calumet & Hecla Conglomerate Lode, Calumet, Houghton County, Michigan**

This small, triple "copper skull" is a fine example of a copper form unique to Michigan's Copper Country. A rare multiple with good convex closure, this is an excellent example of copper mineralization of a felsite conglomerate. The "skull" resulted when copper encased a felsite clast (cobble) within the conglomerate, with subsequent alteration of the clast to an aggregate of clay-sized chlorite minerals. Later removal of the altered material resulted in the hollow, skull-like form so closely associated with the district.

*Plate #11-11, Specimen #1425* **COPPER – Wolverine Mine, Kearsarge Amygdaloid Lode, near Calumet, Houghton County, Michigan. (MH)**

**#1425 6.2×4.5×2.0cm Plate #11-11**

**Wolverine Mine, Kearsarge Amygdaloid Lode, near Calumet, Houghton County, Michigan**

A superb, arborescent grouping of sharp and complexly twinned crystals, this specimen originates from the famous Wolverine Mine. Dark copper in color, this aesthetic grouping of copper crystals was acquired in 1974 when John and his wife Dorothy were in England on business for the United States government. The intrepid pair took the opportunity to visit with famed English mineral dealer Richard Barstow, who brought this small specimen to John's attention and related its history.

The specimen was collected at the turn of the century by a Cornish miner working in the district. Cornish miners were called "Cousin Jacks" because they always had a Cousin Jack at home willing to come over to work. This Cousin Jack had an eye for quality and kept this specimen, of all the copper he had come across in the course of his labors, as a souvenir when he returned home. For John, it was love at first sight. He decided that such a gem deserved to be returned to the Midwest and promptly purchased it. It has had a happy home ever since.

**#1572 12.0×6.7×4.2cm**

**Kearsarge Amygdaloid Lode, north of Calumet, Houghton County, Michigan**

This slightly tarnished grouping of sharp copper crystals resembles a bunch of grapes. The individual crystals, to 1.7cm in diameter, are complex and distorted with somewhat curved faces and exhibit the forms of the dodecahedron and one or more tetrahexahedra, with minor modifications by the octahedron, the trapezohedron *n* {112}, and, possibly, a hexoctahedron. The specimen has been etched from a basalt matrix and has a sawn base. Etched areas exhibit a hackly texture with casts of small quartz crystals. This specimen could have originated from anywhere along the most productive portion of the Kearsarge lode — six miles long. All we can say is that it comes from Calumet or north of there; it may have come from any of Calumet, Copper City, Allouez, Ahmeek, Mohawk, etc. mines.

**#1720 11.0×7.2×5.8cm**
**Plate #11-12**

**Keweenaw County, Michigan**

Slightly tarnished from its pristine copper color, this is a superior example of copper casts after quartz. Well formed and unusually large for casts of this type, this is a classic form for copper from the district, as well as an interesting scientific oddity.

*Plate #11-12, Specimen #1720* **COPPER – Keweenaw County, Michigan. (MH)**

**#2140 5.2×2.8cm**

**Keweenaw County, Michigan**

This specimen consists of a single 2.1×1.5×1.0cm copper crystal on a matrix of epidotized basalt with drusy epidote and adularia feldspar crystals. The crystallization is classic Michigan in that the crystal exhibits a steep tetrahexahedral form easily mistaken for a dodecahedron.

**#2232 4.9×3.3×2.3cm**

**LaSalle Mine, Kearsarge Amygdaloid Lode, south of Calumet, Houghton County, Michigan**

Typical of the open-space-filling nature of copper mineralization in basalts of the district, this specimen consists of rounded amygdule fillings of native copper, to 2.1cm. This is a classic Michigan specimen.

***Plate #11-13, Specimen #2233*** **COPPER – Ojibway Mine, Kearsarge Amygdaloid Lode, Keweenaw County, Michigan. (VP)**

**#2233 8.5×6.0×3.6cm Plate #11-13**

**Ojibway Mine, Kearsarge Amygdaloid Lode, Keweenaw County, Michigan**

This piece is an outstanding and aesthetic arborescent cluster of sharp, cubic copper crystals to 7mm. This is one of the finest and largest of the exceptional specimens from the Ojibway Mine. Originally part of one large specimen that separated during cleaning, the superb specimens from this lot were first brought to the market by Richard Whiteman in 1982. Cubic copper crystals of this size and perfection are rare from the district, and the abundance of crystals modified by the even rarer octahedral form, and those exhibiting superbly developed hoppering, as well as some complex twins similar to those of the famous specimens from the Ray Mine in Arizona, make this one of the most important finds of cubic copper crystals in Michigan. Sharp, microscopic, cubic crystals of silver in an epitaxial relationship to the copper crystals are an association typical of these specimens.

*This is one of the finest and largest of the exceptional specimens from the Ojibway Mine.*

**#2235 5.3×2.6×1.3cm Plate #11-14**

**Ojibway Mine, Kearsarge Amygdaloid Lode, Keweenaw County, Michigan**

Similar to specimen #2233 above, this is an especially aesthetic specimen from the classic Ojibway Mine, which produced some of the finest cubic copper crystals in the district. This rare specimen exhibits the arborescent grouping of sharp, cubic copper crystals, some with octahedral modifications and others with well-developed hoppering, typical of this find. Sharp, microscopic, cubic crystals of silver in epitaxial relationship to the copper crystals occur as an associated species.

**#2292 7.5×4.5×2.5cm**

**Cliff Mine, Keweenaw County, Michigan**

This specimen consists of a single, twinned copper crystal exhibiting a step development of the tetrahexahedral crystal form *l* {035}. It originates from the Cliff Mine, one of the first mines in the district, and exhibits a dark brown color resulting from a thin partial coating of tenorite on the crystal surfaces. While not exceptionally aesthetic, this specimen is a superb example of a crystal form rarely seen well developed in copper.

***Plate #11-14, Specimen #2235*** **COPPER – Ojibway Mine, Kearsarge Amygdaloid Lode, Keweenaw County, Michigan. (MH)**

**#2355 16.0×14.5×9.1cm**
**Plate #11-15**

**Quincy Mine, Pewabic Amygdaloid Lode, Hancock, Houghton County, Michigan**

This is the type of classic, large specimen that museum collections are built upon. The specimen consists of an aesthetic mass of distorted and twinned copper crystals to 5.0cm with drusy quartz crystals on small pieces of attached matrix. It is dark brown in color.

***Plate #11-15, Specimen #2355*** **COPPER – Quincy Mine, Pewabic Amygdaloid Lode, Hancock, Houghton County, Michigan. (MH)**

**#2635 16.5×12.1×5.5cm**

**Iroquois Mine, Iroquois Amygdaloid Lode, Keweenaw County, Michigan**

This specimen from the Iroquois Mine represents another classic form for copper in the district: copper casts after analcime, a zeolite mineral. The aesthetic mass of casts is red-brown in color due to a patina of cryptocrystalline cuprite. Some partial copper crystals exhibiting a tetrahexahedral form are also present.

*Plate #11-16, Specimen #2356* **COPPER – Quincy Mine, Pewabic Amygdaloid Lode, Hancock, Houghton County, Michigan. (MH)**

**#2356 23.5×16.0×9.7cm Plate #11-16**

**Quincy Mine, Pewabic Amygdaloid Lode, Hancock, Houghton County, Michigan**

This is another large, classic specimen of complex, twinned tetrahexahedral copper crystals forming a compact mass with individual crystals up to 3.5cm in size. This aesthetic, dark brown specimen is of the type unique to Michigan and would complement any museum collection.

**#2633 8.0×3.2×2.7cm**

**Centennial Mine, Kearsarge Amygdaloid Lode, Calumet, Houghton County, Michigan**

This specimen consists of a grouping of several large, incomplete, dark copper crystals to 4.0×3.0cm, with minor quartz and epidote from the Centennial #1 or #2 Mine on the Kearsarge lode. The crystals are fairly sharp and exhibit the forms of the tetrahexahedron *f* {013} modified by the trapezohedron *n* {112}. While not a particularly aesthetic specimen, it deserves note for an unusual crystal habit that is typical of Michigan.

*Plate #11-17, Specimen #2634* **COPPER – Champion Mine, Baltic Amygdaloid Lode, Painesdale, Houghton County, Michigan. (MH)**

**#2634 7.0×4.5×2.5cm Plate #11-17**

**Champion Mine, Baltic Amygdaloid Lode, Painesdale, Houghton County, Michigan**

A classic specimen from the Champion Mine, this compact mass of bright, arborescent crystal groups has curved individual branches that reach a maximum length of 3.4cm. Crystals exhibit the forms of the tetrahexahedron and dodecahedron and rarely exceed 3mm in diameter. The grouping is of the form described by Dana in 1886, illustrated in his figure 49, and later coined "Type 1" by Wilson and Dyl (1992). This is typical of arborescent groupings from the Champion and other mines on the Baltic lode.

***Plate #11-18, Specimen #5817*** **COPPER – Lake Superior District (probably Keweenaw County), Michigan. (MH)**

***Plate #11-19, Specimen #2844*** **COPPER – Quincy Mine, Pewabic Amygdaloid Lode, Hancock, Houghton County, Michigan. (MH)**

**#5817 10.5×4.4×4.2cm Plate #11-18**

**Lake Superior District (probably Keweenaw County), Michigan**

This dark brown specimen exhibits the classic herringbone structure of elongated, twinned crystals usually associated with the fissure mines of Keweenaw County (in this case, probably the Phoenix Mine or the Cliff Mine). Individual crystal groupings reach lengths of 6.8cm with "branches" up to 3cm in length. Although the pitted surface of the crystals generally precludes identification of individual crystal forms, the development of at least one type of tetrahexahedron is evident. The crystal groupings resemble those of Dana's (1886) figure 37 and, possibly, figure 54. In all, it is an unusually aesthetic example of an unusual but classic form for crystalline copper from Michigan. This piece was originally owned by Charles Key, then sold to Bryan Lees, from whom Barlow acquired it.

**#2844 23.4×15.1×8.7cm Plate #11-19**

**Quincy Mine, Pewabic Amygdaloid Lode, Hancock, Houghton County, Michigan**

An outstanding museum piece from the Quincy Mine, this specimen consists of a large, arborescent crystal grouping of flattened, distorted, and twinned copper crystals to 5.6cm with a partial coating of cryptocrystalline tenorite. To paraphrase Sam Spade, "this is the stuff museums are made of."

**#3716 13.3×9.5×5.7cm Plate #11-20**

**Quincy Mine, Pewabic Amygdaloid Lode, Hancock, Houghton County, Michigan**

This specimen, also from the famous Quincy Mine, consists of a solid mass of sharp, twinned, tetrahexahedral copper crystals in the arborescent grouping described by Dana in 1886, illustrated in his figure 50, and later coined "Type 2" by Wilson and Dyl (1992).The specimen is a fine representation of an unusual form of twinning in copper.

**#5050 9.0×8.0×5.5cm**

**Refer to Chapter 3, Plate #3-23**

**Point Prospect, Keweenaw County, Michigan**

This is arguably the best specimen from the recent find of large dodecahedral or, more probably, steep tetrahexahedral copper crystals from an outcropping prehnite vein in basalt, east of Copper Harbor. Typical for copper crystals from this locality, the crystals exhibit rough and pitted surfaces and curved faces, complicating the exact determination of crystal habit. The specimen consists of a large grouping of well-defined crystals, the largest of which measures 6.0×5.3×5.0cm, perched on a matrix of basalt and prehnite. Noteworthy for its exceptional single crystal development, unusually high aesthetics, and attached matrix, this is a truly world-class copper crystal specimen. John was alerted to the specimens from this locality shortly after their discovery in 1992 and declined opportunities to purchase several at what he considered to be unreasonably high prices. Demonstrating the truth in the old adage "patience is a virtue," he bided his time, continued to research the discovery, and, within a period of three months, located the best of the lot and secured it for his collection.

***Plate #11-20, Specimen #3716*** **COPPER – Quincy Mine, Pewabic Amygdaloid Lode, Hancock, Houghton County, Michigan. (MH)**

**#4949 8.6×6.4×3.1cm Plate #11-21**

**Quincy Mine, Pewabic Amygdaloid Lode, Hancock, Houghton County, Michigan**

An exceptional and aesthetic example of a rare crystal development from a classic and desirable locality, this specimen demonstrates repetitive stacking of cubic copper crystals along an octahedral axis as first described by Dana in 1886. Other crystal forms present in the specimen include the tetrahexahedra *h* {014} and *l* {035}. The specimen is blackish-brown in color due to a thin tenorite patina. Microcrystals of adularia feldspar are an associated species. This rare classic was previously in the collection of Lazard Cahn.

***Plate #11-21, Specimen #4949*** **COPPER – Quincy Mine, Pewabic Amygdaloid Lode, Hancock, Houghton County, Michigan. (JS)**

**#5178 18.4×12.7×4.7cm Plate #11-22**

**North Kearsarge Mine, Kearsarge Amygdaloid Lode, near Copper City, Keweenaw County, Michigan**

This superb copper specimen originates from the famous North Kearsarge Mine. Dark copper in color, it consists of an arborescent grouping of the type referred to as "Type 2" by Wilson and Dyl (1992) with individual crystals exhibiting the form of the twinned tetrahexahedron *e* {012}. An unusually aesthetic and large piece, this is a fine example of sharply formed crystals from a classic and important locality.

***Plate #11-22, Specimen #5178*** **COPPER – North Kearsarge Mine, Kearsarge Amygdaloid Lode, near Copper City, Keweenaw County, Michigan. (MH)**

***Plate #11-23, Specimen #1590*** **COPPER – Kearsarge Amygdaloid Lode, near Calumet, Houghton County, Michigan. (MH)**

**#1590 16.5×14.0×8.0cm Plate #11-23**

**Kearsarge Amygdaloid Lode, near Calumet, Houghton County, Michigan**

This is another specimen that may be more properly classified as an artifact than as a mineral. It is a fine example of a drill scar in native copper, from the Kearsarge lode. Native copper is quite malleable, and drilling equipment designed to cut through brittle rock will slow or bog down completely when copper is encountered. Such an encounter would leave a drill imprint in the softer copper. It is easy to imagine that specimens such as this represent delays in mining operations of the sort that gave miners headaches and cut into production bonuses.

**#5310 6.6×6.5×3.7cm**

**Quincy Mine, Pewabic Amygdaloid Lode, Hancock, Houghton County, Michigan**

This specimen, originating from the Quincy Mine, consists of an attractive grouping of distorted, twinned tetrahexahedral copper crystals to 1.6cm. The crystals are slightly to moderately tarnished with a cuprite patina that gives a pleasing red-brown to bright copper color.

**#5208 25.0×12.5×12.5cm**
**Refer to Chapter 3, Plate #3-22**

**Ahmeek No. 2 Mine, Kearsarge Amygdaloid Lode, near Ahmeek, Keweenaw County, Michigan**

Another classic specimen from an important Michigan locality, this consists of an aesthetic group of large, flattened, and distorted twinned crystals up to 5.5cm from the Ahmeek No. 2 Mine. The crystals are dark copper in color and face inward from a copper vug in a matrix of massive epidotized basalt. The piece reportedly was collected during the early history of the Ahmeek No. 2 Mine by the mine captain, who later presented it as a gift to one of the clerks in the accounting department of the Calumet & Hecla Consolidated Copper Company. It recently resurfaced as part of an old collection acquired by mineral dealer Robert W. Seasor and quickly found its way to John's collection.

**#5302 5.3×3.7×2.9cm**

**Phoenix Mine, Phoenix, Keweenaw County, Michigan**

Dipyramidal apophyllite crystals, many with copper inclusions, coat this small copper specimen from the Phoenix Mine. The copper forms a mass of small, intergrown crystals exhibiting a sharp but complex habit that includes the rare forms of the cube and octahedron. A black coating of tenorite on the copper is in turn coated by a sprinkling of apophyllite crystals to 4 mm. Apophyllite is quite rare in the Michigan copper deposits, and the dipyramidal form exhibited here is far less common than the platy habit usually observed.

**#5303 18.7×15.1×2.0cm**

**Bohemian Mine, Evergreen Series, Ontonagon County, Michigan**

Typical of Michigan copper originating from narrow veins, this aesthetic, complexly crystallized specimen is a flattened arborescent crystal grouping. A pleasing dark red-brown in color, this classic piece was found in the Bohemian Mine.

**#5324 9.5×4.0×1.4cm**

**White Pine Mine, Nonesuch Shale, White Pine, Ontonagon County, Michigan**

An aesthetic, flattened grouping of copper crystals from the White Pine Mine, this specimen is typical of those etched from calcite veins in the Nonesuch Shale. The crystals are highly distorted, slightly tarnished, and form a branching arborescent group. Minor shale matrix and remnants of incompletely etched calcite are evident.

***Plate #11-24, Specimen #5304*** **COPPER – Phoenix Mine, Phoenix, Keweenaw County, Michigan. (MH)**

**#5304 16.4×11.0×8.6cm Plate #11-24**

**Phoenix Mine, Phoenix, Keweenaw County, Michigan**

A classic Michigan copper, this specimen exhibits a complex crystal habit of distorted dodecahedron modified by trapezohedron and octahedron with additional minor forms. The crystals reach sizes of 2.8cm, are dark copper in color with a golden patina, and form an aesthetic intergrown mass. Minor matrix on the back of the specimen hosts broken crystals of calcite enclosing copper and microscopic crystals of pumpellyite, quartz, and rare white prehnite — a suite more typical of the Quincy Mine in Houghton County than the reported locality of the Phoenix Mine near the old town of Phoenix. Obtained by Barlow from the Paul Heise Collection.

**#5321 3.5×3.0×2.0cm**

**Evergreen Series, Ontonagon County, Michigan**

The tetrahexahedron *l* {035} dominates the habit of copper crystals on this small specimen. Individual crystals are up to 8mm. A lustrous, black tenorite coating with secondary malachite staining on the copper and the association of small crystals of twinned, orange adularia feldspar, quartz with minute copper inclusions, and lime-green epidote with minor calcite are indicative of an origin in the Evergreen Series of Ontonagon County at the southern end of the district. This piece is most likely from the Mass or Caledonia Mine on Mass Hill near the town of Mass.

***Plate #11-25, Specimen #5615*** **COPPER – Mohawk Mine, Kearsarge Amygdaloid Lode, Mohawk, Keweenaw County, Michigan. (MH)**

**#5615   5.0×2.0×1.0cm   Plate #11-25**

**Mohawk Mine, Kearsarge Amygdaloid Lode, Mohawk, Keweenaw County, Michigan**

This superb cluster of several dozen intergrown crystals of copper has a sharp cubic habit. The crystals range from about 2.0 to 7.0mm and are prominently displayed on all sides of the specimen. Minor gray quartz and white calcite matrix are present. Most of the cube faces show steplike growth patterns. This is a very attractive miniature copper.

**#5325   8.7×2.2×1.9cm**

**Champion Mine, Baltic Amygdaloid Lode, Painesdale, Houghton County, Michigan**

Similar to #2634, this aesthetic grouping consists of two long, curved copper branches, 7.2 and 5.0cm in length, which dominate the specimen, with smaller branches to 1.4cm protruding at right angles. Individual crystals show the forms of the tetrahexahedron and dodecahedron and rarely exceed 1 mm in diameter. The whole forms an attractive arborescent grouping of the type described by Dana in 1886, illustrated in his figure 49, and later coined "Type 1" by Wilson and Dyl (1992).

**#5422   13.5×7.0cm**

**Kearsarge Lode, Houghton County, Michigan**

This is a superb cluster of crystals, many of which show a sharp cubic habit. Others are elongate octahedra. A trace of blue-green copper oxide is also evident. Notably, bright, sharp, dodecahedral silver crystals to 4mm are randomly scattered along one edge of the copper crystal cluster. (See Chapter 9, page 235.)

## Copper in Calcite

No discussion of Michigan copper can be complete without considering an association that represents some of the most striking occurrences of copper in the world: bright native copper enclosed in transparent calcite crystals. Following are three examples of special note in the collection.

*Plate #11-26, Specimen #1848* **COPPER – Hancock Mine, Pewabic Amygdaloid Lode, Hancock, Houghton County, Michigan. (VP)**

**#1848 11.0×11.0cm Plate #11-26**

**Hancock Mine, Pewabic Amygdaloid Lode, Hancock, Houghton County, Michigan**

This superb specimen consists of an outstanding, transparent, colorless calcite crystal on minor matrix from the Hancock Mine. The large crystal is complex and twinned scalenohedral in habit, with a cleaved upper termination. It exhibits double refraction of the epidotized matrix as viewed through the crystal. The specimen is enhanced by a smaller calcite crystal, 1.7cm in diameter, filled with bright copper inclusions and by a dusting of copper crystals to 1mm on one side of the larger crystal. Formerly in the collection of the American Museum of Natural History, this is a fine example of the aesthetic and complex copper-calcite specimens that are considered to be classics of Michigan's Copper Country.

*Plate #11-27, Specimen #2109* **COPPER – Quincy Mine, Pewabic Amygdaloid Lode, Hancock, Houghton County, Michigan. (MH)**

**#2109 11.0×10.0×9.0cm Plate #11-27**

**Quincy Mine, Pewabic Amygdaloid Lode, Hancock, Houghton County, Michigan**

This is a small specimen of lustrous, scalenohedral calcite crystals to 7.0cm with bright copper inclusions on a matrix of epidotized basalt. The aesthetics of the specimen are enhanced by the tarnished nature of the native copper where it extrudes from the calcite crystal surfaces and by the association of small crystals of quartz and pumpellyite, as well as some minute copper crystals. The specimen originates from the Quincy Mine on the Pewabic lode in Hancock, Houghton County and represents one of the treasures of the district avidly sought by collectors of both copper and calcite.

*Plate #11-28, Specimen #5617* **COPPER – Caledonia Mine, Knowleton Lode, Mass, Ontonagon County, Michigan. (MH)**

## COPPER IN CALCITE

**#5617 6.1×5.1×4.5cm Plate #11-28**

**Caledonia Mine, Knowleton Lode, Mass, Ontonagon County, Michigan**

An outstanding specimen, especially for the locality, this piece consists of cloudy, grayish white, scalenohedral crystals of calcite in parallel orientation with abundant inclusions of bright native copper. The largest calcite crystal is doubly terminated and measures 6.0cm in length. Minor copper protruding from the calcite is crystallized as euhedral, microscopic cubes, some of which are modified by the octahedron, and rounded crystals of more complex form. Minor quartz, with microscopic inclusions of copper, microcrystals of epidote and pumpellyite, and minute crystals of a white unidentified mineral are associated species. This is the finest miniature specimen of the pocket discovered in the Caledonia Mine by Richard Whiteman in December of 1994.

## DOMEYKITE (COPPER ARSENIDE)

**#5628 10.8×7.6×2.8cm**
**Plate #11-29**

**Mohawk Mine, Kearsarge Amygdaloid Lode, Mohawk, Keweenaw County, Michigan**

This specimen is unique in the author's experience due to the presence of subhedral crystals of domeykite in thick, reniform masses that virtually encompass the copper matrix. Although the Mohawk Mine is famous for its historical production of domeykite ($Cu_3As$), algodonite ($Cu_6As$), and various copper-arsenic mixtures, all collectively known as "mohawkite," specimens are generally massive or consist of small stringers in a quartz matrix. Rarely does domeykite occur as well-developed reniform masses as in this specimen. (Identification of domeykite in specimen #5628 was confirmed by Dr. Terry C. Wallace via X-ray diffraction and laser ablation spectroscopy analyses.) This mass of dull silver-brown, rounded crystal aggregates is truly exceptional and is a fitting addition to the superb copper suites of the collection.

*Plate #11-29, Specimen #5628* **DOMEYKITE – Mohawk Mine, Kearsarge Amygdaloid Lode, Mohawk, Keweenaw County, Michigan. (MH)**

*This mass of dull silver-brown, rounded crystal aggregates is truly exceptional and is a fitting addition to the superb copper suites of the Barlow Collection.*

## NEVADA

**#5309 12.2×10.1×5.2cm**

**Battle Mountain, Lander County, Nevada**

The importance of this specimen lies in its origin — a relatively obscure locality. The specimen is coated with a thick, light green product of oxidation and clearly shows its fracture-filling genesis. Crystals are generally small and clustered in arborescent groups to 5.6cm in length, although some rough, equant crystals to 1.6cm are evident. Some minor matrix is present.

## ZAIRE

A nearly 300-mile-long region of polymetallic copper and cobalt mineralization stretches across south-central Africa from northwest to southeast. Known as the Shaba Crescent in Zaire and the Copper Belt in Zambia, this linear zone of mineralization consists of isolated patches of stratiform sulfide deposits in Upper Proterozoic metasedimentary rocks. Various sections of the deposits were mined at primitive levels since ancient times; large-scale mining developed after the arrival of European explorers in the late 19th and early 20th centuries. Rich oxidized zones of the sulfide orebodies have produced many of the most spectacular examples of secondary copper, uranium, cobalt, and related minerals known. This is particularly true in the Shaba Crescent to the north, where intense structural deformation of the host rocks has resulted in the formation of well-oxidized secondary ores of great extent with diverse mineral suites of outstanding beauty and rarity.

***Plate #11-30, Specimen #5081*** **COPPER – Mashamba West Mine, Shaba Province, Zaire. (MH)**

**#5081 6.0×5.7×5.0cm Plate #11-30**

**Mashamba West Mine, Shaba Province, Zaire**

This rare specimen from the Mashamba West Mine consists of bright copper crystals to 8mm on a matrix coated with cobaltian calcite rhombs 1mm across. Rough tetrahexahedral copper crystals form arborescent groupings on the bright pink matrix of calcite. The stunning aesthetics resulting from the combination of untarnished copper with bright rose-pink calcite make this the best of four specimens that appeared on the market in 1992. This piece is a noteworthy specimen by any standard.

*This piece is a noteworthy specimen by any standard.*

*Plate #11-31, Specimen #4952* **COPPER – Broken Hill, Zambia. (MH)**

## ZAMBIA

**#4952 8.8×3.9×2.2cm Plate #11-31**
**Refer to Chapter 12, Plate#12-35**

**Broken Hill, Zambia**

The Broken Hill Lead-Zinc Mine is located near the town of Kabwe, about 30 miles south of the famous Copper Belt of Zambia. Discovered in 1902 by an engineer searching for ancient copper mines (Notebaart and Korowski, 1980), it was named after the great Broken Hill Mine of New South Wales, Australia. The primary ores consist of various sulfides of lead, zinc, and minor copper hosted in Precambrian metasediments. The sulfides are heavily oxidized, resulting in a beautiful suite of lead, zinc, and copper secondary minerals. Native copper occurs in small amounts with malachite, cuprite, and chalcocite.

This is a superb, arborescent grouping of elongated and twinned copper crystals with an associated growth of fine and slightly rounded dendritic crystals, slightly tarnished. Exquisite aesthetics and the development of twinning, popular with copper collectors the world over, make this an exceptional specimen from a classic locality.

## ARTIFACTS

**#1714 11.5×9.0×2.5cm - fan**
**Plate #11-32**

**Houghton County**

**#CA-1 20.7×2.0cm - spike**

**Florida Ship Wreck**

This is a "chisel-chip" fan of the type produced by miners of the district as gifts for visiting dignitaries. Made by hand with chisel and sledge hammer from solid masses of copper, such fans are highly prized by collectors of Michigan memorabilia, and this is an excellent and aesthetic example.

This spike, another historic artifact, was found at the *Spring of Whitby* which was shipwrecked off the coast of Florida in 1715 and carries Salvors, Inc. certificate of authenticity no. 11516.

*Plate #11-32, Specimen #1714* **COPPER FAN – Houghton County, Michigan.**
*Specimen #CA-1* **COPPER SPIKE – Florida ship wreck. (MH)**

*Plate #11-33, Specimen #5261* **COPPER – Houghton County, Michigan. (MH)**

### Prehistoric Artifacts

**#5261 From 10.0×2.3cm to 3.7×0.8cm**
**Plate # 11-33**

**Northern Michigan/Keweenaw Peninsula**

These nine prehistoric Indian copper tools, including three awls or needles, one arrowhead, one spearhead, one uncompleted knife, and several other partially completed items were produced by Native Americans from surface copper deposits in the Keweenaw Peninsula.

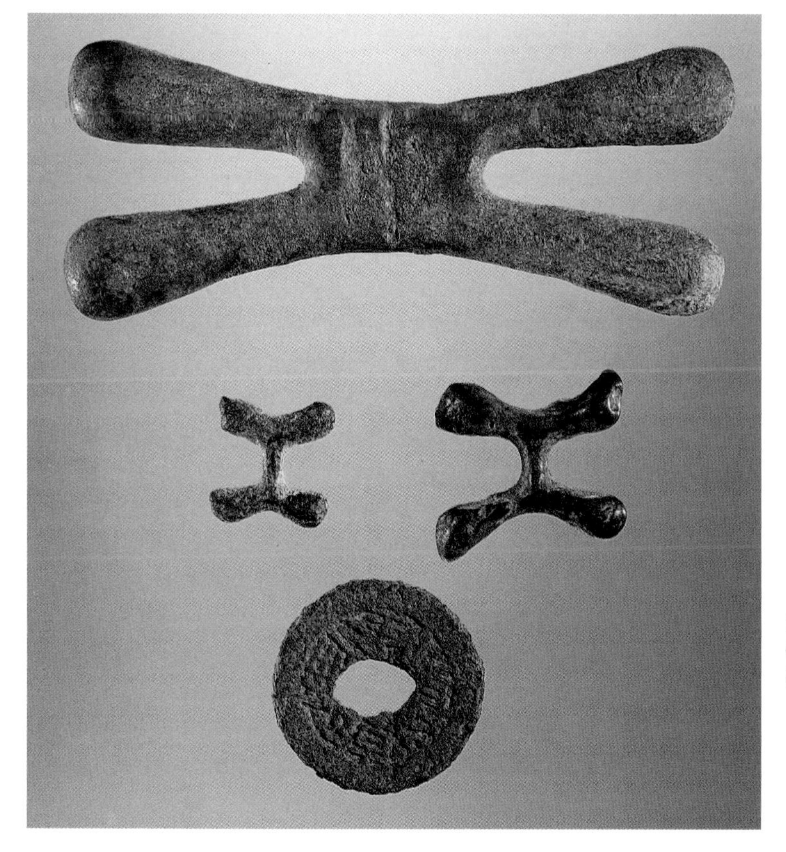

*Plate #11-34, Specimen #5260*
**COPPER – California to Zaire. (MH)**

### Early Copper Money

**#5260**
**Croisettes from 7.3×3.0cm to 1.5×1.5cm**
**Coin, 2.4cm diameter**
**Plate #11-34**

**California to Zaire**

The three croisettes were used as trading chips (money) in the 18th and 19th centuries in Zaire. The "donut," called a Chinese "coolie" copper coin, 2.4cm in diameter, was discovered in a "Frenchman's" grave in Mariposa County, California, probably dating to the Gold Rush of the mid 1800s.

***Plate #11-35, Specimen #1754*** **COPPER – Calumet & Hecla smelter plant, Houghton, Michigan. (MH)**

### Recent Manmade Copper Crystals

**#1754 18.0×18.0×3.0cm Plate #11-35**

**Houghton, Michigan**

This copper specimen was taken from an electrolytic tank at the Calumet & Hecla smelter plant, where small-scale experimental electrolytic refining was being done in 1952 and 1953. Over time, the copper solution in the tank deposited copper crystals on the inside walls, creating this spectacular crystal structure. Purchased from Don Olson in 1978.

## REFERENCES


Bancroft, P. (1984) *Gem and Crystal Treasures*, Western Enterprises, Fallbrook, California, and Mineralogical Record, Tucson, Arizona, 488 pp.

Dana, E.S. (1866) On the crystallization of native copper, *American Journal of Science*, 32:413-429.

Jones, R.W. and W.E. Wilson (1983) The Ray mine, *Mineralogical Record*, 14(5):311-322.

Joralemon, I. B. (1973) *Copper*, Howell-North Books, Berkeley, California, 407 pp.

Murdoch, A. (1964) *Boom Copper; The Story of the First U.S. Mining Boom*, Drier and Koepel, Calumet, Michigan, 255 pp.

Notebaart, C.W. and S.P. Korowski (1980) The Broken Hill mine, Zambia, *Mineralogical Record*, 11(6):339-348.

Wilson, M.L. and S.J. Dyl II (1992) The Michigan copper country, *Mineralogical Record*, 23(2), 104 pp.

# PART V

## *IMPORTANT LOCALITY SUITES*

### *Introduction*

A natural outgrowth of mineral collecting is recognition that certain localities provide exceptional variety or rare and unusual types of species. John's collecting interests have led him to collect not only mineral suites from famous localities like Tsumeb, Namibia, but also suites of rare species like the Tip Top Mine phosphate suite. In another case, as in the Montreal Mine, John's interest is colloquial. Other localities, like the Harris and Flambeau Mines, reveal John's recognition of something unique in the mineral world.

> *The collector has a museum of his own,*
> *and he is the curator.*
>
> *-Douglas and Elizabeth Rigby*

# Chapter 12

# Africa

*by Charles L. Key*

Africa, perhaps more than any other place, was the focus of John Barlow's journey into mineral collecting. As it happened, John's daughter Grace, while in Morocco in 1971, acquired an amethyst geode for her dad. Split in two, it is now catalog #02 and #03. Specimen #01, a small pyrite purchased even before the amethyst, seemed insufficient for "lift off."

As it also happens, Africa is the home of "The Big 5" in big game hunting — cape buffalo, elephant, lion, leopard, and rhino. Bagging these animals would be a challenge to eclipse the "Grand Slam" of North American trophy rams already collected by John. Ergo, the game trophies, the amethyst geode, the African continent, and John's unique DNA all seem to have coalesced into the creation of what has become one of the world's great mineral collections.

Whether or not you are a hunter, you cannot escape the fact that it takes rarefied attributes to conquer the "Everest" of big game trophies. Surely, no less determination is required to build a great mineral collection. So, whatever gene or synergy was operative, Africa was the crucible, Barlow was the force, and — the rest is history.

*An amethyst geode and Barlow's unique DNA seem to have combined to play a pivotal role in creating what has become one of the world's great mineral collections.*

Ten African countries are prominently represented with world-class gem and mineral specimens in the Barlow Collection. A particular wealth of superb specimen material can be found in the locality suites from Tsumeb, Namibia; Kalahari Manganese Fields, Republic of South Africa; Shaba Province, Zaire; Touissit, Morocco; and Jos Plateau, Nigeria. A surprising number of isolated, small mineral finds represented in the collection, such as a super-perfect octahedron of gem white diamond, #T228, weighing 6.4ct from Sierra Leone, or a large spectacular matrix francevillite, #1840, from Mouanana, Gabon, are further testimony to the extraordinary scope of what is certainly one of the great collections of our time.

This author has had a deep personal involvement in Africa over a 30-year interval, visiting countless mines, prospects, and collections, as well as most localities reported here (Tsumeb and Kalahari dozens of times). This involvement has resulted in, among many other things, two mineral species namesaked: keyite and ludlockite. It is from this perspective that the following selections are set forth. Tsumeb is treated first because of its significance, followed by the Kalahari Manganese Fields, then other African localities.

## Tsumeb, Namibia

Tsumeb ranks as one of the great specimen deposits of the world. The abundance and variety of minerals found here are reflected well in the Barlow Collection. The geology of Tsumeb is well explained in what has become known as "The Tsumeb Issue" of *Mineralogical Record*, released as Volume 8, Number 3, May-June, 1977. Reference to that issue, and subsequent articles in that magazine, is suggested.

Simply stated, the Tsumeb deposit is a chimney or pipe that has intruded into the country rock to be later followed by ore-bearing solutions. These solutions more or less followed the periphery of the pipe, creating lenses, pods, mantos, and veins rich in minerals. The pipe is exceptional in its length, extending down to nearly 5,000 feet. It cuts through zones of dolomite and chert, along with smaller zones of limestone and other rocks.

The most remarkable feature of the deposit is the depth of the oxide zones, source of Tsumeb's most famous minerals. There were, in fact, two oxide zones: the upper zone due to normal weathering, and a lower zone due to a fault, known as the North Break Fracture Zone. This fracture zone provided an avenue for meteoric waters to intrude on the lower reaches of the deposit, creating an oxide zone even more profound than the upper zone. Much fracturing, coupled with dissolution of some host rock, created openings in which an exceptional suite of minerals could form.

Tsumeb is best known to the average collector for its azurite, malachite, cerussite, mimetite, dioptase, smithsonite, calcite, anglesite, and the like. But to advanced collectors like John Barlow, the glory of Tsumeb lies in its rare minerals, still being identified through study. The rare mineral suite formed because of the presence of important minor elements including germanium, gallium, cadmium, and cobalt. These joined with the more common metals copper, zinc, lead, silver, arsenic, vanadium, tungsten, iron, tin, barium, magnesium, and more to make Tsumeb one of the world's great deposits and provide collectors with a unique opportunity to gather and study an extensive suite of superb and rare minerals from one locality. *R.W.J.*

Ref.: Wilson, Wendell, et al. *Mineralogical Record*, Volume 8, Number 3, May-June, 1977.

*The wonder of Tsumeb lies in the amazing diversity of truly beautiful minerals there.*

## TSUMEB

### AGARDITE (CALCIUM YTTRIUM COPPER ARSENATE HYDROXIDE HYDRATE)

**#5243 6.0×3.0cm**

These small 3mm radial sprays on brown dolomite crystals, with malachite and conichalcite, are examples of the only rare-earth-bearing mineral found in the Tsumeb Mine. The color is a very unusual shade of pistachio-green, and the crystals are often curved. This extremely rare mineral was found only once, and a couple of dozen pieces resulted.

### ANGLESITE (LEAD SULFATE)

Anglesite from Tsumeb occurs in many guises and, at its best, exceeds anglesite from anywhere else in the world save for Touissit, Morocco.

At Tsumeb, anglesite is most often found as white, colorless or yellow crystals from microscopic size to over 50cm on a face. Rarely, it occurs as lustrous pale blue or pale green crystals commonly overgrown with cerussite, making it nearly impossible to visually distinguish either species.

One curious aspect of Tsumeb anglesite is that when it forms the matrix for azurite crystals, as has been seen in numerous finds, the azurite is invariably the most brilliant and intense blue manifestation of that copper species. I have no idea why this is true. Good examples of this phenomenon may be seen in the Harvard University Collection.

As for crystal form, the anglesite from Tsumeb is most commonly seen as sharp, angular diamond-shaped crystals. Less common are excellent equant bipyramidal crystals. Dozens of other forms have been encountered here in studies of the full range of sizes that occur. A study of anglesite forms from Tsumeb would be an interesting project for a student of crystallography.

**#3301 8.0×6.0cm**

This transparent, colorless gem crystal to 3.0cm is set in a matrix of silicified chert and duftite.

**#2590 18.0×12.5×1.5cm**

This enormous transparent, zoned, colorless, gray-to-pale-yellow single tabular crystal is exceedingly sharp and, while repaired, a most impressive large anglesite. (See Chapter 3, page 38.)

## TSUMEB, NAMIBIA

### ARSENDESCLOIZITE (LEAD ZINC ARSENATE HYDROXIDE)

**#5778 5.0×4.0cm**

Superb gemmy tan/colorless prismatic crystals to 3.5mm in radiating groups lie on brilliant black mimetite crystals with arseniosiderite in vuggy chalcocite ore. Arsendescloizite is extremely rare (whereas descloizite is common by comparison), and these crystal groups are remarkable for their size and quality. The specimen measures 5.0×4.0cm and is distinctive for its black mimetite, which has been found only once in Tsumeb, during the mid 1970s. This is the best arsendescloizite specimen ever recovered from Tsumeb in my experience.

*Plate #12-1, Specimen #852* **AZURITE – Tsumeb, Namibia. A classic crystal of deep blue-black azurite. (VP)**

There are a number of reasons why Tsumeb azurite may well be considered the ultimate as far as competition is concerned. Competition with azurite from other localities is indeed stiff, with sources such as Chessy, Bisbee, Broken Hill, Zacatecas, Touissit and the like producing superb specimens at one time or another. Yet, Tsumeb azurite stands alone for large sharp prisms, some of which reach 25cm in length and display superb steep angular terminations. For dramatic impact, such blades are difficult to beat. Combine such large size with brilliant luster showing strong glints of rich sapphire-blue flashing through and the Tsumeb crystals are unrivaled. As if this were not enough, consider the rich green "bull's eyes" of malachite that sometimes occur, as if perfectly placed, on the prism faces of some Tsumeb azurites, and you have a combination of factors that make Tsumeb azurites unequaled by those from any other azurite source.

### AZURITE
(COPPER CARBONATE HYDROXIDE)

**#852 4.5×4.5×5.2cm Plate #12-1**

Standing upright on a matrix of pinkish quartz and dolomite is a classic crystal of deep blue-black azurite measuring 5.0×2.7×1.8cm. With brilliant luster and steep pyramidal termination, this beautiful cabinet piece represents one reason collectors want a Tsumeb azurite over azurites from other localities.

*Plate #12-2, Specimen #956* **AZURITE – Tsumeb, Namibia. (MH)**

### AZURITE

**#956 7.0×3.5cm Plate #12-2**

This is a superb group of essentially three rectangular, bright blue-black crystals, each about 1.0cm thick, arranged in a dramatic form.

*Plate #12-3, Specimen #2971* **AZURITE with malachite, calcite, cerussite, and duftite – Tsumeb, Namibia. (MH)**

### TSUMEB, NAMIBIA

### AZURITE WITH MALACHITE, CALCITE, CERUSSITE, AND DUFTITE

**#2971 17.5×11.0cm Plate #12-3**

Only Tsumeb produces such a "grand slam" of beautiful copper and lead minerals. This large plate of calcite crystals has 4.0cm bright blue-black azurites positioned off-center with sharp malachite pseudomorph after azurite crystals to 3.5cm. Cerussite crystals, to 1cm, are sprinkled about on one corner. Large areas of small calcite crystals, colored green with duftite, complete the setting. This is a stunning and wonderful specimen.

*Plate #12-4, Specimen #5135* **AZURITE with tsumebite – Tsumeb, Namibia. (VP)**

### AZURITE WITH TSUMEBITE

**#5135 11.0×7.5cm**
**Plate #12-4**

Brilliant blue intergrown aggregates 4.0cm thick form a spectacular mass of crystals associated with minor tsumebite. This azurite is typical of but one form seen here, though there are dozens of forms exhibited in Tsumeb azurites. This piece is a unique and stunningly aesthetic specimen.

*Plate #12-5, Specimen #5289* **BAYLDONITE replacing mimetite – Tsumeb, Namibia. (MH)**

## TSUMEB, NAMIBIA

### BAYLDONITE REPLACING MIMETITE (LEAD COPPER ARSENATE HYDROXIDE REPLACING LEAD ARSENATE CHLORIDE)

**#5289 4.7×2.5cm Plate #12-5**

This specimen essentially consists of two enor mous, stout prisms of sharp, well-terminated mimetite crystals whose perimeter is altering to bayldonite (to a depth of 1mm) with a scintillating surface due to microcrystalline growth. Perched on the prisms are about a dozen well-defined malachite pseudomorph-after-azurite crystals to 0.5cm. All in all, this is a miniature to duel for and a classic Tsumeb specimen, more spectacular than similar material extant. This is an "old-timer," upper level find that found its way to the Barlow Collection via dealers Martin Ehrmann and Richard Kosnar, not to mention the unknown African dealers.

The old upper level workings are notable for their great treasure of large and spectacular finds of carbonates and sulfates. These so-called "German workings" were opened early on when the territory was known as Southwest Africa. Huge crystals and groups of calcite, azurite, cerussite, and malachite pseudomorphs after azurite, smithsonite, anglesite, and gypsum were the most notable species recovered during this time. The pockets were cavernous and frequent. Not many rarities are recognized from this time period, but this may well be a function of lack of awareness coupled with mining incentives rather than a lack of scientific skill or interest.

### BRIARTITE WITH GERMANITE (COPPER IRON ZINC GERMANIUM SULFIDE WITH COPPER IRON GERMANIUM SULFIDE)

**#5250 3.0×3.5cm**

This rich mass of gray-black briartite and copper red germanite is associated with minor galena, chalcocite, and bornite. Tsumeb is the only place in the world to produce this unique sulfide assemblage, and it is rare there.

### BROCHANTITE (COPPER SULFATE HYDROXIDE)

**#5249 4.5×4.0cm**

Blue-to-green prisms to 1.0cm occur in radiating masses associated with anglesite and cerussite. This mineral occasionally gets misidentified as malachite, but the color is different — a more subtle, blue to green similar to blue-green Chivor emerald as opposed to the yellowish green of a Muzo emerald. Quite rare at Tsumeb, but of superb quality when found, this brochantite, while small, is exemplary.

### CARMINITE (LEAD IRON ARSENATE HYDROXIDE)

**#5288 3.7×3.2×3.0cm**

This red mineral is associated with water-clear anglesite crystals, brilliant tabular beaudantite crystals, and unknown mineral T.K. (possibly gartrellite), on a silicified chert matrix. This is one of the most significant Tsumeb specimens in the collection, for both size and quality of the carminite crystals, which reach a maximum length of 0.5cm. The fact that they are much thicker prisms than the norm greatly enhances their beauty. These carminite crystals are flawlessly transparent and saturated with deep carmine-red color and, additionally, have a brilliant luster. Some are included in the anglesite crystals as floating euhedra, rare and breathtaking. This spectacular, thrilling specimen was a gift from Richard Thomssen.

## TSUMEB, NAMIBIA

### CERUSSITE (LEAD CARBONATE)

Tsumeb is world famous for the great variety of mineral species it has delivered, including new and rare species. But what has brought this African locality to the attention of collectors all over the world is the great abundance of some of its species, cerussite being one of note.

As a late-forming secondary lead carbonate, cerussite is found in large amounts throughout the deposit, sometimes varying greatly in form. Cerussite from here ranges from colorless, often gemmy, to white, red, and brown caused by inclusions. Specimen #5253, for example, is a brilliant red as a result of enclosed copper oxide. Other cerussites are coal black due to heavy inclusions of a black sulfide mineral.

Most notable, and best known, are the superb twin forms seen in cerussite here. Most sought are the reticulated twins that form a boxlike network of thick crystals in groups exceeding 20.0cm or more. Often glassy, colorless, or snow white, as seen in specimen #938, these cerussites set the standard for collector excellence the world over. Simple "V" twins are common here as well. Together, these cerussite twin forms found in such large size and great abundance have made Tsumeb famous.

***Plate #12-6, Specimen #938*** **CERUSSITE – Tsumeb, Namibia. (UN)**

### CERUSSITE

**#938 6.5×6.5cm Plate #12-6**

**Tsumeb, Namibia**

This specimen is a beautiful white multiple twin reticulation to 1.5cm thick in a circular grouping, typical for this locality. Cerussite showing this multiple twinning is often referred to as "snowflake" cerussite.

***Plate #12-7, Specimen #5253*** **CERUSSITE – Tsumeb, Namibia. (MH)**

### CERUSSITE

**#5253 1.5×1.5cm Plate #12-7**

Rarely does a thumbnail-sized mineral, much less anything smaller, get any notice in a book intended for a wide audience. This remarkable specimen demands notice, however. This cerussite is two carmine-red prisms, doubly terminated. The color is unique. One crystal is about 1.0cm in length and intersects the other at about 100°. The two are perched on calcite rhombs, which are centered on a crust of green duftite crystals. The specimen is quite fantastic enough to transcend its diminutive size. The color of this cerussite, unique to Tsumeb, is due to copper oxide inclusions. Such cerussite has appeared a few times in the mine but rarely so aesthetically.

## TSUMEB, NAMIBIA

### CUPROADAMITE

**(COPPER ZINC ARSENATE HYDROXIDE)**

**#5132 6.1×6.9×4.2cm**

**Refer to Chapter 3, Plate #3-32**

These radiating aggregates of lustrous dark emerald-green crystals grow to 2.3cm in length. These particular crystals are unique for sheer beauty and size, from Tsumeb or any of the several other localities worldwide where cuproadamite has been found. This specimen's rich color is exceptional. These crystals are sandwiched in a breccia zone of quartz with rare powdery yellow tsumcorite. This was found in the mid 1980s, on the 28th or 30th level, in a small pocket zone that produced a few dozen, mostly small, specimens.

### DIOPTASE

**(COPPER SILICATE HYDROXIDE)**

While Russia can boast of its historical importance as a source of dioptase, Tsumeb certainly can be considered the most prolific. The rich, brilliant green dioptase crystals from here are universally known. Huge amounts of this showy copper silicate have come to light at Tsumeb. Specimens range from a sprinkling of light green microcrystals on white calcite to solid masses of crystals often 15.0cm or more across with crystals to 2.0 or more centimeters. Almost always sharp and lustrous, the dioptase crystal groups from here are universally popular and prized.

### DIOPTASE

**#2194 9.0×4.0cm**

This specimen of brilliant intergrown crystals to 3.2cm in length is an outstanding example and beautiful cabinet piece of what is, by any measure, one of Tsumeb's most sought after minerals.

***Plate #12-8, Specimen #826*** **DIOPTASE – Tsumeb, Namibia. (MH)**

### DIOPTASE

**#826 14.5×10.0cm Plate #12-8**

A matrix of white calcite and dolomite crystals is peppered with superb dioptase crystals to 2.4cm. On the 30th level in the mine at Tsumeb, miners encountered numerous pockets, some very large, that contained thousands of specimens of breathtakingly beautiful, dark emerald-green dioptase. The two specimens described here are among the finest from that find. Dioptase made some miners rich (and crazy) since it is easily sold for high prices, and also because it was (past tense) abundant. It exceeded in value all of the other minerals found at Tsumeb, with fine azurite a distant second.

*Plate #12-9, Specimen #2194* **DIOPTASE – Tsumeb, Namibia. (MH)**

### TSUMEB, NAMIBIA

### DIOPTASE

**#2194 9.0×4.0×3.0cm Plate #12-9**

A matrix of quartz provides the base for this brilliant green covering of dioptase crystals. The large crystal measures 3.0×2.0cm. This specimen was brought to John's home by African pilot Richard Stacpoole.

### FERRILOTHARMEYERITE

**(IRON CALCIUM ZINC COPPER ARSENATE HYDROXIDE HYDRATE)**

**#5245 8.0×4.7cm**

Crusts of yellow-brown microcrystals of this mineral cover rich arsenate ore, the surface of which is sprinkled with lustrous emerald-green crystals of cuproadamite to 0.8cm. Ferrilotharmeyerite was newly described in Germany in 1992, and this is perhaps the most impressive specimen from the find.

*Plate #12-10, Specimen #5039* **HIDALGOITE – Tsumeb, Namibia. (MH)**

### HIDALGOITE

**(LEAD ALUMINUM SULFATE ARSENATE HYDROXIDE)**

**#5039 5.0×4.0cm Plate #12-10**

Masses of botryoidal green aggregates are intergrown with calcite crystals. This is new (in 1991) material from Tsumeb. It was on the 43rd level, and perhaps 44th level also. Tsumeb hidalgoite is easily the finest of the species, and this specimen is one of the best found at Tsumeb.

## TSUMEB, NAMIBIA

### LEADHILLITE (LEAD SULFATE CARBONATE HYDROXIDE)

**#2496 4.5×4.0×0.8cm** **Refer to Chapter 3, Plate #3-59**

This specimen is a sharp tabular gem crystal, extremely rare. This colorless, nearly complete crystal actually has large areas of flawless material that, in the hands of a lapidary genius, could provide the only known faceted gem leadhillite — not that any sane person would attempt to cut what is a "one-of-a-kind," even among leadhillites. Leadhillite is rare from anywhere and extremely rare in Tsumeb, where perhaps fewer than 20 great specimens have ever been saved. Some large and fine ones exist, but such high transparency (read purity) is unrecorded elsewhere.

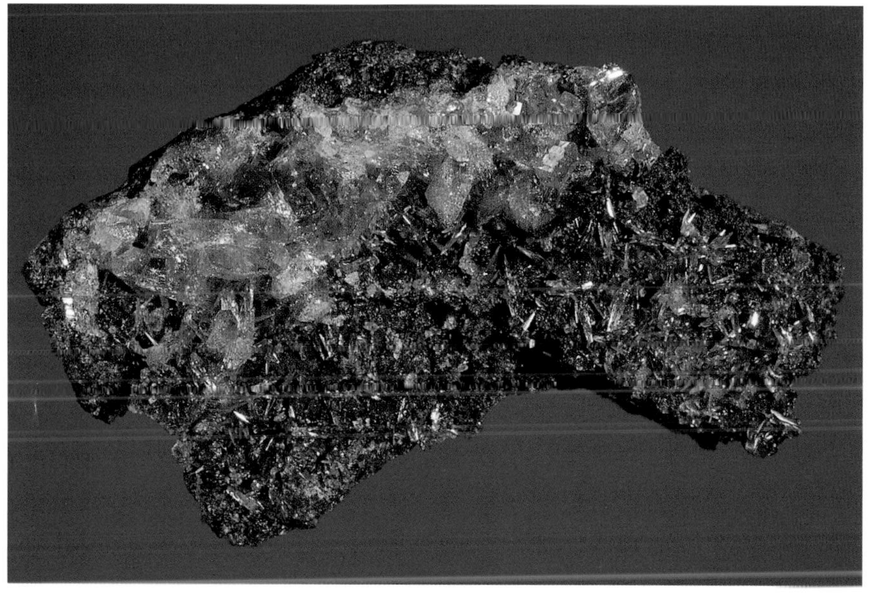

*Plate #12-11, Specimen #5247* **LEGRANDITE – Tsumeb, Namibia. (MH)**

### LEGRANDITE (ZINC ARSENATE HYDROXIDE HYDRATE)

**#5247 3.5×1.7cm Plate #12-11**

Terminated, 3mm, transparent yellow crystals of legrandite rest on tennantite ore with smithsonite needles. Found on the 43rd level, this mineral is quite rare at Tsumeb. It is known in less than a dozen specimens, and this specimen, while small, is of top quality.

### LEITEITE (ZINC ARSENITE) WITH LUDLOCKITE, CHALCOCITE, AND QUARTZ

**#5133 10.5×6.0cm**

**Refer to Chapter 3, Plate #3-61**

This enormous composite crystal mass of diverging gemmy leiteite is intergrown with small, bright sprays of distinct crystals of ludlockite to 1.0cm and crystal quartz and chalcocite. The strange pearly luster of the leiteite, coupled with its thickness, gives it an almost opalescent look. This is surely one of the most important Tsumeb specimens around, and one of the best of this species. It was found in the mine's deep workings, the 44th level, in 1992. Tsumeb is the type locality for leiteite, and only a few examples are extant.

### LUDLOCKITE (IRON LEAD ARSENATE)

**#2589 8.0×7.0cm**

Numerous masses of acicular ludlockite crystals to 0.5cm fill vuggy tennantite-chalcocite ore. The mineral is associated with 0.5cm wulfenite crystals. This rare specimen is one of perhaps six large matrix ludlockites known.

Ludlockite first appeared in the early 1970s. It was acquired from a miner who thought he had "red copper hair." The original discovery was one large vug in a large chunk of massive chalcocite. The piece was trimmed diligently and yielded a mere five specimens, hardly enough to go around. Later mining turned up more ludlockite in minor amounts in the deeper workings of the mine at the 44th level. This later material was not quite up to the standard of the original discovery.

## TSUMEB, NAMIBIA

### MACPHERSONITE (LEAD SULFATE CARBONATE HYDROXIDE)

**#3308 5.0×4.0cm**

These large corroded crystals are a polymorph of leadhillite, wherein much of the entire mass appears to be three intergrown crystals with minor etch cavities containing an unidentified white microcrystal crust. This is another super-rarity from Tsumeb in an impressively large specimen.

### MIMETITE (LEAD ARSENATE CHLORIDE)

**#1277 5.0×4.0×4.5cm** **Refer to Chapter 3, Plate #3-67**

Translucent, yellow to orange, bright, giant crystals rest on oxide matrix. The largest crystal is nearly 4.0×2.0cm. This is easily one of the finest mimetite specimens ever found in Tsumeb.

### MIMETITE

**#0980 4.5×4.5cm**

Transparent yellow crystals to 1.5cm are clustered on matrix. This specimen is from a one-time find in 1974 when a large pocket containing over 100 such mimetites was found. For sheer quality and perfection this find was without equal. Indeed, these were gem mimetites and set new standards for the species. This specimen, while not the biggest, is a perfect jewel.

***Plate #12-12, Specimen #2501*** **ROSASITE – Tsumeb, Namibia. (MH)**

### ROSASITE (COPPER ZINC CARBONATE HYDROXIDE)

**#2501 8.0×5.0cm Plate #12-12**

On this specimen, rosasite is encrusting a malachite pseudomorph after azurite. The crystal is 7.5×5.0×2.8cm on a small oxide matrix. This is most unique and aesthetic. The blue-green velvet rosasite formed over a very sharp pseudomorph produces a wonderful effect. And, while the sequence of rosasite after malachite after azurite is well known, this particular piece stands alone. Obtained from the Sid Pieters Collection.

## TSUMEB, NAMIBIA

### SCHNEIDERHÖHNITE (IRON ARSENITE)

**#2800 6.5×6.5cm Plate #12-13**

Sharp brown to black interpenetrating crystals to 1.0cm protrude from a 3.0cm vug in a mass of chalcocite. This is surely one of the best, if not the best, example of this very rare arsenate, and it ranks as a major Tsumeb specimen.

*A major Tsumeb specimen.*

Plate #12-13, Specimen #2800 SCHNEIDERHÖHNITE – Tsumeb, Namibia. (VP)

### SCORODITE (IRON ARSENATE HYDRATE)

**#T144 2.5×2.0cm**

This is a small superfine group of brilliant blue scorodites to 1.0cm length on tennantite pseudomorph after azurite crystals. The scorodite is selectively coated with silvery-brown crystals of powellite microcrystals. This is probably from the 29th level of the Tsumeb Mine.

Plate #12-14, Specimen #2608 SCORODITE – Tsumeb, Namibia. (MH)

### SCORODITE

**#2608 10.0×12.0cm**
**Plate #12-14**

A mass of tennantite and chalcocite contains a 5.0cm vug of brilliant, sapphire-blue (in incandescent light) gemmy crystals up to 2.8cm in length of scorodite, a rare and beautiful arsenate. Beautiful scorodite like this from Tsumeb is rivaled only by a small find from Zacatecas, Mexico and remains one of the most sought after of all the minerals from Tsumeb. The mineral becomes green in daylight as do most scorodites.

Plate #12-15, Specimen #5134 **SILVER amalgam – Tsumeb, Namibia. (MH)**

*These are among the best silvers ever found in Tsumeb.*

## TSUMEB, NAMIBIA
## SILVER AMALGAM
**#5134 9.0×6.0cm Plate #12-15**

Sharp intergrown silver octahedra (about 2-3mm on face) aggregate into discrete masses of up to 1.5cm in diameter on a cavity surface with malachite crystals. This cavity is in altered cuprite ore found in the deepest workings — 43rd and 44th levels of the mine. The relationship of highly oxidized ore and native silver at great depth (44th level) is unique in the mining world. These are super sharp, albeit small, crystals, associated with arsenates and without equal in mining history. These are among the best silvers ever found in Tsumeb, this specimen perhaps being number one.

### Silver – Eugenite – Silver Amalgam

1992 through 1995 has been a very interesting period of identification for me. You now know about "the jalpaite that wasn't" (#5610). That piece went from being perhaps the number one jalpaite to being one of the top Cornwall chalcocites with the help of the Smithsonian. (Refer to Chapter 3, page 46.)

But you may not know about my proving the existence of crystallized domeykite from Michigan (#5628) with the help of Dr. Terry Wallace of the University of Arizona.

You may also be unaware of my determining the unknown green coating on a multicrystallized nickel-skutterudite specimen from Erzgebirge, Germany with the help of Bill Metropolis of Harvard.

And, now, after acquiring perhaps the number one silver specimen (#5134), I have changed its "color" with the efforts of several Canadian mineralogists to learn that it is identified as the rare eugenite. This discovery caused a stir with several Germans, particularly Dr. Georg Gebhard, who stated that laboratories in Germany found that the Tsumeb silver was actually silver amalgam. With the help of Dr. Terry Wallace, Gebhard's decision after 25 hours of study, was confirmed to be correct.

## TSUMEB, NAMIBIA

### SMITHSONITE (ZINC CARBONATE)

**#5438 9.5×6.0×8.5cm**

**Refer to Chapter 3, Plate #3-85**

This superb specimen is highlighted by a tapering yellow-amber crystal of exceptional size. It ranks among the finest of Tsumeb's smithsonites and has been long considered such. It is pictured in *The World's Finest Minerals and Crystals* by Peter Bancroft, page 51. Procured from the Dr. Webster Collection, and formerly in the David Wilber Collection.

***Plate #12-16, Specimen #4969*** **SMITHSONITE – Tsumeb, Namibia. (MH)**

### SMITHSONITE

**#4969 10.5×7.0cm Plate #12-16**

This mass of 2.0cm crystals of cubelike rhombs has a coffee color. These sharp, well-segregated crystals are zoned with an unusual black inclusion on one plane.

**#2321 4.5×3.0cm**

These are sharply zoned pink rhombs to 1.5cm exhibiting a unique, highly translucent, silvery pink color. Tsumeb is noted for the large number of pink smithsonite specimens produced years ago.

### TENNANTITE (COPPER IRON ARSENIC SULFIDE)

**#984 7.0×7.0cm**

The specimen is composed of three crystals up to 1.7cm, intergrown on a quartz crystal matrix. It is sharp and well defined but lightly surface altered.

### TSUMCORITE (LEAD ZINC IRON ARSENATE HYDRATE)

**#5248 3.0×2.0cm**

This piece consists of botryoidal and cavernous masses of solid, intergrown, radiating 2mm crystals of a mustard-yellow tsumcorite with some massive quartz matrix. A namesake Tsumeb arsenate rarity, this is an exceptionally rich example of the species.

### TSUMEBITE (LEAD COPPER PHOSPHATE SULFATE HYDROXIDE)

**#5246 4.0×3.0cm**

Brilliant emerald-green crystals to 1.2mm are associated with malachite, brochantite, anglesite, and cerussite. These are the finest quality tsumebite crystals I've seen from Tsumeb, particularly with regard to color, brilliance, and perfection of form.

### UNKNOWN 3C-L-K. "SOAP"

**#5252**

This specimen consists of dark chrysoprase/green, massive pieces of a new mineral investigated by both Harvard University and mineralogists in Germany. This material is found on the 43 and 44 levels as pocket fillings resembling green opal but with a greasy "feel" — thus the name "soap" by miners. It is a very hydrated zinc arsenite with a zeolite-like structure! The Barlow specimen is perhaps one of ten in private hands. Analysis is ongoing in Germany; this is definitely a new species as yet unnamed.

### WILLEMITE (ZINC SILICATE)

**#2480 8.0×9.0cm**

Teal-blue, lustrous crystal aggregates of willemite, as epimorphs of azurite and mimetite, fill a 6.0cm vug, with minor white dolomite crystals and a 3.0cm reticulated cerussite, an unusual association. Beautiful and choice, this species has been unavailable for many years, having been restricted to a period of finds during the mid 1970s.

In the arcane world of rare minerals, some minerals hold more stature than others. Tsumeb rarities are a case in point for several reasons. They are often well crystallized and colorful, unlike many rarities that are indistinguishable blebs in a rock. They are often possessed of rare chemistry such as gallium in söhngeite, germanium in stottite, or cadmium in keyite. Such unusual chemical combinations, coupled with their properties, not only attracted the collector world, but ultimately drew the scientific world's attention to that bizarre freak of nature wherein lies concentrated a thimbleful of the planet's rarest juices in a microspeck of mineralogical and geological wonder now known simply as Tsumeb.

## North Cape Province
## Kalahari Manganese Fields

*Plate #12-17, Specimen #5812* **BRAUNITE – N'Chwaning Mine, Kalahari Manganese Fields, South Africa. (MH)**

### BRAUNITE (MANGANESE SILICATE)

**#5812 1.6×0.5×0.5cm Plate #12-17**

**N'Chwaning Mine**

Implanted on a pyramid face of a 2.5cm hausmannite is a doubly terminated crystal of bright-black braunite measuring 1.6×0.5×0.5cm. Braunite appears to be quite rare, at least in free-growing crystals, in the Kalahari manganese mines, although it certainly exists in tonnages within these vast deposits as massive oxide ores. I personally have seen only a few well-formed crystals of any size. This particular specimen is unique in that the braunite crystal is aesthetically perched on another well-crystallized manganese oxide crystal (hausmannite), which is associated with andradite. Purchased from Clive Queit.

*Plate #12-18, Specimen #5045* **CHARLESITE – N'Chwaning Mine, Kalahari Manganese Fields, South Africa. (VP)**

### CHARLESITE
### (CALCIUM ALUMINUM SILICON SULFATE BORON HYDROXIDE HYDRATE)

**#5045 5.0×4.5cm Plate #12-18**

**N'Chwaning Mine**

A nearly complete giant single crystal which exceeds the limits of language for accurate description of its color. To say that it is between chartreuse and lemon-yellow is approximate. No other mineral in nature shares this shocking hue. There is some difficulty distinguishing charlesite as a separate species from sturmanite and ettringite. This specimen is incredible, wherever it may fit. It is from the N'Chwaning Mine #2. Collected in 1989, it formerly belonged to the Bezing Collection.

## KALAHARI MANGANESE FIELDS, SOUTH AFRICA

### GAUDEFROYITE

**(CALCIUM MANGANESE BORATE CARBONATE OXIDE HYDROXIDE)**

**#2975 4.0×2.6cm**

**Wessels Mine**

This specimen consists of lustrous, black, hexagonal crystals to 2.5×1.0cm. This is one of the larger crystal groups found for this mineral, and these are by far the best gaudefroyite crystals found anywhere. Obtained from the Wessels Mine in the early 1990s.

***Plate #12-19, Specimen #5041*** **HAUSMANNITE – Wessels Mine, Kalahari Manganese Fields, South Africa. (MH)**

### HAUSMANNITE

**(MANGANESE OXIDE)**

**#5041 5.5×5.5cm Plate #12-19**

**Wessels Mine**

This large composite crystal made up of many stacked, pagoda-like pseudo-octahedra produces a remarkable effect. Some garnet is attached at the base. Found in 1992 at the Wessels Mine, this may be the largest hausmannite crystal recovered.

*This may be the largest hausmannite crystal recovered.*

### North Cape Province
### Kalahari Manganese Fields

The Kalahari Manganese Fields of South Africa produce some of the world's most spectacular rhodochrosite. The deposit has had such an impact on the mineral and gem collecting world that some note of it should be made here.

The deposits were first discovered in 1922, but mining did not commence until the 1950s. The deposits extend over an area of nearly 60 km and are located north of Sishen, west-northwest of Kuruman in the Northern Cape Province of South Africa. The Hotazel Mine, whose name reflects British humor regarding the climate there, was the first source of superb rhodochrosites. Beginning in the early 1970s, however, other mines, particularly the N'Chwaning, were opened, and a flood of superb rhodochrosites soon followed.

The rhodochrosites from here come in a variety of forms: rhombohedra, scalenohedra, and an odd "wheat sheaf"-like habit. This latter looks like bundles of varicolored pale pink to good red needlelike crystals pinched or tied in the middle so the terminations flair out. The scalenohedron (dogtooth) crystals are the most popular. They are often totally gemmy with a rich red color, many being suitable for faceting. These crystals average 2-3cm long, but superb 6-7cm crystals, in plates 10cm or more across, were common in the halcyon days. Most crystals rest on a black manganite matrix, which provides a stark contrast for the bright red crystals.

Other manganese minerals found here include the ore minerals manganite, hausmannite, and bixbyite. Minor amounts of inesite and other minerals intrigue collectors. For the lapidary, an exceptional occurrence of sugilite, previously a little-known mineral, has lately come from here. The sugilite is massive and shows a rich royal violet color. It is now a very popular and valued gem material. When infused in chalcedony, the resulting material shows a glowing violet color and commands very high prices in the marketplace. Mine owners, recognizing the mineral's value, even trained workers to spot and sort the sugilite, with the resulting product being fenced and guarded with both weapons and noisy geese until it could be offered for sale.

Remarkably, the ore face worked at both the Wessels and N'Chwaning mines is a solid wall of black oxide about 6×14 meters in one vast stratum. *R.W.J.*

Ref.: Wilson, Wendell, and Dunn, Pete J., *Mineralogical Record*, Volume 9, Number 3, May-June, 1978.

## KALAHARI MANGANESE FIELDS, SOUTH AFRICA

### HAUSMANNITE

**#5364 15.0×9.0×4.0cm Plate #12-20**

**Wessels Mine**

This remarkable assemblage of sharp pseudo-octahedral hausmannite crystals ranging in size from 1.0cm to 2.0cm encrusts a plate of reddish brown crystallized garnet. The hausmannite crystals show the typical triangular growth hillocks. Some of the crystals are interpenetrating. Certainly a world-class example of an uncommon mineral.

***Plate #12-20, Specimen #5364*** **HAUSMANNITE– Wessels Mine, Kalahari Manganese Fields, South Africa. (MH)**

### HEMATITE (IRON OXIDE)

**#5233 9.0×6.0×5.0cm Plate #12-21**

**Wessels Mine**

These intensely brilliant, complex, tabular crystals to 3.5cm are associated with large blades of todorokite to 6mm (large for the species) on a matrix of quartz and wad (an indistinct manganese oxide). This is one of the most spectacular arrangements of luster and form imaginable, seemingly designed to intrigue the most discerning collector. Found in the Wessels Mine in 1993, these are arguably the finest hematite specimens from any location — particularly from the standpoint of unequaled brilliance.

***Plate #12-21, Specimen #5233*** **HEMATITE – Wessels Mine, Kalahari Manganese Fields, South Africa. (MH)**

### HEMATITE

**#4671 10.5×8.5×8.0cm Plate #12-22**

**N'Chwaning Mine**

This piece is a brilliantly lustrous group of sharp crystals to 4.0cm across, tightly intergrown and welded together by brilliant microcrystals of garnet. This is an exceptional aggregate of choice tabular hematite crystals.

***Plate #12-22, Specimen #4671*** **HEMATITE – N'Chwaning Mine, Kalahari Manganese Fields, South Africa. (MH)**

*Plate #12-23, Specimen #5163* **INESITE – Wessels Mine, Kalahari Manganese Fields, South Africa. (MH)**

## KALAHARI MANGANESE FIELDS, SOUTH AFRICA

### INESITE (CALCIUM MAGNESIUM SILICATE HYDROXIDE HYDRATE)

**#5163 3.7×3.5cm Plate #12-23**

**Wessels Mine**

A nearly complete 3.0cm ball of red inesite is aesthetically framed with 0.4cm datolite crystals. The rich color and form make this as desirable a specimen as the more famous rhodochrosites from this mining area.

### INESITE

**#5035 30.0×16.5×6.0cm** **Refer to Chapter 3, Plate #3-55**

**Wessels Mine**

Acquired in South Africa and brought to the United States by this writer, the piece is a large matrix specimen almost completely covered by a dense carpet of reddish orange, bladed inesite crystals up to 2.0cm in length.

### RHODOCHROSITE (MANGANESE CARBONATE)

**#2144 4.5×4.0cm Plate #12-24**

**N'Chwaning Mine**

This specimen shows the classic wheat-sheaf bundles of bicolored (light to dark pink) crystal groups to 1.8cm, on contrasting black manganese oxides. This is intensely beautiful material from the early find in 1978 at the N'Chwaning Mine.

*Plate #12-24, Specimen #2144* **RHODOCHROSITE – N'Chwaning Mine, Kalahari Manganese Fields, South Africa. (VP)**

*Plate #12-25, Specimen #1617* **RHODOCHROSITE – N'Chwaning Mine, Kalahari Manganese Fields, South Africa. (VP)**

## KALAHARI MANGANESE FIELDS, SOUTH AFRICA

### RHODOCHROSITE

**#1617 5.0×3.0cm Plate #12-25**

**N'Chwaning Mine**

Pink-red, doubly terminated scalenohedra to 4.0cm rest on manganese oxide matrix. This is a superb small specimen from the 1978 find.

**#1595 12.0×7.0cm**

**N'Chwaning Mine**

Dark red-orange, gemmy scalenohedra to 1.4cm form a solid curving mass of intergrown crystals. The dark red-orange is a unique color for this 1978 find. Most rhodochrosite from here is far lighter in color.

**#3309 10.7×8.0cm**

**N'Chwaning Mine**

Embedded near the center of a brilliant crystal druse of intergrown black manganite covering the matrix is a 3.4cm sphere of alternately pink and dark reddish brown rhodochrosite. This is a dramatic and unusual specimen from Kalahari.

*Plate #12-26, Specimen #5234* **SHIGAITE – N'Chwaning Mine, Kalahari Manganese Fields, South Africa. (MH)**

### SHIGAITE (MANGANESE ALUMINUM SULFATE HYDROXIDE HYDRATE)

**#5234 13.0×9.0cm Plate #12-26**

**N'Chwaning Mine**

This specimen consists of intergrown rosettes to 1.0cm of clove-brown color (yellow in thin sections), overgrown on manganese oxide crystals associated with rosy pink hemispheres of rhodochrosite crystals and minor barite. All these minerals rest on a banded ironstone matrix. This specimen is one of a mere few from the most outstanding find ever of this extremely rare manganese mineral. Found and described from the Shiga Mine in Japan originally, this species was later found at the Iron Monarch Mine in South Australia. Both those occurrences are inconsequential in terms of aesthetic specimens. This find from the N'Chwaning Mine in Kalahari occurred in mid 1993 and is the world's best.

## KALAHARI MANGANESE FIELDS, SOUTH AFRICA

### THAUMASITE
**(CALCIUM SILICON CARBONATE SULFATE HYDROXIDE HYDRATE)**
**#4670 7.8×6.0cm Plate #12-27**

**Wessels Mine**

Occasionally in mineral collecting, a well-known, normally uninspiring species will turn up in such vastly superior form as to leave the collector speechless. Such a find is the thaumasite from the Wessels Mine, Kalahari. These crystals are transparent, pale yellow hexagonal prisms, simply terminated by the basal pinacoid, up to 1.7cm in length by 1.0cm in diameter. Perhaps a thousand or more specimens came to light from the Wessels Mine about 1991, most smaller than Barlow's miniature. The Barlow specimen ranks in the top percentile of what was found and is quite an amazing thaumasite.

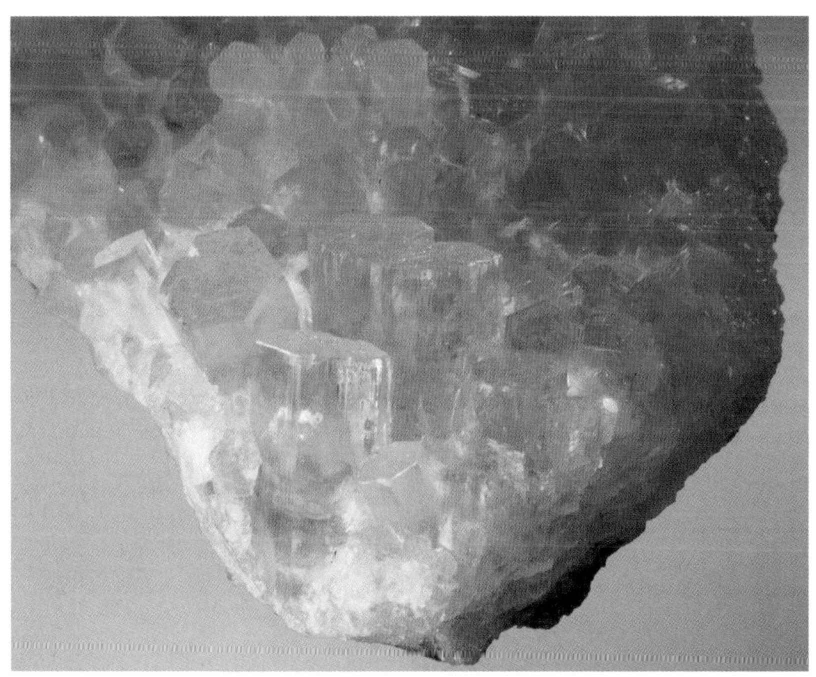

***Plate #12-27, Specimen #4670*** **THAUMASITE – Wessels Mine, Kalahari Manganese Fields, South Africa. (MH)**

***Plate #12-28, Specimen #5235*** **XONOTLITE – Wessels Mine, Kalahari Manganese Fields, South Africa. (MH)**

### XONOTLITE (CALCIUM SILICATE HYDROXIDE)
**#5235 5.7×4.0cm Plate #12-28**

**Wessels Mine**

Three 2.0cm balls of white, acicular xonotlite crystals are perched on a matrix of crystallized datolite. The specimen was recovered from the Wessels Mine in 1992. This is a choice small cabinet piece, the best of the best xonotolite ever found.

## Other African Localities

### ANGLESITE (LEAD SULFATE)

**#2544 4.5×3.5cm Plate #12-29**

**Touissit, Morocco**

Deep yellow, transparent, sharp crystals to 3.5×2.0×1.5cm on galena epitomize the most wonderful anglesites ever found. These most desirable of all anglesites have been coming from the mines in Touissit quite abundantly over the last 10 years. This specimen typifies the top quality from there.

*Plate #12-29, Specimen #2544* **ANGLESITE – Touissit, Morocco. (MH)**

### AZURITE AND MALACHITE (COPPER CARBONATE HYDROXIDES)

**#4742 8.0×7.0cm Plate #12-30**

**Bou Becker Mine, Morocco**

While at first glance this may not seem to be an important azurite, it is truly aesthetic and unique — consisting of radial aggregates that are "scooped out" in organic patterns. It is, in places, bicolored in various shades of blue with some malachite present.

**#5238 15.0×9.0cm**

**Tschouti, Namibia**

Bright, intense blue rosettes in stringers, reminiscent of old Bisbee azurite, partially coat a fracture seam in sandstone, which is also partially coated with malachite. This is one of the very few specimens recovered to date from the new mine at Tschouti — a mine located 22 km from the famous Tsumeb Mine. This deposit has the potential of producing fine mineral specimens.

*Plate #12-30, Specimen #4742* **AZURITE and MALACHITE – Bou Becker Mine, Morocco. (VP)**

### BERYL, VAR. AQUAMARINE (BERYLLIUM ALUMINUM SILICATE)

**#5131 25.0×3.7×3.5cm** **Refer to Chapter 3, Plate #3-6**

**Near Jos Plateau, Nigeria**

This blue beryl is more than an ordinary aquamarine, for it is a complete and perfect crystal of extremely large size and one of the best found here in 1991. It is described in detail in Chapter 3, page 41 and in Chapter 5, page 129.

## BERYL, VAR. EMERALD

**#5020 9.5×1.0×1.3cm**

**Refer to Chapter 3, Plate #3-7**

**Jos Plateau, Nigeria**

This is a doubly terminated crystal of amazing beauty, possessing a true light-emerald color shading rather toward the blue. Quite a large number of these crystals were found near Jos a few years ago. In spite of very high prices, they were bought up instantly because they are so desirable and from a rare locality. This specimen is in the top echelon of the find.

*Plate #12-31, Specimen #5813* **BERYL, var. heliodor – Golden Ridge, Miami District, Zimbabwe. (MH)**

## BERYL, VAR. HELIODOR

**#5813 8.2×3.4cm Plate #12-31**

**Golden Ridge, Miami District, Zimbabwe**

This most unusual euhedral beryl crystal is approximately zoned in thirds, grading from a distinct orange-yellow into yellow into yellowish green. It is quite gemmy albeit separated into a few "silky" zones. Although quite bright, it exhibits some exsolution or etch features on prism faces. This crystal was mined in 1968 in a small pegmatite, quickly worked out, from a district notorious for its many-hued beryls. It was purchased from Clive Queit at Tucson in 1996.

*Plate #12-32, Specimen #4730* **BETAFITE – Ambolotara, near Betafo, Madagascar. (MH)**

## BETAFITE

**(CALCIUM TITANIUM OXIDE HYDROXIDE)**

**#4730 13.7×11.5×5.7cm Plate #12-32**

**Ambolotara, near Betafo, Madagascar**

This specimen is an enormous (to 8.0cm) intergrowth of dull black, low-relief octahedra. While not pretty, this piece is a near-best of this species of niobate/titanate. Obtained from the Sorbonne Collection, Paris. (See Chapter 5, page 138.)

## CHRYSOBERYL, VAR. ALEXANDRITE (BERYLLIUM ALUMINUM OXIDE)

**#1008 4.0×3.5cm**

**Near Fort Victoria, Zimbabwe**

This perfect 2.0cm sixling crystal is well exposed in its schist matrix to reveal high symmetry, twinning planes, and, with strong back lighting, the classic violet-red color in incandescent light that exemplifies alexandrite's "night color."

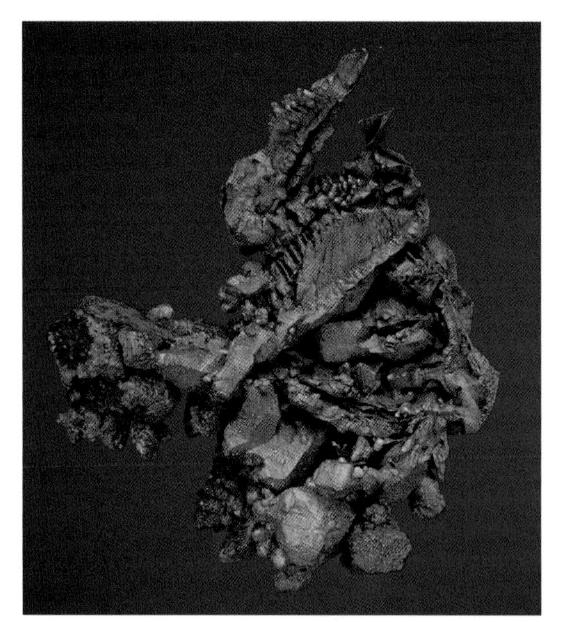

**Plate #12-33, Specimen #2346** COPPER – Onganja Mine, Onganja Mining District, Namibia. (MH)

## COPPER

**#2346 6.0×5.0cm Plate #12-33**

**Onganja Mine, Onganja Mining District, Namibia**

This superior specimen from the mine is an aesthetic combination of twinned and untwinned copper crystals of dark brown color. Individual crystals exhibit the form of the steep tetrahexahedron *l* {035} in a development that could easily be mistaken for the dodecahedron. Untwinned crystals are up to 1.1cm, while twins of the platy, complex type famous from Ray, Arizona reach lengths of 2.0cm. This piece was obtained from the Tom Gressman Collection.

**Plate #12-34, Specimen #5081** COPPER with colbaltian calcite – Mashamba West Mine, Shaba Province, Zaire. (MH)

## COPPER WITH COLBALTIAN CALCITE

**#5081 6.0×5.7×5.0cm Plate #12-34**

**Mashamba West Mine, Shaba Province, Zaire**

This rare specimen consists of bright copper crystals to 8mm on a matrix coated with cobaltian calcite rhombs 1mm across. Rough tetrahexahedral copper crystals form arborescent groupings on the bright pink matrix of calcite. The stunning aesthetics resulting from the combination of untarnished copper with bright rose-pink calcite make this the best of four specimens that appeared on the market in 1992. A noteworthy specimen by any standard.

**Plate #12-35, Specimen #4952** COPPER – Broken Hill, Zambia. (MH)

## COPPER

**#4952 8.8×3.9×2.2cm**
**Plate #12-35**

**Refer to Chapter 11, Plate #11-31**

**Broken Hill, Zambia**

The Broken Hill lead-zinc mine is located near the town of Kabwe, about 30 miles south of the famous Copper Belt of Zambia. Discovered in 1902 by an engineer searching for ancient copper mines (Notebaart and Korowski, 1980), it was named after the great Broken Hill Mine of New South Wales, Australia. The primary ores consist of various sulfides of lead, zinc, and minor copper hosted in Precambrian metasediments. The sulfides are heavily oxidized, resulting in a beautiful suite of lead, zinc, and copper secondary minerals. Native copper occurs in small amounts with malachite, cuprite, and chalcocite. *M.L.W.*

## CUPRITE (COPPER OXIDE) WITH MALACHITE ON CALCITE

**#895 9.5×9.0×7.0cm** **Refer to Chapter 3 Plate #3-31**

**Onganja Mine, Okahandja Mining District, Namibia**

Centered on a calcite matrix sits a superb modified octahedron, 4.7×4.0cm, of gem cuprite uniformly coated by a thin crust of green malachite. This specimen epitomizes the huge, sharp, gemmy crystals that are unique among all cuprites in the world.

*Plate #12-36, Specimen #5025* **CUPRITE – Mashamba West Mine, Shaba Province, Zaire. (VP)**

## CUPRITE

**#5025 8.0×7.0cm Plate #12-36**

**Mashamba West Mine, Shaba Province, Zaire**

This crystal is a perfect, sharp, red/black octahedron to 7.5mm nearly floating over a matrix of gray chert. The matrix is coated with pale blue contrasting chrysocolla. The piece is an exceedingly aesthetic specimen from a large find in Zaire.

*An exceedingly aesthetic specimen from a large find in Zaire.*

I first became aware of Onganja when a contact sent me a sample through the mail identified as "cooperite." The fragment was small, just part of a crystal, but gem flawless and shockingly red.

In a few days I was at the mine and went underground to see the purported pocket zone. At the time, Onganja was a small mine, having been developed only down some 60 m. Since there was no mechanical hoisting machinery, we had to use a jeep to back down an inclined shaft to get to the working level. Everything in sight was massive calcite with occasional splotches of bright green malachite showing randomly about the walls of a large stope.

The exception was an intriguing series of small pockets located about 5-6 m up one wall. These pockets were the source of the now famous giant malachite-coated gem red cuprite crystals so highly prized by collectors. From that moment on, Onganja joined the ranks of noteworthy collector deposits.

## CUPRITE

**#5080 4.0×3.5×2.5cm**

**Refer to Chapter 3, Plate #3-30**

**Mashamba West Mine, Shaba Province, Zaire**

This is a single, sharp, red, off-matrix crystal. That this material could be found in this size and quality would have been unthinkable not so many years ago, in spite of thousands of years of copper mining and hundreds of copper mines.

## DIAMOND (CARBON)

**#1852 4.5×4.5cm**

**New Vaal River Mine, Kimberley Diamond District, South Africa**

A sharp 5mm octahedron of gray-white still in its natural kimberlite matrix. Matrix diamonds are uncommon, and this specimen is notable for being in its original matrix rather than repaired or fabricated.

## DIAMOND

**#4942 1.0×1.0cm**

**Vaal River, Kimberley Diamond District, South Africa**

This octahedron weighs 5.49ct and is 1.0cm square! It is a gemmy white crystal with a typical greenish yellow "skin" exhibiting attractive etch/growth forms. Labeled Cape Province, it appears to be a river stone, perhaps from the Vaal River diggings.

*Plate #12-37, Specimen #T228* **DIAMOND – Tongoma, Sierra Leone. (VP)**

## DIAMOND

**#T228 8.0mm square Plate #12-37**

**Tongoma, Sierra Leone**

A razor sharp, white gem octahedron, this absolutely magnificent single crystal exhibits perfect transparency and "eye-clean" internal clarity.

*Plate #12-38, Specimen #5701* **ERYTHRITE – Bou Azzer, Morocco. (MH)**

## ERYTHRITE (COBALT ARSENATE HYDRATE)

**#5701 7.0×5.2cm Plate #12-38**

**Bou Azzer, Morocco**

Lining a quartz vug in cobalt ore are sharp 2.5cm crystals of magenta-red color. This erythrite is unusually bright. It is so dramatic in form and color that it should be considered unique. The crystal sizes are exceptional, and they are richly colored, sharp, and undamaged.

*Plate #12-39, Specimen #1840* **FRANCEVILLITE – Near Mouanana, Gabon. (JS)**

## FRANCEVILLITE
**(BARIUM LEAD URANYL VANADATE HYDRATE)**
**#1840 18.0×11.5cm Plate #12-39**
**Near Mouanana, Gabon**

This is a solid crust of bright orange-yellow francevillite crystal aggregates to several millimeters, coating the matrix. The crystals are in the typical radial clusters. The rarer mineral curienite is seen as earthy yellow patches. This piece comes from another great find, now long gone, of a very rare uranium vanadate. This specimen, particularly given its large size, ranks high among a small number. The deposit was found near Mouanana in 1956 as a sandstone formation on the road between Franceville and Lasturville. Some nine new species occurred in this remarkable uranium-vanadium deposit.

This deposit is a mineralogical oddity in that it was so rich in uranium that it went through a natural, self-triggered, chain reaction millions of years ago. This phenomenon was discovered when analysis of the mine's uranium output was found to be depleted of fissionable uranium isotope U238. The specimen was secured from Bill Larson via Gilbert Gauthier from the Sorbonne, Paris in the 1970s.

*Plate #12-40, Specimen #5151* **SPESSARTINE – Marienflüss, Namibia. (MH)**

## SPESSARTINE
**(MANGANESE ALUMINUM SILICATE)**
**#5151 8.0×5.0cm Plate #12-40**
**Marienflüss, Namibia**

In a matrix of gray-brown, fine-grained biotite schist, a gem garnet crystal, measuring 1.5×1.2cm, of the most intense and pure orange, sits within a reaction halo of white quartz about 0.5cm thick. The halo grades imperceptibly into the schist. This specimen is one of perhaps six of the finest gem spessartines from anywhere. It was discovered in a remote outcrop near the Namibian border with Angola. The locality in the northern Kaokoveld is known as Marienflüss.

*This specimen is one of perhaps six of the finest gem spessartines from anywhere.*

## SPESSARTINE
**#5150 3.0×2.5cm**
**Marienflüss, Namibia**

This 1.3cm gem crystal is tightly embedded in a silicified biotite schist, producing a magnificent miniature of the same occurrence described above, sans the white rim around the garnet crystal.

*Plate #12-41, Specimen #3302* **GOLD – Sheba Mine, Transvaal, South Africa. (MH)**

## GOLD (NATIVE ELEMENT)

**#3302 7.0×4.0cm Plate #12-41**

**Sheba Mine, Transvaal, South Africa**

Native gold in large masses is rare in this locality. This is a 2.0×1.5cm exposure of sheet gold sandwiched in pyrrhotite/quartz vein rock. For all the gold out of the Republic of South Africa, precious little of it has any specimen merit. This specimen is a noteworthy exception and ranks as a major South African gold specimen.

## JEREMEJEVITE (ALUMINUM BORATE FLUORIDE HYDROXIDE)

**#2470 5.5×3.5×1.0cm**

**Refer to Chapter 3, Plate #3-56**

**Mile 72, near Cape Cross, Namibia**

This is a spray of three intergrown blue, transparent, terminated, gem crystals, the largest group mined near Cape Cross from a small pegmatite by dealer Sid Pieters of Windhoek. This is perhaps the very best specimen of this super-rare borate, known only from here and the Ural Mountains.

## LIBETHENITE (COPPER PHOSPHATE HYDROXIDE)

**#3282 3.6×3.6×2.0cm**

**Refer to Chapter 3, Plate #3-62**

**Rokana Mine, Broken Hill, Zambia**

This is essentially a complete single crystal, forming what appears as a large octahedron but is in the orthorhombic system, with just a bit of matrix attached. The color is deep olive-green with high luster. The crystal faces show deep parallel growth lines. This is the best libethenite I know of by a large margin. It is pictured in *Mineralogical Record*, November-December, 1978, page 344.

*Plate #12-42, Specimen #5024* **MALACHITE – Mashamba West Mine, Shaba Province, Zaire. (VP)**

## MALACHITE (COPPER CARBONATE HYDROXIDE)

**#5024 6.0×5.0cm Plate #12-42**

**Mashamba West Mine, Shaba Province, Zaire**

The world's most amazing crystals of malachite are from Zaire. This specimen consists of huge, sharp crystals to 3.0cm on a face. Found abundantly in 1991, these malachites are now all but gone from the market. This specimen is a prime example and ranks well up in the top percentile.

*Plate #12-43, Specimen #5594* **MALACHITE – Shangulowe, Shaba Province, Zaire. (MH)**

## MALACHITE

**#5594 9.0×6.4cm Plate #12-43**

**Shangulowe, Shaba Province, Zaire**

This crystal aggregate is essentially one huge composite crystal of a size and quality unknown to the mineral world until just recently when this locality produced hundreds of amazing specimens — nearly all of which were snapped up by collectors as fast as they were offered.The vast majority of these wonderful malachite crystal groups were 10–20mm on a face — still enormous by former standards, but this specimen is off the scale for size and virtually perfect in all respects.There are several small octahedral cuprite crystal groups implanted on the surface, but most intriguing is the way the curved faces of the malachite crystals ascend in ellipsoidal lock-step to a common peak, producing a dramatic aesthetic effect.This piece is surely one of the world's great malachite specimens.

## MOLYBDENITE (MOLYBDENUM SULFIDE)

**#5267 5.4×4.0cm**

**Emke Mine, Onganja, Okahandja District, Namibia**

This nearly complete euhedron exhibits classic "beehive" form. It is exceptional for its brilliance and its thicker than normal tabular molybdenite morphology. Molybdenite crystals from this locality, although they are not the largest known in the world nor as well-known as the cuprites from here, arguably rival in quality the finest molybdenites ever found.

## NAMBULITE (LITHIUM SODIUM MANGANESE SILICATE HYDROXIDE)

**#3138 5.0×4.0×3.0cm**

**Refer to Chapter 3, Plate #3-69**

**Kombat Mine, Namibia**

Consisting of red transparent-to-translucent tabular crystals to 2.0×1.1×0.4cm, this jumbled mass of large and small crystals of nambulite is embedded in selenite.The Kombat Mine was the source of what many collectors consider to be one of the most desirable of all Namibian minerals, gem nambulite. Trouble is, only about four really fine specimens were encountered, and only once, in the early 1970s. The Barlow specimen would be about number four in rank while the number one piece is so exceptional as to be off the scale altogether.

*Plate #12-44, Specimen #5021* **ORTHOCLASE – Itrongay, Madagascar. (VP)**

## ORTHOCLASE (POTASSIUM ALUMINUM SILICATE)

**#5021 9.0×4.5×3.0cm Plate #12-44**

**Itrongay, Madagascar**

This spectacular gem crystal is well formed and complete, of a straw-yellow color and virtually flawless internally.This is an excellent example of the best orthoclase feldspar crystals known, exceeded only by crystals in the Sorbonne and the Los Angeles County Museum.

## PARALAURIONITE
### (LEAD CHLORIDE HYDROXIDE)
**#5099 6.0×5.5cm Plate #12-45**

**Pit #9, Touissit, Oujda, Morocco**

This paralaurionite is undoubtedly the best of the species, which was originally known as mere specklike crystals in lead-bearing slags dumped into the ocean near Laurium, Greece during ancient times. Then came the Moroccan discovery, and paralaurionite has subsequently been found in only a few specimens visible to the naked eye. This nearly pure grouping of somewhat pearly-lustered smoky-gray to smoky-gray-brown crystals is on gray oxide ore. The largest terminated crystal is 2.1×1.1×0.4cm and adjoins an immense 4.8×1.7×0.6cm section. Each is transparent to translucent. A large vug on the side of the specimen contains several rosettes of well-formed crystals to almost a centimeter. This piece is extremely rare and exceptionally large for the species. Considered the best ever found.

***Plate #12-45, Specimen #5099*** **PARALAURIONITE – Pit #9, Touissit, Oujda, Morocco. (MH)**

## PREHNITE (CALCIUM ALUMINOSILICATE HYDROXIDE)
**#5088 7.5×5.0cm**

**Tafelkop, Gobbobisberg Mountains, Namibia**

The specimen is 4.5cm crystal "ball" of high luster. The color is medium green, and the ball sits on a matrix of calcite and quartz crystals. The prehnite from this locality at its best is probably the most desirable found anywhere by virtue of exceptional form and color combined with unique high luster. This specimen is a superb example of the "high-end" material in a cabinet size piece.

***Plate #12-46, Specimen #5237*** **QUARTZ, var. amethyst – Brandberg West, Gobbobisberg Mountains, Namibia. (MH)**

## QUARTZ, VAR. AMETHYST
### (SILICON DIOXIDE)
**#5237 11.8×7.8cm Plate #12-46**

**Brandberg West, Gobbobisberg Mountains,Namibia**

This incredible, perfect, bright, gemmy, doubly terminated crystal rests on a matrix of calcite on basalt. This piece, recovered in July, 1993, is one of the finest amethyst specimens ever found at Brandberg West.

*One of the finest amethyst specimens ever found at Brandberg West.*

*Plate #12-47, Specimen #5082* **QUARTZ, var. amethyst – West of Brandberg, Gobbobisberg Mountains, Namibia. (MH)**

## QUARTZ, VAR. AMETHYST

**#5082 24.0×9.5×5.5cm Plate #12-47**

**West of Brandberg, Gobbobisberg Mountains, Namibia**

An outstanding specimen from this locality, this piece consists of two doubly terminated crystals, one piercing the other, ranging from light purple to deep purple in color. Both crystals are perfectly undamaged and show no point of attachment, making this specimen, which was acquired by Barlow from this writer, essentially a "floater," meaning it may have formed suspended in the pocket. The main crystal lies flat with the second jutting from it at a 30° angle. The faces are lustrous and all terminations are perfect. The crystals contain liquid inclusions and phantoms. Mined by G. Cloete in August, 1992.

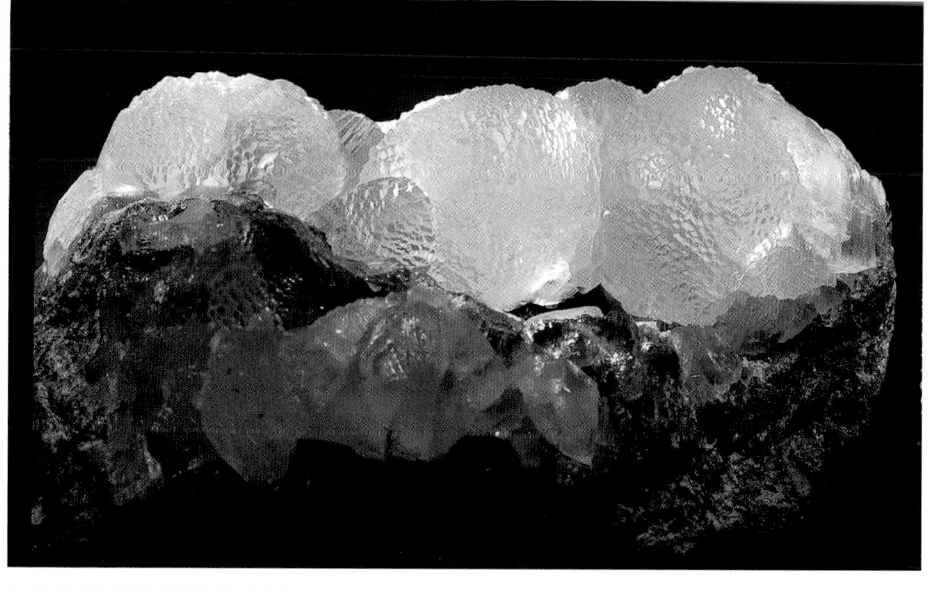

## RUTILE (TITANIUM OXIDE)

**#3307 5.5×2.7cm**

**Gamsberg Mountains, Rehobeth District, Namibia**

One of the finest crystals of rutile from Namibia, this piece consists of two resplendent black prisms in parallel position, terminated and partly obscured by a calcrete matrix encrustation.

*Plate #12-48, Specimen #5136* **SMITHSONITE – Berg Aukus, Namibia. (MH)**

## SMITHSONITE (ZINC CARBONATE)

**#5136 12.5×9.0cm Plate #12-48**

**Berg Aukus, Namibia**

This specimen is one of the greatest Namibian smithsonites in any collection. It demonstrates, contrary to general perception, that Berg Aukus was an important specimen locality on a par with Tsumeb, albeit with far fewer species. Three hemispheres of smithsonite, about 3.0+cm each, fill a vug in massive ore. They are limpid, glistening orbs of glacier-green luminosity. The piece is an ex-Soltau Collection specimen via dealers Jurgen Henn and Mike Ridding.

## TORBERNITE (COPPER URANYL PHOSPHATE HYDRATE)

**#4680 15.5×7.0×6.0cm** **Refer to Chapter 3, Plate #3-91**

**Musonoi Mine, Shaba, Zaire**

This amazing specimen is one of the collection's top prizes. From the Musonoi Mine, Shaba, this torbernite with crystals to 2.0cm is a solid overgrowth on a black-stained matrix rock. While torbernite still trickles out of Zaire, nothing approaching this quality has been seen in decades.

***Plate #12-49, Specimen #2118*** **VANADINITE – Acif Vein, Mibladen, Atlas Mountains, Morocco. (MH )**

## VANADINITE (LEAD VANADATE CHLORIDE)

**#2118 8.5×7.0cm Plate #12-49**

**Acif Vein, Mibladen, Atlas Mountains, Morocco**

This beautiful specimen consists of deep red crystals (measuring up to 1.7cm diameter × 1.0cm thick) perched on opaque white parallel crystals of barite. The piece sets the highest standards for this abundant mineral. Out of the thousands of specimens mined, this one is in the top 2% or 3% for beauty and perfection.

## ZIRCON (ZIRCONIUM SILICATE)

**#5022 2.3×2.0cm Plate #12-50**

**Fatacirero, Betroka, Madagascar**

This loose gem crystal is actually a small 1.0cm crystal intergrown into a larger 1.8×1.0cm crystal. Its color is light yellow-brown to deep red-brown; it is nearly flawless and exhibits good tetragonal symmetry. While small for a world-class zircon crystal, it is superb nonetheless.

*Nearly flawless, sharp, and exhibits good tetragonal symmetry.*

***Plate #12-50, Specimen #5022*** **ZIRCON – Fatacirero, Betroka, Madagascar. (VP)**

The vanadinites from Morocco are exceptional in variety, size, color, and quantity. In the late 1960s we collectors had been used to seeing and collecting beer flat after beer flat of bright red, simple hexagons of vanadinite from the Apache and Old Yuma Mines in Arizona. These sources set the standard, particularly as dealer stock. Crystals did not exceed a quarter of an inch, but they were brilliant red and sharp.

Then along came vanadinite from Mibladen, a desperate desert place. Victor Yount, one of the earlier dealers on the scene, has produced a video of his dozens of trips to Morocco. It is a remarkable record of his perserverance, dedication, and stamina.

Perhaps the most beautiful pieces are rich red vanadinites perched on a black mottramite coating, providing a plush and elegant contrast. Still others are brown and barrel-shaped and appear to need membership in a health club. Anemic in color, fat of waistline, cracked of surface, and listlessly arranged on matrix, they are still worth seeking out for sheer size if nothing else. Some exceed 3.0cm in length

Other fine pieces from here are short, of prism length, almost disclike red crystals that have somehow arranged themselves in an overlapping pattern like flower petals. Perhaps the best, surely the most popular with show judges, advanced collectors, and dealers who find superb minerals in short supply, are those brilliant masses of red intergrown vanadinites with many individual crystals exceeding 1.5cm across, packed and stacked together as though the god of mineral specimens had grabbed a handful of loose crystals and squeezed them just enough to weld them without doing a speck of damage to their coiffure. Such fine specimens fairly leap out at you as you walk by a display case full of fine minerals.

Locals still dig a bit, so fine vanadinites may still be yours for a price not excessive when you consider the rigors of desert mining!

*R.W.J.*

**ZOISITE, VAR. TANZANITE (CALCIUM ALUMINUM SILICATE HYDROXIDE)**
**#4937 5.5×4.4×3.0cm** **Refer to Chapter 3, Plate #3-100**
**71.0g**
**Merelani Hills, Near Arusha, Umba Valley, Tanzania**

## The Sleeping Beauty of Tanzania

Occasionally, nature gives up a treasure, something so unexpected and wonderful that "speechless" is the appropriate response. Such a prize is the crystal known as the "Sleeping Beauty of Tanzania," a giant gem zoisite (tanzanite) crystal uncovered by native miners working the still productive mines of the Merelani Hills, near Arusha in Tanzania. The naming of unforgettable gemstones has long been a tradition in the gem world, and this piece is no exception.

From the miner, this crystal went by way of a local dealer, known as the "Jungle Master of Tanzania," to a gem merchant in Mombasa, Kenya. Sent to a large cutting firm in Bangkok, Thailand, it was rescued from lapidary doom by an astute dealer, Jurgen Henn, of Idar-Oberstein. From Germany it went to dealer Bill Larson of California and, finally, to its just repose in the Barlow Collection, where it has a secure place of honor. Barlow appropriately named this prize gem mineral find the "Sleeping Beauty of Tanzania."

Although its size and clarity are sufficient to set it off from the vast majority of tanzanite mined, its overwhelming attraction is its phenomenal degree of trichroism — virtually off scale by any mineralogical comparison. Trichroism is a gem's ability to affect light, displaying different colors when viewed from different directions. In this case, the "Sleeping Beauty" displays red, blue, and burgundy in each of three axial viewing positions. As each view is so uniquely colored, few viewers of a photomontage of this crystal from three axially differentiated light-angles, such as pictured on the cover of *Mineralogical Record* Volume 24, Number 3, 1993, would initially suspect that they were looking at one and the same crystal. Considering not only its trichroism, but also its size, unusual termination, unique morphology, and intensely saturated color, this crystal properly belongs in the rarefied category of named gems.

## ZOISITE, VAR. TANZANITE

**#5810 6.0×5.0×3.5cm** **Refer to Chapter 3, Plate #3-99**

**Merelani Hills, near Arusha, Umba Valley, Tanzania**

The usual litany of superlatives fail to convey the jolt that even jaded collectors of superb minerals receive when they first lay eyes on this specimen. Clearly, this matrix specimen grew unimpeded in a vug, which is most unusual, as the vast majority of tanzanites are imbedded completely in matrix. Here we have two brilliant, perfectly formed, gem crystals of tanzanite measuring 5×3.7×1cm and 4.5×3.5×1.5cm aesthetically perched on a mass of bright, metallic black crystals of graphite! Associated with the crystals of graphite are minor, albeit choice, free-growing crystals of light green diopside and a large twin carbonate crystal, probably dolomite or possibly ankerite (ergo the evidence of a vug as opposed to an etched piece). Either of these tanzanite crystals would rank as a major specimen by itself, but together in this grouping with these wonderful associated species, in such size and perfection, not to mention the very serious gem values therein, well . . . any adjective you choose would seriously diminish the reality of what truly qualifies as one of the world's great mineral specimens. It was purchased from Tuckman Mines and Minerals Ltd., which secured it from Husseinali Manti.

***Plate #12-51, Specimen #5834*** **ZOISITE, var. tanzanite – Merelani Hills, near Arusha, Umba Valley, Tanzania. (MH)**

**#5834 7.0×5.0×2.5cm Plate #12-51**

**Merelani Hills, near Arusha, Umba Valley, Tanzania**

This specimen is a small cabinet size single crystal with a perfect termination and spectacular color.

***Plate #12-52, Specimen #5833*** **ZOISITE, var. tanzanite – Merelani Hills, near Arusha, Umba Valley, Tanzania. (MH)**

**#5833 4.5×4.0×3.5cm Plate #12-52**

**Merelani Hills, near Arusha, Umba Valley, Tanzania**

A superb small cabinet matrix crystal with perfect termination and excellent color change.

## Tanzanite

In recent years, the gem deposits of Africa have been vigorously explored. Diamond, tourmaline, ruby, sapphire, emerald, alexandrite, garnet — including the superb green gem tsavorite — plus a host of rare gems are currently being extracted.

Yet, of all of these valuable gems, the mineral that has stirred the gem world the most is tanzanite, the blue gem variety of the common mineral zoisite.

You may well recall zoisite as that black-mottled green rock enclosing large hexagonal crystals of silky-red ruby corundum. This massive carving material is found in the Matabatu Mountains near Longido, North Province, Tanzania.

1967 is the year the gem world focused in on Tanzania, for it was then that tanzanite came to light. The exact location, revealed some time later, is the Merelani Hills on the slopes of the Lelatema Mountains, near Shambarai, Laletema District. The closest important town (by Tanzanian standards) is Arusha. Remarkably, it is only about 10 miles south-southwest of the Kilimanjaro International Airport, according to Keller (1992).

The shenanigans associated with a gem discovery of importance are well documented in Keller's text. The maneuvering, intrigue, even robbery and murder, usually associated with Colombian emeralds pale when compared to the Merelani Hills debacle.

Fortunately, a number of major stones, including Barlow's "Sleeping Beauty of Tanzania" (#4937), have survived, and the gem is a viable and avidly sought species in the gem marketplace.

The latest theory on the formation of these gem tanzanites credits hydrothermal alteration of the host metamorphic rocks as the source. Crystals tested show an age close to 600 million years. As it happens, about 550 million years ago a major geologic event occurred. As a result, the rocks were folded into the Lelatema anticline. The rocks were originally two-billion-year-old sediments.

Nearby pegmatite action may have been part of the hydrothermal action that commenced. In any event, heated ground waters circulated through the folded rock, dissolving the necessary ingredients to later form the tanzanites in openings along the upper hinges of the folded rock. The actual host rock of the tanzanite is a bytownite-grossular-diopside-zoisite rock that seems to have provided the necessary elements.

***Plate #12-53, Specimen #5832*** **ZOISITE, var. tanzanite – Merelani Hills, near Arusha, Umba Valley, Tanzania. (MH)**

### ZOISITE, VAR. TANZANITE

**#5832 4.5×4.0×3.5cm Plate #12-53**

**Merelani Hills, near Arusha, Umba Valley, Tanzania**

This specimen is the best miniature size matrix sample seen to date. Excellent termination and superb color change.

*Plate #12-54, Specimen #5838* **ZOISITE, var. tanzanite – Merelani Hills, near Arusha, Umba Valley, Tanzania. (MH)**

**ZOISITE, VAR. TANZANITE**
**#5838 10.5×8.0×8.5cm**
**Plate #12-54**

**Merelani Hills, near Arusha, Umba Valley, Tanzania**

This large cabinet matrix specimen is the second largest seen by Barlow. There are five tanzanite crystals on calcite with laumontite, crystallized graphite, and terminated crystals of green diopside. The large gem tanzanite crystal with excellent color change measures 5.5×2.5cm. This is a prize matrix tanzanite.

## REFERENCES

Bancroft, P. (1973) *The World's Finest Minerals and Crystals*, 1973, Viking Press, New York, 176 pp.

Keller, P.C. (1992) Gemstones of East Africa, Geoscience Press, Tucson, Arizona, 144 pp.

Wilson, W.E., et al. (1977, May-June) Tsumeb!, *Mineralogical Record*, 8(3), 128 pp.

Wilson, W.E., and P.J. Dunn (1978, May June) Famous Mineral Localities: The Kalahari Manganese Field, *Mineralogical Record*, 9(3), 137-153.

CHAPTER 13

# Mexico

*by Peter K.M. Megaw*

Mexico is one of the world's great treasure houses of mines and mineral specimens. The lure of her silver and gold deposits precipitated the Spanish conquest of the Aztec empire, possibly the greatest mining rush in history. Since then, mining has figured prominently in both the economic development of Mexico and in the periodic social upheavals that have arisen against both foreign and domestic developers. Thus mining and minerals are an integral part of the cultural and historical fabric of Mexico. Unlike many places in the so-called "developed world," public perception of mining is very positive in Mexico, creating a favorable environment for continued production and exploration for new ore deposits.

Mexico's documented total silver production exceeds eight billion ounces, with gold production in excess of 50 million ounces. Much additional production has gone unreported. The vast bulk of this came from veins and replacement deposits. Two of Mexico's innumerable silver-gold vein deposits, Pachuca and Guanajuato, rank first and third respectively, in the world in overall silver production. Mexico is also one of the world's largest silver-lead-zinc-copper-gold replacement provinces, containing six of the world's largest districts: Santa Eulalia, San Martin, Concepcion del Oro, Naica, Charcas, and Mapimi. The production history for many of these deposits spans nearly 500 years, beginning with the extraction, by the Spaniards, of enriched oxidized ores in the near-surface environment. Oxide production continued through the early 20th century until new metallurgical techniques allowed exploitation of deeper primary sulfide ores. Fortunately for mineral collectors, both the oxide and sulfide portions of both these deposit types tend to be prolific producers of fine collector specimens. Most major world collections, including the John Barlow Collection, contain excellent examples of them.

*Unlike many places in the so-called "developed world," public perception of mining is very positive in Mexico, creating a favorable environment for continued production and exploration for new ore deposits.*

### Vein Deposits

Vein deposits form from hot, mineral-laden waters, called hydrothermal fluids, derived from buried heat sources. These fluids migrate away from the heat source along open fractures and faults until, near the surface of the earth's crust, temperature and confining pressure drop sufficiently to allow the minerals dissolved in the fluids, chiefly quartz and calcite with lesser metallic minerals, to precipitate in the openings. Eventually the fractures become completely sealed and mineralization ceases until new open spaces are created when crustal forces cause renewed movement along a fault/fracture or, when enough pressure builds up in the system to cause explosive rupturing of the rock.

The largest systems undergo multiple repetitions of this process, creating very wide and long veins containing enormous amounts of metals. At Guanajuato and Pachuca, geologists have recognized as many as 10 separate stages of breakage and mineralization. The largest of these veins exceed 20 m in width and extend for over 10 km. The repeated formation of open spaces in which minerals can grow means that many vein deposits contain large numbers of pockets partially filled with well-formed crystals.

Many exquisite crystals of silver minerals are found in these pockets, commonly on attractive matrices composed of quartz and/or calcite crystals. Miners have long noted that the quartz in the gold-rich portions of many veins is amethystine.

Vein deposits form close to the surface and are therefore found mostly in relatively young rocks that haven't had time to be deeply eroded. In Mexico, the largest vein systems occur in mid-Tertiary-aged [<45 Ma (million years)] volcanic rocks that compose the Sierra Madre Occidental, which runs like a spine down the western length of Mexico, and the Trans-Mexican Volcanic Belt, which is an east-west belt of younger volcanoes (many of which are still active) that cuts across the center of the country.

For specimen collectors, the Fresnillo, Guanajuato, Taxco, Arizpe, Batopilas, and Parral Vein Districts have produced the finest silver specimens. The first three are still doing so.

Descriptions of outstanding silver minerals from these deposits residing in the Barlow Collection can be found in Chapters 9 and 10.

### Replacement Deposits

Replacement deposits form where hydrothermal fluids filter through limestone or dolomite, causing dissolution of the host rocks with nearly immediate precipitation of sulfide or silicate minerals in place of the original carbonate components. Delicate textures of the original rocks are often faithfully preserved in the resulting ores. Replacement deposits form at significantly greater depths than do vein deposits and are often found in direct contact with intrusive stocks, dikes, and sills. These deposits are commonly zoned from the intrusive bodies outwards. Calc-silicate skarns, largely composed of garnet and pyroxenes accompanied by varying amounts of pyrite, pyrrhotite, galena, sphalerite, chalcopyrite, arsenopyrite, and silver sulfides or sulfosalts, occur along and near the intrusive contacts.

Podlike and elongate bodies composed almost exclusively of these same sulfides occur outboard of the skarn and may extend for kilometers from the intrusive rocks. These massive sulfide bodies are called *mantos* (Spanish for blanket or cloak), when roughly horizontal and conformable with sedimentary bedding; they are called chimneys when steeply inclined and cutting across bedding. Both the skarn and massive sulfide ores contain numerous open voids locally lined with well-crystallized sulfides and gangue minerals including fluorite, barite, calcite, dolomite, and quartz. Many of these minerals grow to exceptional size.

In Mexico, the major replacement deposits occur in a 2200km long belt of folded Mesozoic (150 to 65 Ma) carbonate rocks called the Mexican Fold Belt (Megaw, et al., 1988). This folding event created subtle structural plumbing systems that were followed by hydrothermal fluids created during younger (45-25 Ma) intrusive activity. Replacement deposits form more than a kilometer below the surface, so they require a combination of uplift and erosion to bring them close enough to the surface to mine.

In Mexico, relatively recent (<20 Ma) regional block faulting has uplifted many replacement deposits into the area of influence of descending surface (meteoric) waters. These waters attack the primary sulfide minerals and oxidize them to a wide variety of secondary minerals. A volume reduction of 20% or more occurs during this process, resulting in abundant, and often very large, open spaces into which these secondary minerals can freely grow. Crystals exceeding 20cm in length are not uncommon in these areas.

A large number of Mexican replacement deposits very well known to mineral collectors include Naica; Mapimi; Santa Eulalia; Charcas; Concepcion del Oro; Los Lamentos; San Pedro Corralitos; San Carlos, Chihuahua; and Guadalupe Victoria, Durango. Less well known, but nonetheless important, specimen producers include San Martin, Zacatecas; Zimapan; Velardena; Bilbao, Zacatecas; and Aurora, Chihuahua. Specimens from many of the latter deposits have been incorrectly attributed to the better known localities because they tend to be sold in the merchandising centers established around the better known sites.

Naica is an excellent example of a largely unoxidized replacement deposit that produces superb sulfides and primary gangue minerals. The Mapimi deposit is one of the world's best examples of an oxidized replacement deposit and is the source of several hundred secondary mineral species, several of which are unique to the district. The largest replacement deposit in the world is Santa Eulalia, which alone produced one quarter of all the silver shipped from the New World to Spain in the period 1703–1739 (Hadley, 1975). Santa Eulalia's mantos and chimneys are so large that an oxide zone over 500 m deep has developed on top of more than 600 m of primary ores. This district has, therefore, produced a wide variety of both primary and secondary minerals in its 300-year history.

## John Barlow and Mexican Minerals

John Barlow's acquisition of minerals from Mexico was not the result of a deliberate decision to accumulate Mexican minerals, but rather evolved from his love for Mexico. John and his wife, Dorothy, vacationed and fished frequently in Mexico for many years before he became a mineral collector. On these trips, they fell in love with the country's beauty and developed a deep affection for the Mexican people.

As John began mineral collecting, he naturally became aware of the existence of fine minerals from Mexico, especially the silvers and sulfosalts from Guanajuato and elsewhere. He almost casually decided to try getting some specimens during these vacation trips. On one early trip to Taxco, a search for silvers resulted in nothing more than a few crude specimens purchased from children in the street. John bought these to help the kids and left the specimens behind in his hotel room when he departed Taxco.

John largely abandoned his efforts to make the necessary contacts for good specimens, partly because he was hampered by a lack of fluency in Spanish, but more because he was averse to altering the relaxed tenor of his vacations. With the exception of his self-collected boleites and cumengites, he never acquired another specimen while in Mexico. However, John's relentless pursuit of silver minerals inevitably kept bringing him into contact with Mexican specimens, and dealers gradually began bringing him Mexican minerals of all kinds. He rapidly became acquainted with the minerals of famous districts such as Fresnillo, Naica, Mapimi, and Santa Eulalia and began to acquire individual specimens from these places that caught his eye.

Several wholesale dealers who specialize in Mexican minerals later began sending John entire lots of specimens, which he bought. John's Mexican collection has, therefore, grown by a combination of single steps and huge leaps and is characterized by either individual specimens or large suites from a single locality or find. These suites are important because they range from virtually complete pockets that show important variations within a cogenetic group to accumulations that display the variations exhibited by examples of the same species extracted from a series of pockets in the same mine or orebody.

John's first "keeper" specimen from Mexico was an unprepossessing pyrite and jamesonite (now gone) from the Concepcion del Oro District, Zacatecas, and one of his latest (February, 1996) is a specimen composed of clumps of white aragonite with offshoots of colorless aragonite needles, liberally covered with balls of lustrous lime-green adamite. This is perhaps the best cabinet-sized sample from this particular find.

These two specimens bracket a collection that includes 468 total pieces. The distribution of samples closely reflects the manner in which John has acquired his Mexican collection: 34 are single specimens of a single species, but 266 of the total make up suites of the same species from individual localities. The largest species suites are creedite from Santa Eulalia, Chihuahua (77); boleite from Boleo, Baja California Sur (51); adamite from Mapimi, Durango (46); legrandite from Mapimi (25); ludlamite from Santa Eulalia (21); and rhodochrosite from Santa Eulalia (11).

John and his wife, Dorothy, adopted a Mexican "daughter," Maria del Bosque, from Aguascalientes, who lived with them in Appleton and went to high school with their daughter, Alice, during 1970. The two families have remained in constant touch ever since.

This chapter departs from the normal format of the book, since so many of the described Mexican specimens come from a single district or orebody. So, specimen descriptions will be grouped alphabetically under a general geologic description of each district to provide a general geologic context for the minerals. The species that come from sources other than the major districts are listed alphabetically at the end of the chapter.

## Boleo, Baja California Sur

The Amelia Mine, Boleo, Baja California Sur, Mexico is a very unusual ore deposit, unlike any other in Mexico. Geologically, the Boleo District consists of an elongate, fault-bounded basin cutting Jurassic-aged granitic and metamorphic rocks and Eocene andesitic volcanic rocks. The basin was filled by a cyclical succession of very young (Pliocene to Pleistocene) volcanic rocks interbedded with gypsiferous sediments and massive bedded gypsum. Each cycle began with fault movements that triggered rapid deposition of a thick layer of coarse sediments, now conglomerate. After each faulting event, the basin experienced slow deposition of airborne volcanic ash from nearby volcanoes. During these "quiet" periods, numerous hot springs along the basin-bounding faults pumped hot sulfate and metals-bearing water into the basin. As these fluids cooled and mixed with seawater, they precipitated abundant gypsum and primary Cu-Co-Zn-Mn-Fe and Ag minerals with the slowly accumulating ash. This continued until the next faulting episode again swamped the system with coarse sediments. At least five separate cycles have been recognized. The tuff beds have been largely converted to clays, and in some the very fine grained primary sulfide mineralization has been oxidized and transported to form secondary native copper, copper oxide, and copper-chloride bodies up to 2.5 m thick. It is these exotic bodies that produce the rare boleites, pseudoboleites, and cumengites that were first identified and described from this locality by French mineralogists in the 1890s. The best boleites and cumengites came from the middle, or Number 3, tuff bed, source of 85% of the ore mined from the district. When fresh, this bed is a very warm and moist clay that dries quickly, releasing the boleites (Panczner, 1987). This characteristic means that the matrix of most specimens must be stabilized to prevent loss of the crystals.

In contrast to his "silver-pick" buying of most of his Mexican minerals, John Barlow self-collected most of his boleites and cumengites. When Bill Larsen and Ed Swoboda were operating the Amelia Mine in the late 1970s, John and his grandson Greg Helein, flew down to the mine after the Tucson Show. The best collecting at the time was on the 920 level, and it was so hot and humid that John was more preoccupied with surviving than with collecting. Nonetheless, he was able to collect a number of good specimens in the mine and managed to get a few more that Bill and Ed had recently mined. He was a bit chagrined, however, when Greg found some excellent cumengites and boleites on the dumps without suffering the heat underground. One of John's best cumengite on matrix is #T366. Other Mexican boleites can be found in Chapter 10, Plate #10-22, and in Chapter 3, Plate #3-29.

The Mexican collection is richly endowed with native silvers from Batopilas, Taxco, and Zacatecas (17 specimens). Refer to Chapter 9 (Silver), pages 221-223.

The collection is also heavily endowed with silver bearing minerals. Refer to Chapter 10, acanthite, aguilarite, pearceite, pyrargyrite and stephanite from Guanajuato; amalgam from Taxco; pearceite and stephanite from Sonora; pyrargyrite from Durango; and boleite and cumengite from the Amelia Mine, Boleo, Baja California Sur, Mexico.

### CUMENGITE
**(COPPER LEAD CHLORIDE HYDROXIDE)**
**#T366 1.8×2.5×2.5cm**

The cumengite is a 0.5cm "star" of indigo blue tetragonal crystal pyramids growing off each of the cube faces of a 0.25cm central boleite crystal. Mined in the late 1970s, this is an excellent example of this much sought after habit of cumengite overgrowing boleite found at the Amelia Mine. Fine examples like this are a rarity.

## Replacement Deposits

### Mapimi, Durango

The Mapimi, or Ojuela, District lies in north-central Durango, near the combined cities of Torreon, Gomez Palacio, and Lerdo. Mapimi was discovered in 1598 and was a significant silver producer for the Spanish until they were expelled from Mexico in the early 1800s (Panczner, 1987). Production sharply diminished during most of the remainder of the 1800s until President Porfirio Diaz opened Mexico to foreign mining investment.

The "Porfiriato" time period witnessed Mapimi's largest boom from 1890 until the 1910 Revolution, during which time the engineering marvel of the Roebling Suspension Bridge was constructed here (Castro-Carrillo, 1980). The Peñoles Company acquired the district after the revolution and mined deep sulfide ores until the end of World War II. Peñoles then turned the mines over to a miners' cooperative, which under took pillar robbing and scavenging operations in the oxide zone until the mid 1980s (Panczner, 1987).

The district produced a total of about five million tons of high-grade silver-rich oxide ores and perhaps as much as a million tons of sulfides (Megaw, et al., 1988). The local miners were educated by American and Mexican mineral dealers and collectors on the value of the spectacular mineral specimens found throughout the mines, so enormous numbers of specimens were produced during the cooperative's tenure. Today, the district is in the hands of a private individual, and specimen mining has slowed to a trickle. However, new minerals are still being discovered by detailed study of specimens mined in the past, and undoubtedly more will be forthcoming.

### Mapimi Deposit

The Mapimi deposit consists of a series of four groups of inclined mantos formed along the crests of a series of parallel tight folds in the local Cretaceous limestones (Prescott, 1926). These mantos branch and intersect complexly and appear rooted in seven major chimneys (Villarello, 1906). The chimneys grade downward into skarns formed in proximity to felsic intrusive rocks (McLeroy, et al., 1986; Megaw, et al., 1988). The oxide level extends to a depth of over 500 m, which encompasses all the mantos and many of the chimneys. The deep primary ores consist mostly of galena, sphalerite, chalcopyrite, and pyrite, but both the skarns and deep sulfide ores locally contain up to 20% arsenopyrite (McLeroy, et al., 1986). Oxidation of this arsenopyrite is why the secondary ores are so rich in arsenates. Probably only the Tsumeb deposit in Namibia has a larger suite of arsenates than Mapimi.

### The Ojuela Mine

The Ojuela Mine is the largest and best known of the dozen or so major mines in the district, so much so that the district is often referred to by that name. Most specimens carry the Ojuela Mine as their location because modern access to the labyrinthine underground workings has largely been through the Ojuela Mine. When given, mine level designations for the district are usually in reference to the Ojuela Shaft. However, this is only part of the story as it is most important to know which manto or chimney a specimen actually came from. In most cases this information is lost. For older specimens, and for many species, only the mine level is widely known.

Fortunately, in recent years several dealers and authors (see Panczner, 1987) have been more diligent in providing stope information, although such information obtained from the specimen miners may not always be accurate because they want to protect their collecting turf.

### ADAMITE (ZINC ARSENATE HYDROXIDE)

Adamite is a secondary zinc arsenate originally described from Chañarcillo, Chile in 1866. Mapimi has yielded excellent specimens of adamite since the 1940s, and the district is probably the world's most prolific producer. Other important localities are Laurium, Greece; Tsumeb, Namibia; and Gold Hill, Utah. Recent discoveries of adamite at Santa Eulalia, San Pedro Corralitos, Adargas, Minillas, and Tres Hermanos, Chihuahua; Asientos, Aguascalientes; and San Martin, Zacatecas indicate that adamite is relatively common in the oxidized portions of many replacement deposits.

Mapimi has produced adamite in a bewildering variety of colors and habits from a number of stopes between the 4th and 15th levels of the Ojuela Mine. The best known type is probably the botryoidal aggregates, to 6.0cm in diameter, with brilliant lime-green color, which come from several different stopes on the 4th level. This material characteristically fluoresces yellow-green under UV light because of trace amounts of uranium (Modreski and Newsome, 1984). Coarse, blocky yellow-green crystals to 3.0cm, often topped with limpid twinned calcite scalenohedra, come primarily from the Esperanza Orebody, 4th level. These are significantly less fluorescent than the botryoidal material.

*John Barlow's collection contains 46 Mapimi adamite specimens, of which 37 are from a single lot of mixed purple and green material (Specimens 2900–2936).*

Small amounts of large, bright purple adamite crystals have also been found in the San Judas Orebody on the 5th level (see below). Coarsely crystalline green crystals with purple tips, green crystals with purple cores, and purple crystals with green tips are known from the upper portions of the purple adamite zone. The purple color of these adamites was originally attributed to the presence of cobalt, and the Mapimi locals still call the purple-tinged adamites "cobaltos." In fact, the purple color is imparted by manganese, not cobalt, and several other manganese species, including chalcophanite, occur with it. The rare species lotharmeyerite was originally identified from the matrix of these purple adamites (Dunn, 1983), but no lotharmeyerite has been noted on Barlow's specimens.

John Barlow's collection contains 46 Mapimi adamite specimens, of which 37 are from a single lot of mixed purple and green material (specimens #2900 -#2936).

Several of these are notable and are described below. The remainder are from diverse sources and include both purple and yellow-green specimens. His best example of the familiar botryoidal habit (#2655) is a 13.0×8.0×3.0cm cluster of intergrown spheres and fans of lime-green radial crystal aggregates to 2.0cm in diameter. His best adamites are two purple specimens acquired in the early 1980s (#2360, Plate #13-4 and #3070, Plate #13-5).

Purple adamite specimen #3070, Plate #13-5, is one of Barlow's four best non-silver-bearing Mexican specimens, and its acquisition reflects John's big game hunting approach to mineral collecting. In 1981, El Paso dealer John Whitmire and several associates undertook to collect at Mapimi. Their work began with mucking out an enormous amount of backfill in long-abandoned stopes and finished with their abrupt expulsion from the district in late 1981, shortly after they hit a superb pocket of purple adamite crystals. John Barlow got wind of the find and set out to track it down at the 1982 Tucson Show. Unfortunately for John, all the best pieces were gone by the time he got in touch with Whitmire and company, so he missed out except for a middling quality piece acquired shortly thereafter (#2360, Plate #13-4).

But John had already opened another avenue of approach: Whitmire's wife Rosa, with whom he got along well. Rosa told him, "Don't worry, John, I'll get you one. What do you want?" John expressed a modest desire for a hand-sized piece with excellent color and crystallinity.

He promptly forgot about it until the following Tucson Show when Rosa sidled up to him saying "Got something for you, John." John's eyes popped out as she handed him a 7.0×5.5×4.0cm plate of well developed individual crystals 2.5×0.8cm across, up to 2.5cm long, with prominent dome terminations and superb saturated color zoning from bluish white at the base to brilliant magenta at the tips (#3070, Plate #13-5). The specimen is one of the best of the find.

Barlow's second best purple adamite is #2360, Plate #13-4 — a 9.0×5.5×4.0cm plate of 2.5 and 1.2cm long sprays of elongate crystals in a vug of botryoidal iron oxides. Their color is very clean and saturated, and individual crystals are zoned from bluish white bases to magenta tips.

In November of 1985 John purchased a lot of 37 fine, coarsely crystalline, green and purple adamites from the Rock Shop of El Paso (specimens #2900-#2936). This lot is important because it clearly shows the habit, color, and zoning variations possible in a single pocket. Many of the specimens are relatively unattractive with muddy purple color (e.g.,#2911, #2914, #2919), but others show good purple (#2925, Plate #13-2) or green color (#2934) or display a showy zoning from green bases to purple tips (#2925, Plate #13-2; #2936) or the reverse (#2918, Plate #13-1). Specimen #2895, Plate #13-3, was acquired shortly before this lot but appears to have come from the same zone if not the same pocket. Descriptions of the better pieces follow.

## ADAMITE

**#2918  2.5×2.5×2.0cm  Plate #13-1**

**Mina Ojuela, Mapimi, Durango, Mexico**

Plate with 2.0cm fan composed of 1.5cm crystals with prominent dome terminations. Color is well saturated and zoned from brown-purple bases to yellow-green tips. This zoning is the reverse of the more common yellow-green base, purple tip specimens (compare with #2925).

***Plate #13-1, Specimen #2918*** **ADAMITE – Mina Ojuela, Mapimi, Durango, Mexico. (MH)**

***Plate #13-2, Specimen #2925*** **ADAMITE – Mina Ojuela, Mapimi, Durango, Mexico. (MH)**

**#2925  6.5×4.5×5.0cm  Plate #13-2**

**Mina Ojuela, Mapimi, Durango, Mexico**

Plate with an exceptionally large 2.0×2.5×1.5cm dominant crystal, with very prominent dome termination, surrounded by six smaller crystals. Color is zoned from green-brown bases to purple tips. Purple is tinged by green.

**#2895  8.0×6.5×5.0cm  Plate #13-3**

**Mina Ojuela, Mapimi, Durango, Mexico**

Rodlike aggregate of intergrown fans composed of wedge-shaped individual olive-green crystals. Points of terminations are purplish indicating transition to manganoan composition. Overall shape is very unusual.

***Plate #13-3, Specimen #2895*** **ADAMITE – Mina Ojuela, Mapimi, Durango, Mexico. (MH)**

*Plate #13-4, Specimen #2360* **ADAMITE – Mina Ojuela, Mapimi, Durango, Mexico. (MH)**

## ADAMITE

**#2360 9.0×5.5×4.0cm Plate #13-4**

**Mina Ojuela, Mapimi, Durango, Mexico**

A solid mass of iron oxides with an open vug 3.0×5.0cm lined with sooty black material provides the setting here. On the rim of the vug is a 2.5×0.8cm spray of diverging adamite crystals. Their color ranges from faint bluish white to brilliant magenta. Canted off at the base of this spray is a slender 0.8cm crystal. All the crystals are very lustrous. Nested deep in the vug is a 1.2cm cluster of slender magenta-tipped adamites. Randomly scattered within the vug are several slender adamite crystals as well.

**#3070 7.0×5.5×4.0cm Plate #13-5**

**Mina Ojuela, Mapimi, Durango, Mexico**

Composed of a mass of randomly intergrown and freestanding adamites, this specimen has a minimum of iron oxide matrix. The adamite crystals range in length from 1.2 to 2.5cm. Their color is zoned from bluish white at the base to bright magenta at the terminations. The specimen shows virtually no damage.

*Plate #13-5, Specimen #3070* **ADAMITE – Mina Ojuela, Mapimi, Durango, Mexico. (MH)**

## KÖTTIGITE

**(ZINC ARSENATE HYDROXIDE)**

**#3340 10.0×5.0×5.0cm Plate#13-6**

**Mina Ojuela, Mapimi, Durango, Mexico**

Triangular banded gossan martrix with 4×2cm vug containing a 2cm long radiating spray of gray-blue elongate thin crystals. This is an excellent example of this species from near the water table at this classic locality.

*Plate #13-6, Specimen #3340* **KÖTTIGITE – Mina Ojuela, Mapimi, Durango, Mexico. (MH)**

## LEGRANDITE

### (ZINC ARSENATE HYDROXIDE HYDRITE)

**Mina Ojuela, Mapimi, Durango, Mexico**

Legrandite is a rare zinc arsenate found in only a few localities worldwide. The type locality is the Lampazos Mine in Nuevo Leon, Mexico, an obscure locality from which little material has been recovered. Very nice legrandites have recently been found at Tsumeb, Namibia, and it is also very sparsely distributed at Santa Eulalia, Chihuahua. Mapimi remains the world's premier legrandite producer and has yielded the largest crystals, to 28.0cm. Legrandite at Mapimi occurs from the 12th to the 17th levels, but stope data are apparently unavailable. The best specimens reportedly come from near the water table, which usually lies around the 16th to 17th level, so only during prolonged droughts is collecting feasible (Panczner, 1987).

In October of 1977, the best legrandites yet found anywhere were unearthed at Mapimi: the "Aztec Sun" (in the Romero Mineral Museum, Téhuacan, Puebla), and "The Club" (at the American Museum of Natural History, New York). El Paso dealer Jack Amsbury brought the pieces out and offered them as a lot to John Barlow. The price was high by 1970s standards, but it was intended to be Jack's retirement fund.

Barlow mulled it over for two days before deciding to pass... he continues to regard this decision as one of the worst mistakes he made in building his collection.

In compensation, he has since acquired 25 legrandites that appear to have come from a series of finds mostly preceding the 1977 pockets.

*Barlow mulled it over for two days before deciding to pass... he continues to regard this decision as one of the worst mistakes he made in building his collection.*

***Plate #13-7, Specimen #1257*** **LEGRANDITE – Mina Ojuela, Mapimi, Durango, Mexico. (MH)**

***Plate #13-8, Specimen #1274*** **LEGRANDITE – Mina Ojuela, Mapimi, Durango, Mexico. (MH)**

## LEGRANDITE

**#1257 4.5×3.0×5.0cm Plate #13-7**

**Mina Ojuela, Mapimi, Durango, Mexico**

Spray of 2.5cm long, freestanding, elongate, well-terminated, brilliant yellow crystals surrounded by 1.5cm crystals at the base.

**#1274 4.5×2.5×1.5cm Plate #13-8**

**Mina Ojuela, Mapimi, Durango, Mexico**

Radiating spray composed of bundled, elongate, terminated, brilliant canary-yellow crystals. The specimen is remarkable for its superb, saturated yellow color.

## LEGRANDITE

**#1797 8.0×7.0×2.6cm**

**Refer to Chapter 3, Plate #3-60**

170° radiating stellate group of 2.5–3cm elongate terminated yellow crystals with platy, pseudohexagonal (?) adamite and rice-grain adamite or smithsonite on an iron-rich hard matrix.

## SANTA EULALIA

The Santa Eulalia district lies in central Chihuahua State, about 20km east of Chihuahua City. Chihuahua City exists because it offers the closest surface water to the district. The district was discovered in 1702, and it rapidly became one of the most important silver mines in Mexico because of the ease of mining and extraction of silver. As at Mapimi, the early Spanish-era mining focused on the near-surface mantos composed of massive argentiferous cerussite laced with silver chlorides and native silver. Grades in the 1700s exceeded 1.5 kg of silver per ton and 25%+ lead. The lead was not recovered until the late 1800s when deeper oxide ores became the focus of modern-style mining during the Porfiriato. A selective flotation plant was constructed in 1925, allowing exploitation of the extensive sulfide ores of the chimneys and deeper mantos. Oxide production tapered off to almost nothing over the next 10 years. Silver grades in the sulfides dropped to about 400 g/ton, but the lead and zinc recovered made for very profitable operations. Mining continues in the eastern part of the district.

Santa Eulalia has been continuously active since 1702 and has produced over 40 million tons of ores averaging 8.8% lead, 330 g/ton silver, and 7.8% zinc (Megaw, 1990). Minor amounts of gold, copper, tin, and vanadium have been produced from the eastern portion of the district.

Santa Eulalia is unique in replacement systems because it exhibits a wide variety of ore types and compositions within a single district (Hewitt, 1968; Megaw, 1990). The district is divided into two parts: the East Camp, dominated by the San Antonio Mine; and the West Camp, dominated by the Potosi and Buena Tierra Mines. No significant mineralization has been found in the intervening, 2km wide, Middle Camp. The East Camp is dominated by tabular garnet-hedenbergite skarn chimneys symmetrically developed on both sides of a series of felsic dikes. Massive sulfide pods flank the skarns, and sulfide mantos locally extend like fingers off the skarn. Discrete sulfide mantos also occur separately from the skarns in the peripheries of the camp. Distinctive, carrot-shaped tin and vanadium-bearing chimneys extend off the top of the skarn bodies in two places (Hewitt, 1943). In addition to the usual sulfide and gangue-lined pockets found in the skarn and massive sulfide ores, a series of open structures cut the skarns and provided space for the growth of very well crystallized siderite, fluorite, ludlamite, and vivianite.

In contrast to the East Camp, the West Camp is characterized by elongate massive sulfide chimneys 1km tall and mantos that exceed 3km in length. These orebodies have yielded enormous quantities of well-crystallized ore minerals from their deep, unoxidized zones and equally important amounts of secondary oxide minerals from the upper zones.

The West Camp contains virtually no East Camp-like skarn; instead, it hosts a series of unusual johanssenitic hedenbergite-manganoan-fayalite ("knebelite") skarn bodies that lie in the upper portions of the system, over 500 m from the nearest ore-related intrusive rocks. These iron-calcic skarns underwent a late ore-stage hydration event that converted the original anhydrous silicates to a mixture of chlorite, ilvaite, quartz, and rhodochrosite. This hydration event, called "retrograde skarn," was contemporaneous with the introduction of ore sulfides and resulted in the creation of numerous irregular voids into which grew crystals of ore sulfides, rhodochrosite, selenite, and creedite. The Barlow Collection contains an excellent representation of creedite and rhodochrosite.

## EAST CAMP, SANTA EULALIA

The central part of the San Antonio Mine skarns, below the 10th level, are known as the R-40 zone. This area consists of typical garnet-hedenbergite skarn with both disseminated and massive sulfide zones. The R-40 skarn is cut by several late vertical fractures that pinch and swell repeatedly, creating small open pockets at irregular intervals along their length. These pockets typically contain greenish yellow-brown siderite, often pseudomorphing scalenohedral calcite. Sparse colorless cuboctahedral fluorite crystals to 3.0cm are also present, typically coated by the siderite. Rarely, the pockets contain bright yellow-green ludlamite and deep blue-green vivianite crystals to 8.0cm in length. Each ludlamite pocket exhibits distinctive crystal morphology including stacks of multiply twinned, bladed crystals; sharp, coffin-shaped crystals; and sheaves, fans, rosettes, and ball-like crystal aggregates. Many have inclusions and partial encrustations of minute pyrite cubes.

The high-grade sulfide zones above the 12th level throughout the San Antonio Mine were mined by underground "glory hole" methods prior to 1968, leaving open stopes extending over 250 m vertically. In 1968, large-scale cut-and-fill trackless mining techniques were adopted allowing exploitation of the virgin ores between the 15th and 12th levels. Mining in the R-40 began in the late 1970s, and because of the cut-and-fill mining techniques, the ludlamite zone was intersected intermittently as mining swept through the zone at successively higher levels.

This activity produced the spectacular ludlamites found in the early 1980s (Wilson, 1982). Mining has consumed almost all of the R-40 zone between the 15 and 12 levels and is now primarily focused on the remnant areas around the old glory hole stopes, a very hazardous mining situation. Small amounts of ludlamite continue to trickle out of the R-40, including a very nice pocket hit in 1991 when a manway was driven from the 14th to the 13th level, and, recently, specimens have been found in the glory hole area. Ludlamites evidently came from the glory hole areas years ago, because old pieces, misidentified as "cuprian hemimorphite" because of the crystal habit, are known (E. J. Huskinson, personal communication). John Barlow has 21 ludlamite specimens representing several of the major finds of the 1980s. He also has an excellent miniature of siderite crystals with ludlamites (#2890) from the zone described.

**Plate #13-9, Specimen #3135 LUDLAMITE – San Antonio Mine, Santa Eulalia, Mexico. (MH)**

### LUDLAMITE
**(IRON MAGNESIUM MANGANESE PHOSPHATE HYDRATE)**

**#3135 4.0×2.5×2.0cm Plate #13-9**

**San Antonio Mine, Santa Eulalia, Mexico**

2.5cm "bow-tie" composed of platy yellow-green crystals perched on subsidiary sheaves of ludlamite crystals. The main bow-tie has a secondary bow-tie extending perpendicularly from its middle, making a very aesthetic shape.

**Plate #13-10, Specimen #2857 LUDLAMITE – San Antonio Mine, Santa Eulalia, Mexico. (MH)**

**#2857 3.5×2.5×2.0cm Plate #13-10**

**San Antonio Mine, Santa Eulalia, Mexico**

This is a very aesthetic group of stacked crystals. The color is vibrant green typical of the R-40 zone.

***Plate #13-11, Specimen #2890*** **SIDERITE – Santa Eulalia, Chihuahua, Mexico. (MH)**

## SIDERITE (IRON CARBONATE)

**#2890 3.5×2.0×1.5cm Plate #13-11**

**Santa Eulalia, Chihuahua, Mexico**

Group of six, 1.5cm rounded olive-green rhombohedral crystals. Rounding is notable, but these are NOT pseudomorphs after fluorite as the original label states. These are some of the best formed siderite crystals known from this zone.

## VIVIANITE (IRON PHOSPHATE HYDRATE)

**#5369 9.5×1.6×1.6cm Plate #13-12**

**Level 13, San Antonio Mine, Santa Eulalia, Mexico**

The pockets that produce ludlamite also produce small amounts of vivianite. Most of the vivianites are small, poorly formed, and commonly badly cleaved. Unlike vivianites from sedimentary deposits, these specimens do not decrepitate quickly and appear to be color stable for at least 20 years. Between 1991 and 1993, several small pockets containing exceptionally fine vivianites were found. Several were doubly terminated, blocky, and elongate crystals to 5.0cm long on siderite matrix. In March, 1994, a small pocket was encountered containing one superb crystal that Barlow was able to acquire. The specimen (#5369) is a single matrixless dark emerald-green elongate crystal 9.5cm long and 1.6cm square with two stepped angled terminations. The crystal shows several incipient cleavages that make it gemmy/transparent parallel to cleavage and translucent perpendicular to cleavage. The entire crystal has a smoothly rounded "melted" appearance similar to many galenas from this district and Naica. This specimen ranks with the two or three finest examples from the locality, and its unusual equant-elongate habit puts it in the ranks of the best vivianites anywhere.

***Plate #13-12, Specimen #5369*** **VIVIANITE – Level 13, San Antonio Mine, Santa Eulalia, Mexico. (MH)**

## WEST CAMP, SANTA EULALIA

### ADAMITE (ZINC ARSENATE HYDROXIDE)

**#5830 12.5×12.0×12.0cm Plate #13-13**

**Levels 8 to 9, Q-Trend, Potosi Mine, Santa Eulalia District, Chihuahua, Mexico**

One of the best specimens, and probably the best cabinet-size sample, from a narrow, aragonite-lined fissure in the oxide zone of the West Camp mined in the summer of 1995. The fissure produced only 15–20 high-quality specimens and several hundred inferior pieces before it pinched down. No other fissures in the area have been found to contain adamite, although many are lined with aragonite needles. This specimen is a roughly equant head composed of clumps of white aragonite with offshoots of colorless aragonite needles to 1.0cm long, liberally covered with balls of lustrous lime-green adamite to 1cm in diameter. Some of the adamite balls are impaled on the aragonite needles. The overall effect is of clumps of fresh green grapes on hoarfrost. The adamite fluoresces yellow-green under both long- and short-wave UV, with a stronger response to short-wave. The fluorescence is brighter than that of adamite balls from Mapimi. This material first appeared at the Denver and Munich Shows in Fall 1995 where it was sold as smithsonite. *Mineralogical Record* (Volume 27, Number 1, 1996, page 61) published a photograph of the best small cabinet specimen from the find and repeated the vendors' misidentification as smithsonite. Other specimens have been seen mislabeled as prehnite. The Barlow specimen was obtained at the Tucson Show in February, 1996 from the Jeffrey Mining Company. It is one of the finest Mexican specimens in the collection.

***Plate #13-13, Specimen #5830*** **ADAMITE – Levels 8 to 9, Q-Trend, Potosi Mine, Santa Eulalia District, Chihuahua, Mexico. (MH)**

### CREEDITE (CALCIUM ALUMINUM SULFATE FLUORIDE HYDROXIDE HYDRATE)

Creedite is an unusual hydrothermal mineral found originally near the Creede District of Colorado, with additional important finds from near Tonopah, Nevada and from Russia. Creedite may be colorless or may range from pale violet or lavender to a deep grape-purple. The color may be related to fluorine content and is very unstable. The purples fade to completely colorless relatively quickly under strong incandescent lights or indirect sunlight. The mineral is best stored in the dark.

Creedite from Santa Eulalia comes from the Piñata and Inglaterra calcic-iron skarn orebodies on the 10th level of the Potosi Mine in the West Camp and on the 5th level of the Inglaterra Mine, respectively. These two orebodies are less than 150 m apart: the Inglaterra 5th level is equivalent to, and connected with, the Potosi Mine, 10th level. Creedite has been known from the Inglaterra orebody, also correctly known as the A-10 orebody (Prescott, 1926) or incorrectly as the Condesa Mine (Panczner, 1987), since the late 1940s. Inglaterra creedites typically form radiating sheaves of cloudy white to pale violet crystals less than 1.0cm in length. Many are coated with gypsum, var. selenite. A few stout prismatic, limpid, doubly terminated, assymetrical crystals perched on quartz have also come from this orebody.

In 1982, creedites with Inglaterra type morphology but deeper color began to be found in the Piñata orebody.

In 1984, a spectacular pocket of deep purple elongate crystals to 8.0cm were encountered, and "creedite fever" began. Over the next two months, several more pockets lined with 1.0-4.0cm elongate crystals and druses of 0.5–1.0cm crystals were found. Gradually, the management noticed a distinct drop in mine production, and rumors of collecting activity on the 10th level filtered up to the mine superintendent. To see for themselves, he and several staff engineers hid away in the area and, as the first few hours of the next shift passed, they were astonished to see miners assigned to stopes as distant as the 22nd level of the mine appear to collect specimens.

When they finally sprang their trap, 34 miners were charged with stealing specimens and summarily fired... all but six were rehired two to three weeks later because good miners are hard to come by. However, the stope was closed for mining and has remained so to the present. Stories describing "pickup truck loads of creedite" that never made it to market are false.

The Barlow Collection contains 77 creedite specimens, all from the Piñata orebody. Barlow's acquisition of the creedites marks the beginning of what was to become a long-term, and voluminous, association with the Rock Shop of El Paso. John saw five or six specimens from the first find at the 1984 Denver Show and bought them all. Shortly thereafter he purchased an entire lot of creedites, and this lot was followed by another. In short order, John found himself in possession of a very significant percentage of the creedites from the Piñata pockets, although, unfortunately, none of the deep purple 8.0cm elongate crystals. His final purchase was a group of slender doubly terminated pale lilac crystals to 3.5cm in length that probably left the mine the day before management lowered the boom. Because he bought entire lots, the quality varies from superb to ordinary, but the completeness of the sampling gives him examples of most of the notable variations of color, habit, and size seen in the pockets.

***Plate #13-14, Specimen #2660*** **CREEDITE – West Camp, Santa Eulalia, Chihuahua, Mexico. (MH)**

### CREEDITE

**West Camp, Santa Eulalia, Chihuahua, Mexico**

**#2659 4.0×2.5×2.5 cm**

"V"-shaped pair of intergrown deep violet crystals. The major crystal is 3.5cm long with a subsidiary crystal 2.5cm long. Crystals are lightly included with iron oxides.

**#2660 5.0×2.5×2.5cm Plate #13-14**

**West Camp, Santa Eulalia, Chihuahua, Mexico**

This well-colored miniature has several crystals prominently displayed. The largest cluster shows a typical "bow-tie" effect and reaches 2.8cm in length. Two single crystals barely touch and are 2.2 and 2.4cm in length. The color of the creedite is a good violet, with some zoning to a lighter color evident. The matrix is coated with brown microcrystallized siderite.

*Plate #13-15, Specimen #2665* **CREEDITE – West Camp, Santa Eulalia, Chihuahua, Mexico. (MH)**

*Plate #13-16, Specimen #2741* **CREEDITE – West Camp, Santa Eulalia, Chihuahua, Mexico. (MH)**

## CREEDITE

**#2665 8.5×6.0cm Plate #13-15**

**West Camp, Santa Eulalia, Chihuahua, Mexico**

This superb specimen is a good violet color, showing virtually none of the paleness common to creedite. The entire face of the specimen is crystallized creedite with the supporting matrix completely hidden. The crystals are typical, sharply terminated, lustrous, and in diverging groups and sprays. They range in size from about 1.0cm up to 2.2cm. Dominating the central area of the specimen is a thick, multicrystal diverging spray containing the major crystals fanning out a remarkable 3.5cm. The good color, the large, prominent, crystal spray, and the completely covered surface make this one of the better creedites extant.

**#2740 12.0×8.0×8.5cm**

**Refer to Chapter 3, Plate #3-28**

Plate of very well formed individual lavender prismatic crystals to 3.5cm long with prominent dome terminations and minor subsidiary terminal faces. Exceptionally well crystallized and damage free. Barlow's best overall. (See Chapter 3, page 55.)

**#2741 10.0×7.0×4.0cm Plate #13-16**

**West Camp, Santa Eulalia, Chihuahua, Mexico**

Radiating groups of 1.2cm vibrant medium purple crystals with rust-colored tips, caused by a partial coating of iron oxide that was common on many specimens from these pockets.

**#2797 5.0×3.0×3.5cm**

**West Camp, Santa Eulalia, Chihuahua, Mexico**

"Bow-tie"-shaped group 2.5cm by 1.3cm composed of deep lavender crystals. The central crystals are 2.0cm long. Matrix is coated by radiating spheres and sprays of golden goethite crystals locally overgrown by limpid doubly terminated quartz.

## RHODOCHROSITE
### (MANGANESE CARBONATE)

The Main Silicate Orebody lies between the 10th and 11th levels of the Potosi Mine and is pierced by the #1 Shaft. This saucer-shaped orebody lies between the Chorro and Potosi Trends and is the largest of the calcic-iron skarn bodies at Santa Eulalia (Hewitt, 1968). It is the sole producer of well-crystallized rhodochrosite. The rhodochrosite occurs in the abundant vugs within the upper portion of the orebody and displays a wide variety of crystal habits and gemminess. The body is best known for quantities of medium-pink botryoidal masses and bright pink cloudy rhombohedra, as well as rarer sharp blood-red gemmy crystals. A wide variety of accessory microminerals occurs with the rhodochrosite including other carbonates, quartz, ore sulfides, huebnerite, helvite, ilvaite, pyrosmalite, argentopyrite, natanite, jeanbandyite, wickmanite, and fluorite. The fluorite displays a wide variety of forms and combinations of forms, making it a micromounter's dream.

John Barlow's Santa Eulalia rhodochrosites include one of the best specimens ever produced from this orebody.

## RHODOCHROSITE

**#1835 5.0×4.0×4.0cm Plate #13-17**

**Santa Eulalia, Chihuahua, Mexico**

This is a blood-red aggregate of transparent scalenohedral crystals with a central 3.5×2.0cm diameter crystal flanked by myriad parallel subsidiary crystals in crystallographic continuity. The specimen is virtually damage free and very lustrous. The terminations are multiple, sharp rhombohedra. The rich color, bright luster, and lack of damage makes this one of the best Santa Eulalia rhodochrosites. It was purchased from Jack Amsbury in 1978.

***Plate #13-17, Specimen #1835*** **RHODOCHROSITE – Santa Eulalia, Chihuahua, Mexico. (MH)**

**#3687 7.0×1.3×1.3cm Plate #13-18**

**Santa Eulalia, Chihuahua, Mexico**

This is a very elongate pinkish red scalenohedral crystal composed of parallel subcrystals with multiple rhombohedral terminations creating an overall trigonal cross-section. There is a dusting of tiny fluorite, calcite, and quartz crystals on the back. This specimen is one of the best described by Art Smith (1987) as "cathedral rhodochrosite." It was purchased from Alfonso Arriga in 1987.

***Plate #13-18, Specimen #3687*** **RHODOCHROSITE – Santa Eulalia, Chihuahua, Mexico. (MH)**

***Plate #13-19, Specimen #1521*** **RHODOCHROSITE – Santa Eulalia, Chihuahua, Mexico. (MH)**

**#1521 8.5×5.0×4.0cm Plate #13-19**

**Santa Eulalia, Chihuahua, Mexico**

Pink, radiating spheres to 2.0cm of rhodochrosite with multiple minute rhombic terminations with 2.5mm semi-equant balls or stout rhombohedra of dark blue-green pyrosmalite. Small, doubly terminated, limpid quartz crystals partially cover the pyrosmalite. These are the largest pyrosmalites known from Santa Eulalia.

## Specimens from Other Replacement Deposits

In addition to his extensive suite collections from the Santa Eulalia and Mapimi Districts, John Barlow has also obtained several excellent individual specimens from some of the other major replacement deposits, in particular, Concepcion del Oro, Zacatecas; Charcas, San Luis Potosi; and Naica and San Carlos, Chihuahua.

### AZURITE (COPPER CARBONATE HYDROXIDE)

**#4962 9.0×9.0×6.5cm Plate #13-20**

**Rancho Santa Rosa, Concepcion del Oro, Zacatecas, Mexico**

Azure-blue rosette of curved individual crystals. Superb example of this species from a replacement deposit.

***Plate #13-20, Specimen #4962*** **AZURITE – Rancho Santa Rosa, Concepcion del Oro, Zacatecas, Mexico. (MH)**

### DANBURITE (CALCIUM BOROSILICATE)

**#5293 9.0×7.0×6.0cm Plate #13-21**

**San Sebastian Mine, Charcas, San Luis Potosi, Mexico**

A divergent spray of milky white, "chisel-point" terminated danburites with a pale amethystine chalcedony coating and sparse, perched, 1-3mm, doubly terminated, twinned, scalenohedral calcites. The amethystine color is very unusual but is imparted by the coating, not the danburite.

***Plate #13-21, Specimen #5293***
**DANBURITE – San Sebastian Mine, Charcas, San Luis Potosi, Mexico. (MH)**

### DANBURITE

**#5370 15.0×13.0×15.0cm Plate #13-22**

**San Sebastian Mine, Charcas, San Luis Potosi, Mexico**

A cluster of about ten pale pink, "chisel-point"-terminated danburite crystals ranging from 3.0 to 6.0cm in length, surrounding a single prominent pale pink crystal 12.0cm long by 3.0cm across. Crystals are zoned from cloudy bases to gemmy tips. Specimen shows a coating of white, doubly terminated scalenohedral calcites to 5mm long. One of the better groups from the large 1993 find of pinkish danburites at the San Sebastian Mine.

***Plate #13-22, Specimen #5370*** **DANBURITE – San Sebastian Mine, Charcas, San Luis Potosi, Mexico. (MH)**

## GALENA (LEAD SULFIDE)

**#2850 4.5×3.0×2.0cm Plate #13-23**

**Naica Mine, Naica, Chihuahua, Mexico**

Flattened pseudohexagonal spinel twinned crystals to 3.0cm from Naica, another replacement deposit. It has excellent luster and none of the back-etching typical of most similar specimens from the deposit. Minor pyrite on matrix.

***Plate #13-23, Specimen #2850*** **GALENA – Naica Mine, Naica, Chihuahua, Mexico. (MH)**

## NATROLITE (SODIUM ALUMINUM SILICATE HYDRATE)

**#2200 8.5×6.5×3.0cm** **Refer Chapter 3, Plate #3-70**

**Surface near Aurora Mine, Charcas, San Luis Potosi, Mexico**

This 3.0cm diameter white natrolite sphere lies on a bed of 3-8mm calcite crystals lining a vesicle in basalt. There is a secondary partial 2.0cm sphere of natrolite. This is one of the best known natrolites from Mexico and is from an outcrop near the entrance to the famous Aurora Mine at Charcas.

***Plate #13-24, Specimen #5048*** **VANADINITE – Apex Mine, San Carlos, Chihuahua, Mexico. (MH)**

## VANADINITE (LEAD VANADATE CHLORIDE)

**#5048 2.5×0.7×0.7cm Plate #13-24**

**Apex Mine, San Carlos, Chihuahua, Mexico**

Typical tannish-orange hexagonal prisms to 1.5cm with white distorted rhombohedral calcite perched on top. Very nice small miniature, classic for the replacement deposit at San Carlos.

## Vein Deposits

### BARITE (BARIUM SULFATE)

**#3027 8.0×7.0×6.0cm Plate #13-25**

**Mina Santa Rita, Nieves District, Zacatecas, Mexico**

Single, simple, diamond-shaped crystal 8.0×7.0×2.5cm with 3.5cm high stack of smaller crystals piled on top. Crystals are basically white but have contrasting irregularly distributed patches of black jamesonite inclusions. A good example of this unusual occurrence.

*Plate #13-25, Specimen #3027* **BARITE – Mina Santa Rita, Nieves District, Zacatecas, Mexico. (MH)**

### GOLD (NATIVE ELEMENT)

**#5705 3.0×3.0×0.6cm**

**Refer to Chapter 7, Plate #7-14**

**San Julian Mine, Ramos, Chihuahua, Mexico**

This roughly square plate weighing 5.5g is composed of radiating fans of finely reticulated dendritic elongate octahedral gold crystals. A fine satiny luster is imparted by parallel growth lines of the dendrites. This piece was clearly removed from selvage within an epithermal vein, and tiny euhedral quartz crystals are locally lodged among the gold crystals. This old classic gold was in the Bally Collection and traded out of a German museum. It carries an old acquisition number (2838) and an exceptional catalog card handwritten in German indicating location, weight, and a general description. The card indicates the specimen was originally obtained in 1914 for 50 currency units (marks?). Although the Ramos District is not a name currently in use, the name "San Julian" and the dates of the specimen match well with the now defunct San Julian high-grade vein mining district that lies almost on the Chihuahua-Durango border southeast of the more famous Guadalupe y Calvo District.

*Plate #13-26, Specimen #2652* **QUARTZ, var. amethyst – Peregrina Mine, Guanajuato, Gto., Mexico. (MH)**

### JALPAITE (SILVER COPPER SULFIDE)

**#5336 5.7×4.7×4.5cm**

**Refer Chapter 10, Plate #10-32**

**Level 407, San Juan de Rayas Mine, Guanajuato, Gto., Mexico**

This is a world-class example of a rare silver-bearing mineral. It is a nearly solid mass of material crowned by several sharp discrete crystals, the largest being 2.5×2.1cm. The crystals are slightly lustrous with numerous growth hillocks on the faces. Certainly a premier example of a rare silver copper sulfide.

### QUARTZ, VAR. AMETHYST (SILICON DIOXIDE)

**#2652 16.0×15.0×6.0cm Plate #13-26**

**Peregrina Mine, Guanajuato, Gto., Mexico**

This is a deep royal purple plate of multiple prismatic 0.5-1.0cm amethyst crystals branching off central single crystals to 4.0cm long. The plate is a full floater with termination overgrowth on the back where it separated from the vein wall. This is a classic example from the district. Specimen #2653 in the collection, not described, is the twin of this group.

*Plate #13-27, Specimen #5808* **QUARTZ, var. amethyst – San Vicente Mine, Veta Madre, Guanajuato District, Guanajuato, Gto., Mexico. (MH)**

## QUARTZ,

**VAR. AMETHYST**

**#5808 6.0×6.0×4.0cm**
**Plate #13-27**

**San Vicente Mine, Veta Madre, Guanajuato District, Guanajuato, Gto., Mexico**

Roughly triangular slab covered by 1.0cm purple crystals with central pair of 3.5cm lavender "rabbit ears." Most of the larger crystals have 1-3mm purple secondary crystals growing perpendicular to the prism faces. This is a very good minature from this prolific amethyst locality.

### Additional Species

## ANDRADITE

**(CALCIUM IRON SILICATE)**

**#5286 8.0×8.0×5.0cm Plate #13-28**

**Mina La Linda Prieta, Julimes District, Chihuahua, Mexico**

Mound-shaped group of extremely lustrous, jet-black sharp dodecahedral crystals ranging from 1.0 to 3.5cm across. Largest crystal faces have well-developed growth spirals and attractive 3mm wide beveled edges that represent negative dodecahedra. This specimen is probably the best ever to come from the small open pit mine operated by Benny Fenn. John bought the specimen from Benny and Elva Fenn at the 1996 Tucson Show. This is an extremely aesthetic, undamaged specimen and ranks as one of the world's best garnets.

*Plate #13-28, Specimen #5286* **ANDRADITE – Mina La Linda Prieta, Julimes District, Chihuahua, Mexico. (MH)**

*Plate #13-29, Specimen #5417* **SANIDINE – Mina El Pili, Sierra Tortuga, Naica Region, Chihuahua, Mexico. (MH)**

## SANIDINE
**(SODIUM POTASSIUM SILICATE)**
**#5417 5.5×5×4.5cm Plate #13-29**

**Mina El Pili, Sierra Tortuga, Naica Region, Chihuahua, Mexico**

Blocky group of champagne-colored thick tabular monoclinic prisms with minor step faces and small hornblende intergrowths. Uppermost faces show remarkably strong blue-white "schiller" or moonstone effect. This effect is visible on all but the side faces. These are some of the largest undamaged crystals recovered from this small underground locality by Benny Fenn in 1994. Purchased from Benny and Elva Fenn at the 1995 Tucson Show. (Dr. David London did the compositional analysis.)

## JAROSITE (POTASSIUM IRON SULFATE HYDRATE)
**#4911 9.0×8.0×3.5cm; 2.5×2.0cm crystal Plate #13-30**

**Sierra Peña Blanca Uranium Mine, near Aldama, Chihuahua, Mexico**

Flat plate of dark amber-colored hexagonal platy crystals forming partial rosettes. Largest crystal is 2.0cm across and nearly 1.0cm thick. These are uncommonly well crystallized and among the largest known examples of this common mineral. This specimen is one of the better specimens collected from dump boulders by Dr. Philip C. Goodell before the dumps were turned. Obtained from Cal Graeber in 1993.

*Plate #13-30, Specimen #4911* **JAROSITE – Sierra Peña Blanca Uranium Mine, near Aldama, Chihuahua, Mexico. (MH)**

## TOPAZ (ALUMINUM SILICATE FLUORIDE HYDROXIDE)

**#2856 4.0×3.0×2.5cm Plate #13-31**

**San Luis Potosi, Mexico**

Excellent example of crystals that occur abundantly in pockets in rhyolite flow dome complexes in west-central San Luis Potosi state and eastern Zacatecas. The specimen is a medium brown, 2.0cm long, by 1.0cm across, prismatic gemmy crystal showing well-developed pinacoidal termination. A 1.6cm "brush" of hyalite opal over rutile needles sprouts off the upper flank of crystal. There is a smaller topaz crystal at the base. This is a good example of the associations here.

***Plate #13-31, Specimen #2856*** **TOPAZ – San Luis Potosi, Mexico. (MH)**

## TOURMALINE, VAR. SCHORL (SODIUM IRON ALUMINUM BOROSILICATE HYDROXIDE)

**#5590 12.0×9.5×5.5cm Plate #13-32**

**Municipio de Santa Cruz, Sonora, Mexico**

On a flat plate of bright smoky quartz crystals to 1.5cm sit two beautiful black schorl crystals. The largest crystal is flat lying and measures 8.0×6.0cm. It is sharply terminated. The termination faces are mirror smooth, while the prism faces are a myriad of fibrous striations giving them a rich, yet lustrous, velvet effect. Such tourmalines were, in fact, called velvet tourmalines when being mined three decades ago. The second crystal is stubby and nearly equant, measuring 3.0×4.0×3.5cm. Its sharp terminations are mirror smooth, while the prism faces show the typical velvet sheen. These tourmalines come from a deposit also noted for gemmy yellow to honey-colored scheelites to 7.0 or 8.0cm in size and cream colored apatites 1cm across and up to 10cm long. It was purchased from Jendon Minerals.

***Plate #13-32, Specimen #5590*** **SCHORL (tourmaline) – Municipio de Santa Cruz, Sonora, Mexico. (MH)**

## SELECTED REFERENCES


Castro-Carillo, M (1980) Monografia del Estado de Durango, *Tipo Gobierno de Durango*, 236 pp.

Dunn, P. J. (1983) Lotharmeyerite, a new mineral from Mapimi, Durango, Mexico, *Mineralogical Record*, 14:35-36.

Hadley, P. L. (1975) Mining and society in the Santa Eulalia mining complex, Chihuahua, Mexico 1709-1750, Ph.D. dissertation, University of Texas, Austin, Texas, 365 pp.

Hewitt, W.P. (1943) Geology and mineralization of the San Antonio mine, Santa Eulalia mistrict, Chihuahua, Mexico, *Geological Society of America Bulletin*, 64:173-204.

Hewitt, W.P (1968) Geology and mineralization of the Main Mineral Zone of the Santa Eulalia district, Chihuahua, Mexico, *American Institute of Mining Engineers, Trans.*, 240:229-260.

McLeroy, D.F., R.F. Franquesa, and M.S. Romero (1986) Origin of the breccia pipes and mantos of the Ojuela lead-silver district, Mapimi, Durango, in K.F. Clark, P.K.M. Megaw, and J. Ruiz (Eds.), *Lead-Zinc-Silver Carbonate-Hosted Deposits of Northern Mexico, Society of Economic Geologists Guidebook, Nov. 13-17, 1986,* pp. 153-168.

Megaw, P.K.M. (1986) *Mineralogy of the Santa Eulalia mining district, Chihuahua, Mexico, Rochester Academy of Science, 13th Mineralogical Symposium 1986*, pp. 21-31.

Megaw, P.K.M. (1990) Geology and geochemistry of the Santa Eulalia mining district, Chihuahua, Mexico, unpubl. Ph.D. dissertation, University of Arizona, Tucson, 463 pp.

Megaw, P.K.M., J. Ruiz, and S.R. Titley (1988) High-temperature, carbonate-hosted, Pb-Zn-Ag massive sulfide deposits of Mexico: An overview, *Econ. Geol.*, 83.

Modreski, P.J. and D. Newsome (1984) Green uranium-activated fluorescence of adamite from the Ojuela mine, Mapimi, Mexico, *FM-MSA-TGMS Joint Symposium, Minerals of Mexico, Tucson, AZ, Feb. 12, 1984*, p. 7.

Panczner, W.D. (1983) Notes from Mexico, *Mineralogical Record*, 14:169-172.

Panczner, W.D. (1987) *Minerals of Mexico*, Van Nostrand Reinhold Co., New York, 459 pp.

Prescott, B. (1916) The Main Mineral Zone of the Santa Eulalia district, *Am. Inst. Mining Eng. Trans.*, 51:57-99.

Prescott, B. (1926) The underlying principles of the limestone replacement deposits of the Mexican province, *Eng. Min. Jour*. 122:246-253.

Smith, A. (1987) New cathedral rhodochrosite from Santa Eulalia, *Mineral News*, 3(11):3.

Stone, J.G (1956) Ore genesis in the Naica district, Chihuahua, Mexico, *Econ. Geol.*, 54:1002-1034.

Villarello, J.D. (1906) Le Mineral de Mapimi, XXth International Geologic Congress, Mexico, Guide to Excursion 27, 18 pp., 2 plates.

Wilson, W.E. (1982) What's new in Mexico?, *Mineralogical Record*, 13:181-183.

Chapter 14

# Red Beryl

## The Harris Mine, Wah Wah Mountains, Beaver County, Utah

*by F. John Barlow*

### The Harris Mine

*(The following is based on an article by F. John Barlow that appeared in* Lapidary Journal, *March 1979; the text has been edited to update it.)*

The Wah Wah Mountains in southwestern Utah are not a well known range, but like all of the western ranges following the Continental Divide from Alaska to Mexico, they have many things in common. Having spent many moons climbing the high country of Alaska, British Columbia, Alberta, Montana, Colorado, Wyoming, California, and Mexico hunting wild game, I've come to appreciate and respect the beauty and adventure the western ranges hold.

To tell the truth, I didn't get to meet the Wah Wahs in person until early in 1978. Of course, I had heard of them. My atlas has a circle around them that dates back to 1974.

It was in 1974 that I bought a suite of exceptionally fine thumbnail-sized minerals from Bill Larson's collection. You guessed it, one specimen was a red beryl, two tiny crystals on gray rhyolite. Nothing unusual, yet two things impressed me. The crystals were very small. They were so small you could hardly see them (a slight exaggeration). If the crystals hadn't been so bright, the specimen might have gotten lost in the rock box. The second thing that impressed me, and the real reason I've still got it, is that it carried Bill's low catalog #47, and was labeled "self-collected in 1968." That made it special.

With this introduction to red beryl, I had to find out the size of the largest known crystal and where it was. My instincts told me not to check with one of the mineral dealers, many of whom I didn't know at that time anyway. So I checked with the Smithsonian. Sure enough, they had perhaps the biggest and best red beryl at that time. It was a single crystal about 1.1×1.6cm on rhyolite, a loose single with fine termination and good color. Now, I felt I was an "expert." I had the red beryl bracketed from the smallest to the largest.

As collectors, we are continually telling each other, "It's a small world." You may also know red beryl is very rare in gemmy, large crystal form. So within months, in the very small world of the red beryl collecting, would you believe the second best crystal on matrix would appear before me? Dave Wilber, noted collector, was willing to part with it, so I acquired it from him for many times what he had paid for it. He was happy and so was I, even though I knew what he paid for it. I was now sure the red beryl portion of my collection was secure and set. I would discover later that part of the learning curve is that we should not jump to such conclusions. It's hard to forget how rare bixbyites, Hotazel rhodochrosites, legrandites, and "BSLB" (before Swoboda, Larson, Baja) boleites had been. I am sure you can add to the list of once rare species that now are available in some abundance. Maybe someone is about to bring out a four or five inch jeremejevite, but don't plan on it.

Be that as it may, 1975, and then 1976, saw a flood of red beryl crystals by the bagful. It looked like Mother Earth was heaving red beryl. But, alas, nothing of any consequence showed up. The crystals were small, few were matrix specimens, and most of all, the color was horrible due to rhyolite inclusions. These were nothing like I had come to know from the Wah Wahs. There was a reason. These new crystals were not from the Wah Wahs, but from the nearby Thomas Range. We were unimpressed and disinterested.

*Thinking I had spotted a faceted red beryl, I asked if they had cut any. They displayed a few very small 1/2 carat and under, very fine red gems. Then Rex said, "Would you like to see the world's largest gem red beryl cut stone?" I said yes, and when I saw it I flipped. You know what happened. Complete strangers in a half lit room, we struck a deal. The specimens and the world's best faceted red beryl were added to the Barlow Collection. Remember that this was in 1977.*

About this same time, in late 1976 and early 1977, things were happening over in the Wah Wahs. A location at the south end of the range, known to only a few, above the junipers and in the enchanting, scented piñons at about 7500 feet, was being investigated. This location is now known as the Harris Mine, source of the world's finest red beryls.

To me, the real excitement of this story started the night before the 1977 Tucson Show in a dealer display room just being set up in the Marriott Hotel, Tucson.

Arriving to check in about 8:00 PM, I ran into my friend, Jim Honert, of Black Hills Minerals. Jim said, "John, have you seen the new Wah Wah red beryls?" I said, "No," so Jim said, "Go up to room 409; the Harris brothers are here from Delta, Utah. Don't waste any time — no one has seen them." Jim didn't know it, but he was reawakening an old love affair. He was lighting the fuse.

My grandson, Greg Helein, was with me, and I said, "Greg, let's go. We'll check in later." When we got there, they were setting up in a half-lighted room. I knew Bob Harris very slightly from his reputation for digging and selling trilobites. He sells, under the name "Bug House," more trilobites internationally than anyone I know. I had never before met Ed or Rex.

When I asked about the red beryls, they looked at me a bit quizzically and dug into a box under the table. Out came three specimens on matrix and out went the lights. Literally. Fuses were blowing and the lights were going on and off while I was trying to see the red beryls. If the lights had stayed on, Ed and Rex might have asked twice the price, and I might not have bought them. I do have to say, though, knowing what I did about red beryl, I didn't need much light to tell me I was looking at the best representative red beryl specimens in existence to that date.

Thinking I had spotted a faceted red beryl, I asked if they had cut any. They displayed a few very small ½ carat and under, very fine red gems. Then Rex said, "Would you like to see the world's largest gem red beryl cut stone?" I said yes, and when I saw it, I flipped. You know what happened. Complete strangers in a half lit room, we struck a deal. The specimens and the world's best faceted red beryl were added to the Barlow Collection. Remember that this was in 1977.

This initial meeting has developed into one of the most pleasant friendships of my rock and mineral experience. The Harris brothers are my kind of guys. They're straight up, smart, honest as the day is long, and, in my book, they have their feet on the ground and their heads screwed down and on straight. I have helped them all I can, and they have been great to me.

All of the fine red beryl specimens and gem cutting material have come from two claims, No. 6-7 and 8, together known as the Violet Mine, Beaver County, Wah Wah Mountains, Utah. The mine name has since been changed to the Harris Mine. My grandson, Greg, and I have mined with Rex Harris and his two sons on several occasions. The boys, Steve and Gary, are fine hardworking kids like their father, and I would classify them as the "All American Boy types." The

Harrises all live in Delta, Utah, an arduous 125 mile trip from the mine.

Driving southwest out of Delta on Highway 275 on the way to Milford, you pass through a sleepy little hollow that you quickly realize from the dilapidated buildings to be an old village. This is the town of Deseret, which is one of the earliest old Mormon towns. It dates back to about 1850. Passing through this southwestern desert country, as you would expect, the sun usually shines, the temperature is warm to hot most of the time, and they say the wind can really rip up a howl. Yet, all our trips have been pleasant, and both Greg and I enjoyed riding to the mine with Rex and Ed that first time.

On the drive to the mine, I was attracted to the great steam clouds rising up from the hills to the east of us. These are geothermal wells that have been drilled by both Getty Oil and Phillips. Presently, they are building a 52 megawatt electric generating plant and have plans for 10 or 12 more. Before turning west into the south end of the Wah Wahs, you pass through Milford, an old mining town, with a population of approximately 1200 today. Milford dates back to 1855, when mining was started in earnest. When the Union Pacific Railroad came through, Milford became a crew change and watering stop. I believe even today approximately 100 crew members live there enjoying a quiet, beautiful little desert town. A warning though — don't break the speed limit.

Out of Milford you travel an unpaved gravel road to the foothills. It's a typical sandy but good (passenger car travel) desert road, what we would call in Wisconsin a third class road. If you know where the mine is, you may be able to see the white diggings as you start into the hills. Once in the foothills, a jeep or a four-wheel drive pickup is a necessity.

Originally, the Harrises were thinking of opening a part of the mine to the public on a daily fee basis, but sale of the property canceled that.

I have heard a great dispute took place as to who owned the Harris Mine, made up of eight claims known as the "Violet Mining Claims 1 through 8," in Beaver County. The dispute has been settled in court without bloodshed. However, unless the records have been purged, one might find a report that one party meant business. I understood he carefully guided two .357 slugs within three inches of a disagreeing partner's ear. This settled a lot of rhyolite dust. The writer, not having been present, reports this heresay, which may not resemble the truth.

If you are going to try a little mining on your own at the Violet (now Harris), you'd better get permission first. It might prevent that sickening feeling one gets when he hears that high changing pitch of a ricocheting slug. That would be the bad news. The good news would be that you heard the zoom.

The claims are in very hard, white to gray rhyolite. With the red beryl seemingly randomly scattered through the rhyolite, it's some hit, but mostly miss, digging. There appears to be no pattern or known method of tracking the deposits. The crystals in the upper mine are more numerous but tend to be darker. They are not, however, as dark or as poor quality as the Thomas Range red beryl I have seen.

The crystals in the lower mine are fewer in number but tend to be larger and exhibit the bright gemmy, raspberry color. This is where the very fine specimens in the Barlow Collection came from. The crystal #1630, Plate #14-9, was pictured on the cover of *Lapidary Journal*, March 1979 and *Mineralogical Record*, September-October 1979, I considered it to be the best specimen in the world, [at the time, 1979].

The largest specimen ever found to date (1979) #1878, Plate #14-5, is a Harris specimen,

*I haven't forgotten the dream in that pick-up on the way to the Wah Wahs. I'll tell you a secret! I had another one, too! (Are you listening, Ed, Rex, Steve, and Gary?). I dreamed it in my mineral room. I dreamed of a rough crystal that cut a superb gem. The crystal was 2 inches down the c-axis and 1/2 inch across, doubly terminated on pure white rhyolite. The cut stone was a 4 1/2 carat as gemmy as my 2.42 carat or just a little cleaner. Did you hear all of this, Rex and Ed? When will you make my dream come true?*

doubly terminated, measuring 13.0×26.0mm, on a small piece of matrix. It is without doubt the best red beryl thumbnail specimen in the world, as of 1979. I have perhaps the second, third, and fourth largest specimens on matrix. The largest faceted stone is a 2.93ct, which went to a collector in Germany. The second largest, 2.42ct, is in my collection [1979].

When Greg and I drove to the mine the first time in 1978 with Ed, Rex, Gary, and Steve, I thought, "Boy, here we are, right in the 'pocket'." On the way down, with Ed telling all the Mormon history of the country, I didn't hear much of what he was saying. My mind was wandering. I was thinking of the monster gemmy crystal on matrix I was going to find. I was thinking of Greg and me mining with Bill Larson in the Amelia Mine in Santa Rosalia, Baja, Mexico, and then to the "Fourth of July Pocket" of rubellite tourmaline in 1974 when I was underground for three days opening pocket after pocket in the Tourmaline Queen Mine with Bill Larson, Ed Swoboda, and Pete Bancroft. That was the second largest rubellite find of the century, at the time. It was the time I found and brought out the big "Barlow Buster," #922, Chapter 4, Plate #4-7. My mind was really turning at high rpm, and I wasn't hearing anything Ed was saying. While I was daydreaming of bringing out the biggest, gemmiest, clear perfect raspberry crystal on matrix sitting up doubly terminated, Ed hollered, "Look at the coyote." I was so far away mentally I just caught a glimpse of a shaggy looking thing hightailing it across the desert. In the confusion, I thought I heard Ed say "Benduki Pace Pace" which in Swahili means gun quick. Ed kept on talking, Greg was falling asleep, and the Great Beryl Hunter was back to reality. We were heading into the hills now and even though I was back to earth, I still believed that dream.

Hours later, when we had to head back, it had been a hard day but one of the great days you never forget. The dream didn't come true, and I'll tell you why. I knew the rhyolite was hard because I tried and did trim that Wilber piece I told you about earlier. But I never realized how really darned hard it was to break up the blasted boulders, darn near killing yourself and then finding nothing or maybe a spot where a crystal could have been formed. Mine production runs about one good quality crystal in each two to three tons of rhyolite and maybe one or two fine collector specimens per month. The rhyolite must be drilled and blasted with small amounts of dynamite. The boulders are then wedged and pried out with bars and then broken up with hammers. We learned in a hurry on that first trip why there are so few good red beryl specimens around and why they cost so much. To me, they are worth it. Very few crystals are faceting quality and most of those found have been cut by owner, Ed Harris, and Rex's daughter, Tina Nielsen.

I haven't forgotten the dream in that pick-up on the way to the Wah Wahs. I'll tell you a secret! I had another one, too! (Are you listening, Ed, Rex, Steve, and Gary?). I dreamed it in my mineral room. I dreamed of a rough crystal that cut a superb gem. The crystal was 2 inches down the *c*-axis and 1/2 inch across, doubly terminated on pure white rhyolite. The cut stone was a 4 1/2ct as gemmy as my 2.42ct or just a little cleaner. Did you hear all of this, Rex and Ed? When will you make my dream come true?

The question is — did Rex Harris, Ed, or Gary ever fulfill John's dream? Not really, but in the 18 years since the "dream," the Harris Mine has produced larger and gemmy crystals and many, by comparison, awesome. Where are these finest specimens today? Many are described on the following pages, for, the main red beryl suite in the Barlow Collection consists of 14 specimens now. Each has a reason for being in the suite. Rex Harris has seen perhaps all the red beryl specimens from the Harris Mine. In his estimation, and in Barlow's eyes, the Barlow suite is the finest anywhere in the world. *R.W.J.*

***Plate #14-1, Specimen #5330*** **BERYL, var. red – Harris Mine, Wah Wah Mountains, Beaver County, Utah. (MH)**

### BERYL, VAR. RED (BERYLLIUM ALUMINUM SILICATE)

**#5330 4.7×2.7cm Plate #14-1**

This is the largest crystal on matrix known to the mine owner, Rex Harris, as of 1994. The crystal is 3.6cm long and 1.5cm wide. The color is good, not gemmy due to included rhyolite. Some of the crystal faces show contact marks from the enclosing, since removed, rhyolite. It is, however, an awesome sized crystal sitting across a small pedestal of matrix.

**#4771 6.0×2.7×2.6cm Plate #14-2**

**Refer to Chapter 3, Plate #3-11**

This is a superb matrix specimen with a dominant cherry-red single crystal, 2.7×1.1cm, interconnecting with a cluster of nine similarly colored red beryl crystals. The color is superb. The specimen is considered the number one red beryl in the world as of 1994.

***Plate #14-2, Specimen #4771*** **BERYL, var. red – Harris Mine, Wah Wah Mountains, Beaver County, Utah. (MH)**

**#5331 7.0×6.5×5.5cm Plate #14-3**

This superb specimen called the "Regent," is the best of the best. It is perhaps the finest large (2.8×1.5×1.5cm) doubly terminated crystal on a rhyolite matrix found to date. Major portions of the crystal are gem clean.

***Plate #14-3, Specimen #5331***
**BERYL, var. red – Harris Mine, Wah Wah Mountains, Beaver County, Utah. (MH)**

**#4772 2.8×1.3cm**
**Plate #14-4**

This choice single crystal, doubly terminated, rests horizontally on a ball of matrix containing minor red beryls. The dominant crystal is 2.8cm long and of a rich violet-red color. Small gemmy areas may be seen at each termination. This specimen is an outstanding crystal superbly displayed, one of the top three in the collection.

***Plate #14-4, Specimen #4772*** **BERYL, var. red – Harris Mine, Wah Wah Mountains, Beaver County, Utah. (MH)**

*Plate #14-5, Specimen #1878* **BERYL, var. red – Harris Mine, Wah Wah Mountains, Beaver County, Utah. (MH)**

### The "Sinkankas Red Beryl"

**#1878 2.6×1.7×2.0cm Plate #14-5** **Refer to Chapter 3, Plate #3-12**

A superb red beryl crystal on matrix, the number one thumbnail in existence as of 1980. A stunning, gemmy, bright, intense fiery red, doubly terminated 2.7×1.3×1.3cm crystal on small ashen gray rhyolite. A very famous specimen.

When John Sinkankas asked if he could paint my red beryl #1878 because he might want to put it in his new book, *Emerald and Other Beryls* (1981), I agreed. When I found he did include it in his book (page 174-2), I was elated.

After the book was published, I asked John each year from 1982 to 1990 if I could acquire his original watercolor painting. Each year he kindly said no, he never sells his paintings. The last time I asked at Tucson, 1990, he said, "Maybe, John, I could do this to keep it with the specimen but under certain conditions." Finally, in early August 1990, I received his bill of sale, Plate #14-6, with his original watercolor, Plate #14-7.

As this book was being considered, I wrote John for his permission to use it. He agreed, and I'm most grateful to him.

JOHN SINKANKAS
CAPTAIN, U.S. NAVY (Ret.), Ph.D.

5372 VAN NUYS COURT
SAN DIEGO, CA 92109
Telephone: (619) 488-6904

BILL OF SALE

August 8, 1990

SOLD TO: F. John Barlow
Earth Resources
P.O.Box 567
Appleton, WI 54912

The original red beryl watercolor painting used as an illustration in my Emerald and Other Beryls, painted upon illustration board, the latter measuring 10 x 7 1/2 inches, and the actual painting measuring approximately 2 3/4 x 2 1/4 inches.

As mutually agreed upon in the future, all rights to reproduction in any literary production are retained by the painter.

The sum of $500.00 in the form of personal check No.1234, 79-86/759, dated August 3, 1990, to pay in full is herewith acknowledged.

John Sinkankas

**Plate #14-6 Sinkankas bill of sale of his watercolor of red beryl, Barlow specimen #1878. (MH)**

***Plate #14-7*** **Sinkankas watercolor of red beryl, Barlow specimen #1878. (MH)**

**#5332 2.8×1.4cm Plate #14-8**

This specimen consists of two nearly parallel growth crystals slightly offset along the long axis. Each crystal is 2.6cm long. A third, small, crystal penetrates from the side. The crystals are 80% exposed, sharp, and a rich violet-red color.

***Plate #14-8, Specimen #5332*** **BERYL, var. red – Harris Mine, Wah Wah Mountains, Beaver County, Utah. (MH)**

***Plate #14-10, Specimen #4787*** **BERYL, var. red – Harris Mine, Wah Wah Mountains, Beaver County, Utah. (MH)**

***Plate #14-9, Specimen #1630*** **BERYL, var. red – Harris Mine, Wah Wah Mountains, Beaver County, Utah. (MH)**

**#1630 1.0×.80cm. Plate #14-9**

Bright, gemmy, excellent colored crystal on 4.2×4.0cm matrix, pictured on *Lapidary Journal* cover, March, 1979 and *Mineralogical Record* cover, September, 1979.

**#4787 12.0×10.5×4.0cm Plate #14-10**

This is the largest matrix specimen in the red beryl collection. It contains seven prominent red beryl crystals up to 2.3cm in length arranged in three distinct groupings, expertly exposed. The main crystal of the central cluster has numerous smaller crystals, intergrown with and attached to it on an enclosing rhyolite matrix. Nearby, two single crystals are well exposed, the entire effect being a very pleasing and an uncommon matrix grouping. The major crystal is doubly terminated, measures 2.4cm, and is of excellent color.

**#4930 2.6×1.3cm Plate #14-11**

This slender, doubly terminated crystal is 2.7cm long and sits right up on a piece of matrix. The color is superb. Near the center of the crystal a collar of rhyolite divides the prism into two sections — clearly a re-healed crystal.

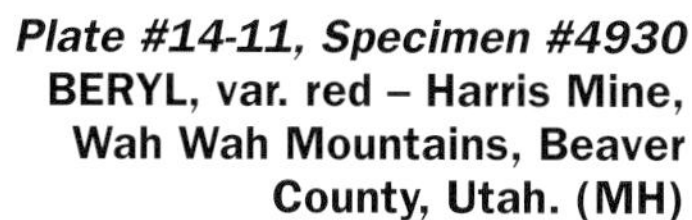

***Plate #14-11, Specimen #4930*** **BERYL, var. red – Harris Mine, Wah Wah Mountains, Beaver County, Utah. (MH)**

**#3121 7.9×6.0×5.0cm Plate #14-12**

A small, solid, hard rhyolite boulder that, when opened, shows a 1.3×1.0cm terminated crystal with a perfect cast of that crystal on the opposite half of the rock. This specimen is scientifically important as it raises the question, "Does red beryl form in solid hard rhyolite or only along soft seams of clay?"

***Plate #14-12, Specimen #3121*** **BERYL, var. red – Harris Mine, Wah Wah Mountains, Beaver County, Utah. (MH)**

As of this writing, Kennecott Copper Co. has acquired a lease on the Harris Mine properties to explore the gem potential of the deposit. The Harrises are still mining though private collecting is not possible.

## REFERENCE

Sinkankas, J. (1981) *Emerald and Other Beryls*, Chilton Books Co., Radnor, Pennsylvania, 664 pp.

# CHAPTER 15

# Phosphates

## The Tip Top Mine, Custer County, South Dakota

*by Martin Jensen*

The Barlow Collection contains a number of exceptionally fine examples of various rare crystallized secondary, hydrated phosphate minerals from the Tip Top Mine, Custer County, South Dakota. This locality, although not extensive, has produced perhaps the finest known examples, both in large crystals and specimens, of minerals such as jahnsite, leucophosphite, montgomeryite, robertsite, and tavorite. In addition, the deposit has produced approximately 90 different mineral species and is the type locality for ten new species (ehrleite, fransoletite, jahnsite, pahasapaite, parafransoletite, pararobertsite, robertsite, segelerite, tinsleyite, and tiptopite). Because of such significant occurrences, the Tip Top Pegmatite must be ranked as a famous and scientifically important mineral locality, even though a large portion of the species occur in small crystals or microcrystals.

Situated in the very scenic southern Black Hills of western South Dakota, the Tip Top Mine is located a short drive south of the town of Custer, in Pleasant Valley, a very aptly named green meadow with a flowing stream flanked by pine forests. The pegmatite proper is one of many large Precambrian pegmatites emplaced within even older (2.5 billion years) schists and phyllites. These pegmatites typically consist of large "rolls" or egg-shaped pods, about 100 m in diameter, connected edge to edge, in an en echelon manner. These rolls are zoned in the classic sense and commonly contain a core enriched with beryl, triphylite, columbite-tantalite, and a few sulfides. The pegmatites have been commercially worked over the last century for products such as beryl, feldspar, scrap mica, columbite-tantalite, spodumene, and quartz. Mines such as the Hugo, Peerless, Etta, Ingersoll, Bull Moose, Helen Beryl, Tin Mountain, White Elephant, and Tip Top have all produced commercial quantities of these types of pegmatitic minerals at one time or another. Today, however, all mining has essentially ceased, due in large part to depletion of the deposits or poor economic conditions.

The Tip Top Mine had been worked by open-pit methods most extensively in the 1950s and 1960s, largely for its very clean, large masses of perthitic microcline feldspar and beryl. In 1981-1982, a local miner by the name of Vern Stratton dewatered the pit and began mining the remaining portion of the beryl-rich core, still exposed at the bottom. It was during this period of renewed activity that some extremely fine mineral specimens were discovered, some of which were acquired by John Barlow.

The author began visiting the mine on a regular basis in October, 1981, when Vern had begun to expose some odd, red, crystallized crusts on intergrown albite-muscovite matrix. These were found to be manganoan montgomeryite and were the finest examples of this species ever seen (see specimen #5191). Shortly thereafter, one of the blasts exposed a corroded and heavily fractured beryl crystal about 20.0×50.0cm in size. It was from this crystal that the secondary hydrated beryllium-bearing species tiptopite, fransoletite, and erhleite were recovered, all as crystals associated with red to yellow montgomeryite, white englishite, orange roscherite, and colorless whitlockite. These specimens were indeed exceptional for their variety of rare crystallized minerals and colorful associations.

The day Vern "loaded heavy" was perhaps the most significant day of the Tip Top Mine's history. He had drilled a long bore hole into the top of the pegmatite core above his cut and loaded it heavily with explosives. Taking his cigarette and climbing up on the boom of his track drill he lit the fuse, and we all stood back behind his white Ford pickup (its windows long ago all shattered and broken . . . ). A huge mass of triphylite had been exposed in this part of the deposit, and the blast brought down a tremendous amount of material (in addition to sending fly rock directly at us and all over the mine). After the fresh exposure stabilized and settled down, the next day proved to be almost unbelievable.

Vern continued to mine sporadically for another year or so, working further into the core. At this time, some very fine specimens were exposed, including columbite crystals to 6.0cm, euhedral vivianite crystals to 4.0cm, and large joint surfaces to 30.0cm heavily covered with crystallized colorless to purple-pink whitlockite.

In the blasted muck, there were literally tons of mineral specimens, consisting for the most part of the leucophosphite-hureaulite-robertsite-jahnsite assemblage, with lesser laueite, tavorite, switzerite, and messelite. While going through only one third of this muck pile (because of time constraints and lack of space in the vehicle to carry any more material), the hureaulite specimen in the Barlow Collection (#5190) was found. The entire muck pile was moved with the loader and buried for backfill elsewhere in the bottom of the mine the next day.

As if this exposure had not provided enough exceptional material, a couple of weeks later in early October, 1982, another blast revealed a large, fractured beryl crystal about 1.0×2.0 m in size. During early collecting in the exposed end of this beryl, I discovered many specimens of joint surfaces coated with olive-green mitridatite and minute, unknown, white crystals and sprays. These turned out to be the new species parafransoletite. The next blast into this beryl crystal yielded another round of exceptional specimens of the tiptopite-roscherite-englishite-montgomeryite assemblage, as well as purple glassy crystals of the new species pahasapaite. Specimens from this discovery are well represented in the collection.

Vern transferred his interest in the mine over to his associate, Dale Ludington, shortly thereafter, and Dale worked the small remaining portion of the core. A few interesting specimens were collected during this period (one day Dale had a 30 gallon drum nearly full of cobbed ferrocolumbite crystals), but for the most part, the best "ore" had already been mined out. Dale ceased pumping and mining after a couple of months in 1982, and the mine, once again, began to fill with water.

Today, the Tip Top is silent, having a still, green pool of water about 30 m deep inside it. The likelihood of renewed mining seems depressingly slim because of both engineering and "environmental" obstacles. On the other hand, unlike most of the other Black Hills complex pegmatites which have two rolls, the Tip Top probably has a third roll which is, as yet, untouched. One can only imagine the potential for beautiful specimens and perhaps other new species that exist in this unexploited part of the deposit.

From this important locality, the following minerals, not all described, are represented in the collection: bermanite, carbonate-hydroxylapatite, collinsite, dufrenite, englishite, ferrisicklerite, ferrocolumbite, fransoletite, frondelite, heterosite, hureaulite, jahnsite, laueite, leucophosphite, messelite, microcline, mitridatite, montgomeryite, muscovite, pyrite, quartz, robertsite, rockbridgeite, roscherite, schorl, sphalerite, switzerite, tavorite, tinsleyite, tiptopite, todorokite, triphylite, vivianite, and whitlockite.

While working on my Master's degree in Geology (1981–1984) at the South Dakota School of Mines in Rapid City, I came into contact with other mineral collectors in the area, including W. L. (Bill) Roberts, Pete and Neal Larson, and Bob Farrar. It was through these friends that I was able to meet John Barlow and converse about the crystallized pegmatitic phosphates from the Tip Top Mine. I had collected extensively at this deposit and was at the locality after almost every blast during its most productive period.

Because of John's interest and his active pursuit of minerals from the mine during that time, he obtained a large proportion of very fine and important specimens now preserved in his collection.

Seeing this material intact ten years later while participating in this project was like taking a step back in time. The Tip Top has long since closed and flooded and there is a total lack of specimens available for collection or field study, thus making the Barlow suite particularly significant.

**BERMANITE** (MANGANESE PHOSPHATE HYDROXIDE HYDRATE)
**#5213 6.0×4.0cm**

A fracture surface on triphylite with an almost complete covering of minute, rust-red scaly bermanite crystals on lustrous, colorless leucophosphite crystals, associated with a flat-lying, 1.0cm radiating spray of black rockbridgeite-frondelite. Impressive for the bermanite quantity and coverage, even though not exceptionally aesthetic.

**HUREAULITE** (MANGANESE PHOSPHATE HYDRATE)
**#5190 12.0×8.0×6.0cm Plate #15-1**

The piece exhibits a lustrous, translucent, deep pink, simple, monoclinic hureaulite crystal 1.4cm long perched on the very front of a vug 3.5cm across of dark brown, lustrous leucophosphite crystals in the center of a mass of greenish black triphylite. This hureaulite crystal is notable for its superb presentation, color, and luster and is probably one of the finest specimens of this species currently known, at least from the United States. From the Martin Jensen Collection.

*Plate #15-1, Specimen #5190* **HUREAULITE – Tip Top Mine, Custer County, South Dakota. (MH)**

**JAHNSITE** (CALCIUM MANGANESE IRON PHOSPHATE HYDROXIDE HYDRATE)
**#5218 7.5×6.0cm**

This hand specimen of massive, glassy quartz has a surface completely covered with a thick druse of lustrous, pale brown jahnsite crystals to 0.5mm. Several very large and lustrous brownish purple leucophosphite crystals (to 5mm) occur on the jahnsite. Although larger crystals of jahnsite were found at the mine, this piece is an excellent example for the deposit and the species.

## JAHNSITE

**#5240 7.0×3.0×3.0cm**

A hand specimen of massive glassy quartz with a flat fracture surface coated with carbonate-hydroxl-apatite and sprinkled with later jahnsite crystals and spherules of crystallized mitridatite. There are local, undamaged, lustrous, nut-brown jahnsite crystals to 5mm on this piece, making it one of the finest examples of this species known from the locality.

## LAUEITE (MANGANESE IRON PHOSPHATE HYDROXIDE HYDRATE)

**#5214 10.0×6.5×4.5cm**

A hand specimen of dark-brown ferrisicklerite/heterosite with local deep, orange-yellow, lustrous laueite crystals to 2mm associated with leucophosphite and botryoidal robertsite/mitridatite.

## LEUCOPHOSPHITE (POTASSIUM IRON PHOSPHATE HYDROXIDE HYDRATE)

**#5219 7.5×6.0×3.0cm Plate #15-2**

At one corner of this hand specimen is a cluster of very slightly dulled, monoclinic, brown leucophosphite crystals to 5mm (very large for the species) resting on a very thick druse of pale yellow-brown jahnsite and associated large clumps to 2.0cm of fine-grained robertsite.

## MESSELITE (CALCIUM IRON MANGANESE PHOSPHATE HYDRATE)

**#5239 3.4×3.0×2.2cm**

Lustrous, bladed, milky-white messelite crystals to 5mm thickly cover one side of the specimen and occur associated with small lustrous jahnsite crystals. An exceptionally fine miniature from the locality.

***Plate #15-2, Specimen #5219*** **LEUCOPHOSPHITE – Tip Top Mine, Custer County, South Dakota. (MH)**

***Plate #15-3, Specimen #5191*** **MONTGOMERYITE – Tip Top Mine, Custer County, South Dakota. (MH)**

## MONTGOMERYITE (CALCIUM MAGNESIUM ALUMINUM PHOSPHATE HYDROXIDE HYDRATE)

**#5191 22.0×13.0cm Plate #15-3**

An extremely rich, deep red crust of lustrous, radiating, montgomeryite crystals to 4mm, associated with crystallized robertsite, englishite, and whitlockite, cover a fracture surface 8.0cm across on intergrown albite-muscovite matrix. I am unaware of any better examples of this species in existence.

*Plate #15-4, Specimen #5192* **MONTGOMERYITE – Tip Top Mine, Custer County, South Dakota. (MH)**

## MONTGOMERYITE

**#5192 10.0×10.0×5.0cm**
**Plate #15-4**

The surface of this hand specimen of albite-quartz is heavily covered with radiating sprays to 4mm of lustrous, deep red montgomeryite crystals associated with black robertsite, colorless whitlockite, and minor white englishite. This is an exceedingly fine specimen, being a close second to #5191 above. From the Martin Jensen Collection.

## ROBERTSITE

**(CALCIUM MANGANESE PHOSPHATE OXIDE HYDRATE)**
**#5220 7.5×6.0×6.5cm**

This hand specimen of dark-brown ferrisicklerite-heterosite-quartz matrix has a large vug 5.0cm across totally lined with a thick crust (5mm) of lustrous, black robertsite crystals. Minor associated leucophosphite and carbonate-hydroxlapatite are present. Certainly one of the top five examples of the species known.

## ROCKBRIDGEITE

**(IRON MANGANESE PHOSPHATE HYDROXIDE)**
**#5221 13.0×9.0×6.0cm**

Noteworthy for the size of its flat-lying, silky, dark green-black, lustrous rockbridgeite crystals, some up to 12.0cm, the piece is a split-apart fracture surface of a matrix of solid triphylite/rockbridgeite.

## ROSCHERITE

**(CALCIUM MANGANESE BERYLLIUM PHOSPHATE HYDROXIDE HYDRATE)**
**#5188 7.0×4.5cm**

This small hand specimen of beryl has a fracture surface about half covered with large, lustrous, black roscherite crystal clusters to 2mm, which are associated with yellow montgomeryite crystals and brownish mitridatite.

## SWITZERITE

**(MANGANESE IRON PHOSPHATE HYDRATE)**
**#5196 8.0×6.0×5.0cm**

This specimen is a large spray 14mm long of silky, cream-white, lath-shaped switzerite crystals in a vug of dark leucophosphite crystals. One of the top three switzerite specimens from the locality.

## TAVORITE

**(LITHIUM IRON PHOSPHATE HYDROXIDE)**
**#5217 8.0×10.0×8.0cm**

A large specimen of solid, green-black massive rockbridgeite and triphylite with a large vug 6.0×6.0cm across containing abundant crystallized pink-brown hureaulite, brownish leucophosphite, and yellow-green tavorite. Crystals of the latter species from anywhere are exceptionally rare; although small and slightly coated with minute, rust-red bermanite, they are abundant on this specimen.

## TIPTOPITE (LITHIUM POTASSIUM SODIUM CALCIUM BERYLLIUM PHOSPHATE HYDRATE)

**#5193 11.0×5.0×7.0cm**

This is a large hand specimen of beryl coated with tiptopite, although the abundant spherules of red montgomeryite are distracting. Radiating white spherules to 8mm of tiptopite needles cover opposite surfaces of the piece in association with red montgomeryite crystals, small white englishite rosettes, orange-brown roscherite crystal clusters, small black todorokite spherules, and druses of colorless rhombohedral whitlockite crystals. This piece was found in October, 1981 and is very likely the best specimen recovered from that discovery.

**#5199 8.0×8.0cm**

Tiptopite crystals predominate on a jointed surface of beryl, although a few clusters of gray-white, bladed fransoletite to 2mm are present. Associated minerals include crystallized orange-brown roscherite, white englishite, whitlockite, pink montgomeryite, spherules of grayish carbonate-hydroxylapatite, and unknown white porcelaneous spherules on tiptopite crystals.

## WHITLOCKITE (CALCIUM MAGNESIUM IRON PHOSPHATE HYDROXIDE)

**#5189 9.0×6.0×4.5cm**

Whitlockite was locally abundant, as druses of lustrous, large crystals lining open fractures. Associated with carbonate-hydroxylapatite crystals, it occurred in blocks of massive gray quartz at the Tip Top. The color varies from colorless to purple-pink. This example exhibits a druse 3.0×4.5cm across of lustrous, purple-pink, simple rhombohedral crystals to 5mm. From the Martin Jensen Collection.

In addition to the extensive Tip Top suite, the collection contains one noteworthy specimen from another South Dakota deposit. The Richmond-Sitting Bull Mine is an old silver-gold property located in the northern Black Hills of South Dakota. The deposit consisted of oxidized ores hosted in flat-lying Cambrian sandstones and was mined through adits by room and pillar methods. At one small area deep in the mine, a vuggy zone was encountered that contained exceptional examples of crystallized secondary arsenate minerals, particularly pharmacosiderite, beudantite, tsumcorite, and scorodite associated with superb pyrolusite, manganite, and other manganese oxide species. Because of the localized nature of this occurrence, very little, if any, material reached the commercial specimen market. The best specimens remained in South Dakota, in the W. L. (Bill) Roberts (now deceased) Collection in Rapid City and in the Black Hills Institute of Geological Research collection, Hill City. The Barlow Collection has fine examples from this locality.

## PHARMACOSIDERITE (POTASSIUM IRON ARSENATE HYDROXIDE HYDRATE)

**#1208 3.0×3.0×2.5cm Plate #15-5**

**Richmond-Sitting Bull Mine, Black Hills, South Dakota**

This is a fine example of gray-green pharmacosiderite crystals locally overgrown with an unknown yellow mineral. A small miniature, but the pharmacosiderite may be one of the largest examples of the species from the locality and exceptional worldwide.

Realizing the temporary nature of active mines and having the enthusiasm to pursue significant specimens during times of operation when they might otherwise be lost, collectors such as John Barlow are able to make available at later dates material for examination, aesthetic appreciation, and/or scientific study. Luckily, a large fraction of the Tip Top mineralogy has been preserved in such a manner as a subcollection of the Barlow Collection.

***Plate #15-5, Specimen #1208*** **PHARMACOSIDERITE – Richmond-Sitting Bull Mine, Black Hills, South Dakota. (MH)**

# CHAPTER 16

# The Wisconsin Collection

*by Gene L. LaBerge*

## Introduction

Although Wisconsin is not widely known as a source of fine mineral specimens, the Barlow Collection contains impressive suites of superb minerals from several mines, as well as fine specimens from a number of localities.

The premier suite of Wisconsin minerals in the collection comes from the newly opened Flambeau copper-gold mine at Ladysmith in the north central part of the state. Serendipity played a major role in assembling this suite; first, an opportunity to visit the mine by Casey Jones and Jane Koepp, dealers and collectors, resulted in a contract that allowed them to save some of the superb crystallized specimens. Second, Barlow's contact with Jones, through his network of friends, enabled John to assemble an outstanding suite from this major new locality.

The other significant locality represented is the Montreal iron mine in far northern Wisconsin, which has been closed for more than 30 years. This suite was obtained from a former miner and collector who had the foresight to save high-quality mineral specimens when they were available from the mine. Minerals from other localities include the now-closed lead-zinc mines in southwestern Wisconsin, the Vulcan dolomite quarry in southeastern Wisconsin and individual specimens from quarries and outcroppings in other parts of the state.

## The Flambeau Mine

The collection of chalcocites and other minerals from the copper-gold mine at Ladysmith that opened in 1993 is the most significant suite of Wisconsin minerals in the Barlow Collection. The mine is located in northwestern Wisconsin about 300 km southwest of the famous Copper Country of northern Michigan. However, the deposit is not related to the Michigan copper deposits.

The mine is operated by the Flambeau Mining Company, a wholly owned subsidiary of Kennecott Copper Corporation, which, in turn, is owned by the RTZ Corporation. The orebody was discovered in 1968 by Bear Creek Mining Company, an exploration branch of Kennecott. The deposit was not exposed; it was discovered by drilling geophysical anomalies as part of a regional exploration program. It is located about 2 km south of Ladysmith adjacent to the Flambeau River, from which it takes its name.

Typical of a major collector, Barlow was first on the scene when specimens from the mine became available in the summer of 1994. World-class chalcocites were recovered by Casey Jones of Burminco, from Monrovia, California, who holds exclusive rights to collect mineral specimens from the mine. The Barlow Collection contains more than 50 outstanding chalcocites from the Flambeau Mine.

The deposit was brought into production in June, 1993, and mining will be completed in 1997.

Flambeau Mining Company spent more than $22 million on environmental safeguards prior to initiating mining operations. This is, perhaps, the most modern, environmentally safe mine in the world. The orebody is being mined as an open pit with a planned depth of about 100 m. Upon completion of mining operations, the pit will be backfilled and the site will be restored to its original configuration, complete with the development of a wetland. Therefore, no additional collecting at the site will be possible after mining operations cease.

The Flambeau Mine is a volcanogenic massive sulfide deposit formed about 1,850 million years ago when hot springs on the flanks of an ancient volcanic island brought metal-rich brines to the sea floor. The metals were precipitated as sulfide minerals including pyrite, pyrrhotite, chalcopyrite, sphalerite, minor galena, and traces of gold in a thin lens about 700m wide and 13.5m thick. The pancake-shaped lens was originally horizontal but was tilted to near vertical by mountain-building forces that occurred after volcanism. In this regard, the deposit is like hundreds of other orebodies from which copper, zinc, lead, and some silver and gold are produced. What makes the Flambeau Mine unique is that it contains a thick accumulation of secondary minerals produced by chemical weathering of the upturned edge of the orebody.

The secondary minerals have an interesting history, for they were formed during Late Precambrian time, some 600 to 1,200 million years ago when the deposit was subjected to deep chemical weathering. During this ancient period of Earth history, the area destined to become Wisconsin was located near the equator. Subsequent plate movements have shifted the continents so that Wisconsin is now in a more temperate climate zone. However, the secondary minerals in the Flambeau Mine almost certainly formed long ago in a subtropical climate with the water table well below the surface, not in the temperate climate that exists now. A major portion of the orebody escaped erosion during the invasion of the Cambrian sea, some 520 million years ago, when a layer of sandstone was deposited on the enriched ore. Much of this sandstone was subsequently eroded, but a small patch remained over the orebody to protect the secondary minerals from erosion when continental glaciers moved over the deposit about 20,000 years ago.

Chalcocite is, by far, the most important secondary mineral recovered from the Flambeau Mine. Like other orthorhombic minerals (such as barite), the chalcocite occurs in a wide variety of crystal forms, including slender or stout tabular crystals that may form singly or in intergrown clusters. Crystals range in length to 6.0cm and may form in distinctive "X" twins, or they may be disclike pseudohexagonal crystals. The most spectacular feature of the Flambeau Mine chalcocites, in addition to their large size, is the purple, gun-metal blue, reddish-bronze, brassy yellow, or iridescent patina that coats nearly all of the crystals. The patina is produced by a thin coating of bornite on the chalcocite. This bornite coating formed much later than the chalcocite, when the water table was much closer to the surface, similar to what it is today. Only rarely are the more typical lead-gray colored chalcocites encountered in the Flambeau Mine.

Several generations of chalcocites are present at the Flambeau. The earliest and most extensive chalcocite is massive fine-grained ore produced by enrichment of the primary chalcopyrite and pyrite. This variety consists of extremely fine grained "sooty" gray powder and somewhat coarser granular crystals that constitute much of the ore. This gray-black sandy chalcocite forms the matrix for the well-crystallized chalcocite. The ore contains numerous relatively open fractures and vugs (or pockets) in which the collector crystals formed. Clearly, the larger crystals have grown on the earlier granular chalcocite. Pockets of various size have yielded some remarkable crystals and crystal groups.

Another significant pocket, encountered on Friday, the 13th of October, 1995, was named by Jones the "Lucky Friday Pocket." Several large three-dimensional clusters along with numerous smaller groupings of brassy yellow chalcocites with a pseudohexagonal habit were recovered.

The most spectacular pocket was encountered in the late afternoon on February 1, 1995 during drilling for blasting on the 1010 level in the mine. The cavity, named the "Drill Pocket," measured 1.5×1.5×2.5 m (4×4×8 feet) and was lined by intergrown and single chalcocite crystals to 6.0cm long mainly with a midnight blue or iridescent patina. This must have been a truly fantastic sight! Because the area was to be blasted the next morning, the mine geologist worked into the frigid night recovering ore bags of crystals. The next morning the area was, indeed, blasted and several more bags of somewhat damaged crystals were removed. A majority of the crystals, however, were destroyed. Such is the fate of crystals in most mines!

Over 1.5 million tons of ore, mainly chalcocite, will be removed from the Flambeau Mine, but only a few hundred *pounds* of crystals will be preserved. When the mine is closed and backfilled, no future collecting will be possible from this unique mine.

Some of the more significant crystals and groups from the Flambeau Mine are described below.

**Plate #16-1, Specimen #5514** BORNITE – Flambeau Mine, Wisconsin. (MH)

## FLAMBEAU MINE, WISCONSIN

### BORNITE (COPPER IRON SULFIDE)

**#5514 3.0×3.0×2.0cm Plate #16-1**

A superb 2.0×2.5cm tabular orthorhombic crystal with pyramidal terminations and a lustrous, blue-purple patina. One side of the large crystal is profusely covered with intergrown tabular crystals to 8.0mm of bornite with chalcocite crystal habit. An excellent, interesting, miniature pseudomorph.

### BORNITE WITH CHALCOCITE (COPPER SULFIDE)

**#5647 3.5×4.0×2.0cm Plate #16-2**

A grouping of fine intergrown tabular crystals in an open network of midnight blue 1.0–4.0mm clusters. The tabular form of the bornite suggests that it formed as pseudomorphs after earlier chalcocite. A rare bornite from the mine.

**Plate #16-2, Specimen #5647** BORNITE with chalcocite – Flambeau Mine, Wisconsin. (MH)

**Plate #16-3, Specimen #5836** CHALCOCITE – Drill Pocket, 1010 Level, Flambeau Mine, Wisconsin. (MH)

### CHALCOCITE (COPPER SULFIDE)

**#5836 5.0×5.0×5.0cm Plate #16-3**

**Drill Pocket, 1010 Level**

A very aesthetic miniature specimen from the spectacular Drill Pocket in the mine. A large, complex, 2.5cm, tabular, "X"-twinned chalcocite, typical of Drill Pocket chalcocites, rises above a group of smaller tabular and untwinned crystals on a base of cellular chalcocite. The surface coating of bornite produces the interesting deep midnight-blue patina.

## FLAMBEAU MINE, WISCONSIN

## CHALCOCITE

**#5835 6.0×6.0×4.5cm Plate #16-4**

**Drill Pocket, 1010 Level**

A superb, aesthetic miniature consisting of a twinned 4.5cm "spear-point" chalcocite crystal with a violet-blue to purple patina rising from a cluster of intergrown twinned 1.0cm chalcocites. One of the larger crystals to come from the Drill Pocket.

***Plate #16-4, Specimen #5835*** **CHALCOCITE – Drill Pocket, 1010 Level, Flambeau Mine, Wisconsin. (MH)**

***Plate #16-5, Specimen #5719*** **CHALCOCITE – Drill Pocket, 1010 Level, Flambeau Mine, Wisconsin. (MH)**

**#5719 6.0×5.3×2.5cm Plate #16-5**

**Drill Pocket, 1010 Level**

A superb cluster of sharp, twinned, 2.0–3.5cm iridescent brownish-bronze crystals on a base of cellular chalcocite. The bornite coating produces a very attractive patina on the specimen. This is a classic example of the Drill Pocket chalcocites from the Flambeau Mine.

***Plate #16-6, Specimen #5726*** **CHALCOCITE – Drill Pocket, 1010 Level, Flambeau Mine, Wisconsin. (MH)**

**#5726 18.0×14.0×8.0cm Plate #16-6**

**Drill Pocket, 1010 Level**

This is one of only a few "plates" of chalcocite crystals recovered from the Drill Pocket at the mine. The surface of the plate is covered with intergrown striated, tabular, twinned 1.0–2.0cm chalcocites with a deep bluish purple patina. The chalcocite crystals are on a 3.0cm thick base of peculiar cellular chalcocite that formed the wall of the large pocket. Several twinned crystal groups to 3.0cm rise from the base. A very unusual piece from the Flambeau Mine.

*Plate #16-7, Specimen #5721* **CHALCOCITE – Drill Pocket, 1010 Level, Flambeau Mine, Wisconsin. (MH)**

*Plate #16-8, Specimen #5745* **CHALCOCITE – Drill Pocket, 1010 Level, Flambeau Mine, Wisconsin. (MH)**

## FLAMBEAU MINE, WISCONSIN

## CHALCOCITE

**#5721 5.0×3.4×4.0cm Plate #16-7**

**Drill Pocket, 1010 Level**

An outstanding miniature with a sharp, twinned, lustrous purplish blue chalcocite crystal aesthetically rising above about six smaller twinned and un-twinned chalcocites. The vivid patina produced by the tarnished bornite coating results in exceptionally distinctive chalcocites. This is one of the premier miniatures from the famous Drill Pocket.

**#5745 6.0×4.5×3.5cm Plate #16-8**

**Drill Pocket, 1010 Level**

A superb cluster of brilliant metallic blue 1.5–2.5cm pseudohexagonal chalcocites in subparallel, fanlike arrangement. Most of the crystals are not twinned and have a brilliant purple patina. This is an excellent miniature from the Drill Pocket.

**#5449 6.5×5.5×4.5cm Plate #16-9**

**Sunrise Pocket, 1010 Level**

This is a superb "spear-point" chalcocite crystal with a brassy yellow to red to blue iridescent patina. Because of its fine crystal form and aesthetic qualities, it has been designated the "Flambeau Warrior." It stands on a base of smaller similarly iridescent chalcocites. The size of the dominant crystal (3.5×2.2×1.2cm) is exceptional. On the major crystals are smaller, to 7.0mm, secondary chalcocites showing a more typical blocky habit. Although this is one of the finest specimens recovered from the Flambeau open pit operation, it comes from a small isolated pocket. The crystal form and iridescence of the specimen are decidedly different from the black, heavily striated chalcocites from Bristol, Connecticut; the Flambeau chalcocites must be considered a very significant American occurrence.

*Plate #16-9, Specimen #5449* **CHALCOCITE – Sunrise Pocket, 1010 Level, Flambeau Mine, Wisconsin. (MH)**

## FLAMBEAU MINE, WISCONSIN

## CHALCOCITE

**#5390 3.2×2.2×1.4cm**

**Refer to Chapter 3, Plate #3-17**

The specimen is a small miniature consisting of two sharp and brilliant crystals arranged in "V" form. The longer crystal is 3.2cm. Both are beautifully and sharply twinned along the vertical *c*-axis. The twins show a classic "X" or cruciform twin pattern when viewed down the termination. Unlike most chalcocites, both have a brilliant bronze patina of chalcopyrite and so have a high metallic luster.

The major crystal has a tight cluster of small sharp chalcocites jutting from the middle of the prism. The base of the crystal pair has another small chalcocite cluster protruding.

Other chalcocites found in the pocket show a range of crystal forms and colors due to the chalcopyrite-colored patina. This find is more than significant — it ranks as having produced the best American chalcocites in nearly 150 years!

**#5450 7.5×5.5×4.5cm**

**Refer to Chapter 3, Plate #3-19**

**Sunrise Pocket, 1010 Level**

This is one of the most spectacular specimens recovered from the mine. Because it is such a superb example of a species not commonly found so well crystallized, it has been named the "Flambeau Chief." Nothing of this quality has been mined in America in this century. Like the "Flambeau Warrior," it was recovered from a relatively small isolated pocket. The specimen is mostly composed of five intergrown sixling chalcocite twins with typical pseudohexagonal form. They range from 1.5 to 3.5cm in diameter. The largest of the group sits in dominant position on the face of the specimen. All the sixlings are made up of tabular chalcocite wafers each about 1.0mm thick, producing a laminated surface along the edges of the twinning. The faces of the sixlings are studded with small, to 5.0mm, tabular, blocky chalcocites.

Along the base of the specimen is a 2.5cm wide "chisel-point" termination of yet another chalcocite. The color of the entire mass of crystals is a dull brassy yellow, with red and blue iridescent highlights.

***Plate #16-10, Specimen #5580***
**CHALCOCITE – 1010 Level, Flambeau Mine, Wisconsin. (MH)**

**#5580 3.5×3.5×2.0cm Plate #16-10**

**1010 Level**

A superb lead-gray chalcocite grouping that has been named the "Flambeau Flower" because of its flowerlike shape. The specimen is a flat disclike cluster of lustrous lead-gray chalcocites with some reddish purple patina. The "petals" of the "flower" are delicately striated, thin tabular chalcocite crystals to about 1.5cm in a fanlike array. A grouping of several intergrown stout chalcocites produce a 1.0cm mound at the center of the flower.

***Plate #16-11, Specimen #5786***
**CHALCOCITE – Lucky Friday Pocket, 1000 Level, Flambeau Mine, Wisconsin. (MH)**

**#5786 3.3×2.5×2.0cm Plate #16-11**

**Lucky Friday Pocket, 1000 Level**

A superb miniature specimen consisting of an intergrowth of two well-formed pseudohexagonal chalcocites about 3.2cm in diameter and 0.8cm thick. The large purple to brassy yellow intergrown crystals have a sunburst appearance produced by radial striations and minor parallel growth on the crystal faces. Several smaller 1.0–2.0cm chalcocite crystals add highlights to the larger crystals. This is an excellent miniature from the Lucky Friday Pocket. The brassy yellow color, blocky pseudohexagonal form, and a lack of "X"-twinning make these specimens very distinctive.

***Plate #16-12, Specimen #5787*** **CHALCOCITE – Lucky Friday Pocket, 1000 Level, Flambeau Mine, Wisconsin. (MH)**

## FLAMBEAU MINE, WISCONSIN

## CHALCOCITE

### #5787 4.3×3.5×2.2cm Plate #16-12

### Lucky Friday Pocket, 1000 Level

A spectacular miniature specimen composed of seven intergrown chalcocite crystals forming a fanlike arrangement with great aesthetics. The larger crystals and smaller accent crystals all have a brassy yellow color with a purple patina produced by the bornite coating. The chalcocites are not twinned or striated. One of the finest miniatures from the Lucky Friday Pocket.

***Plate #16-14, Specimen #5800*** **CHALCOCITE – Lucky Friday Pocket, 1000 Level, Flambeau Mine, Wisconsin. (MH)**

***Plate #16-13, Specimen #5790*** **CHALCOCITE – Lucky Friday Pocket, 1000 Level, Flambeau Mine, Wisconsin. (MH)**

### #5790 4.7×4.3×3.2cm Plate #16-13

### Lucky Friday Pocket, 1000 Level

A superb miniature composed of a group of intergrown, subparallel platelike crystals of lustrous brassy yellow to violet-colored chalcocite. The faces of the main 2.5cm crystal have triangular striations due to crystal growth. A very fine druse of tiny bornite crystals gives some faces of the chalcocite crystals a dark gray coloration when viewed from certain directions. A classic miniature from the Lucky Friday Pocket.

### #5800 17.0×18.0×9.0cm Plate #16-14

### Lucky Friday Pocket, 1000 Level

This is undoubtedly the finest large grouping of chalcocites recovered from the Flambeau Mine. It consists of a circular mound of numerous 2.0 to 3.0×1.0cm disclike pseudohexagonal chalcocites that stand upright on the specimen. The specimen is almost completely covered with the disclike crystals, with only very minor attachment. It was enclosed in loose, black, sandy chalcocite and may have been essentially a "floater," without real attachment. The crystals have a lustrous bronze-yellow color that changes to purple when the specimen is viewed from a slightly different angle. Some crystals have a vivid blue patch on the patina that highlights the overall specimen. Some of the crystals have a brownish black rind produced by slight etching of these crystals. The specimen is very different from Drill Pocket specimens in crystal form and color. It is truly a unique specimen of chalcocite, not only from the Flambeau Mine, but from anywhere.

***Plate #16-15, Specimen #5828*** **CHALCOCITE – 990 Level, Flambeau Mine, Wisconsin. (MH)**

## FLAMBEAU MINE, WISCONSIN

## CHALCOCITE

**#5828 13.0×9.5×5.0cm Plate #16-15**

**990 Level**

A superb specimen of closely intergrown 1.0 to 1.5×0.5cm sharp, wedge-shaped chalcocites with a lustrous gun-metal blue patina. Cellular chalcocite and pyrite form the base of the specimen. The sharp, wedge-shaped, untwinned, gun-metal blue crystals are quite distinct from those in the Lucky Friday Pocket, about ten feet above it. The cause for these variations in crystal forms and coloration is not known; however, it certainly results in spectacular variations in the Flambeau Mine chalcocites.

In addition to these spectacular chalcocite specimens, the Flambeau Mine has produced a variety of other interesting specimens. Superb masses of botryoidal chalcopyrite, bornite, and chalcocite have been recovered, along with specimens of native gold on chalcocite, sheets of native silver, and sheets of native copper. Some of the native copper sheets on a matrix of limonite-stained schist are attractive. Plates of native copper to 10cm or more with a druse of fine crystals are also present. Minor chalcotrichite (fibrous, red copper oxide), as well as some malachite and azurite, was also recovered. Lustrous crystals of arsenopyrite to more than 2.0cm are present in the ore, along with ankerite and dolomite crystals to about 1.0cm that line cavities.

Excellent samples of all of these minerals reside in the collection, making it an important reference suite from this short-lived mine.

## THE MONTREAL MINE

Iron mines are not noted for yielding mineral specimens that are prized by mineral collectors because iron species have a very limited range in composition and often lack spectacular color and crystal form. However, a notable exception to this general rule is the ore of the Montreal Mine, Montreal, Wisconsin, near the Wisconsin-Michigan border in the Gogebic Iron Range. With 56 cataloged specimens, the Barlow Collection contains not just a representative suite of this mine's minerals, but the finest Montreal Mine suite yet assembled. Fortunately, the specimens survived the rigors of mining and the uncertain postmining conditions attendant in many mineral collections.

The Gogebic Iron Range is a 120km long belt of two-billion-year-old sedimentary rocks in northern Wisconsin and Michigan in which iron ores were formed by surface weathering, although a vast majority of the rock remained unenriched. The largest deposit of ore on the Gogebic Range was encountered in the Montreal Mine.

Located in the Snowbelt along the south shore of Lake Superior, the mine site was discovered in the forested wilderness in the early 1880s. The Oglebay Norton Company commenced mining at the Montreal Mine site in 1886. More than 45 million tons of iron ore were mined before the mine closed in 1962. The Montreal Mine was a huge complex of six shafts with a labyrinth of drifts, stopes, inclines, and haulage ways interconnected on many levels throughout the mine. Mining reached a remarkable depth of 4,335 feet (1,321 m) and, as such, was the deepest underground iron mine in the world.

Most of the "collector" minerals at the Montreal Mine formed after the iron ores, as these minerals form cement for brecciated ore or occur as encrustations on the hematite, goethite, and "psilomelane" of the ores.

What are the "collector" minerals from the mine? How did they form? And how did they end up in the Barlow Collection?

The iron ores were produced by oxygen-rich ground water percolating downward through the original rock, oxidizing the iron to hematite or goethite, while removing nearly 50% of the original rock. This resulted in very porous ore bodies that were rather unstable and required extensive timbering during mining operations. This extensive dissolution of the original rock also produced

**Mining is always unpredictable.**

One of John Brottlund's more interesting recollections is that of miners drilling blast holes on the 4,000-foot level. They broke into a huge cavity filled with water under high pressure. The pressure shoved the drill out the hole, and water shot from the hole as if from a water cannon. As the water began filling the drift, the miners quickly set up pumps to stop the flooding. The water was soon so deep that the men brought a rowboat down into the mine to get to the hole. Watching the miners row a boat along the water-filled drift 4,000 feet below ground level must have been a unique sight! When pumps were installed and the water was under control, the pocket was opened. It contained thousands of spectacular barite crystals! Some of the crystals in the Barlow Collection are presumed to relate to this "rowboat" event.

numerous cavities in which crystals and botryoidal masses formed. The major ingredients in the rock, hematite and goethite, along with minor components were concentrated by the circulating waters. Among the minor ingredients were barite with numerous crystal habits, globular masses of manganocalcite, crystals of manganite, botryoidal masses of "psilomelane," minor pyrite, pyrolusite, and quartz, and occasional crystal groups of rhodochrosite. No systematic study of the mineralogy of the Montreal Mine has ever been undertaken, so composition of the several minerals is rather imprecisely known.

Mineral specimens were casually collected by a number of the more than 1000 miners who worked in the mine during its lifetime. Of these, one man, John Brottlund, made a concerted effort to recover and preserve mineral specimens from the mine. Brottlund worked in the Montreal Mine from about 1951 until it closed in 1962. After working his eight-hour shift, he would walk through the various levels of the mine in search of pockets of minerals. He recalls seeing masses of selenite filling cavities as big as a house that were removed by sawing the selenite into blocks with cross-cut saws. Curiously, almost none of the selenite survived to be in collections. Brottlund carried the specimens he collected in his coat pockets and in the bib of his overalls. While collecting, he once got stuck in a large crystal-lined cavity for hours. He was rescued when miners on the next shift heard his call for help.

Brottlund recalls that the shallower levels in the mine yielded mainly botryoidal masses and needle ores of hematite and goethite. Hematite was dominant in the interior of the orebody, whereas goethite was more abundant near its margins. The more "exotic" minerals were mainly from the deeper levels in the mine. Brottlund amassed a large collection (probably over 1000 specimens) of Montreal Mine minerals that he kept at his home. His favorites were housed in a cabinet in his living room. Others soon occupied a large part of his basement, and still others were housed in Brottlund's antique shop in his backyard. He continually upgraded the collection in his living room where his finest pieces were housed. In addition to his own collecting, he obtained fine specimens from several other miners.

Enter John Barlow. During the early years of his collecting, the early 1970s, Barlow was chided about the lack of Wisconsin minerals in his exhibits. Early on, John had obtained an outstanding millerite specimen from Estabrook Park near his childhood home in Milwaukee. But one specimen does not make a collection, so John actively sought other Wisconsin minerals.

In 1978, Don Olson, a midwest dealer who had spent much time searching out collections in northern Michigan and Wisconsin, told Barlow about the minerals of the Montreal Mine and the Brottlund Collection. Barlow contacted Brottlund, fought his way through a fall snowstorm to Brottlund's home some ten miles north of Ironwood, Michigan, and purchased a group of minerals displayed in Brottlund's antique shop.

John mentioned his purchase to Don Olson, who asked him how he liked Brottlund's collection. John replied, "What collection?" Don responded, "The one in his living room." Barlow replied, "I never saw that." Don told Barlow that the collection in Brottlund's living room was outstanding, and that they might work together to acquire it. In the spring of 1979, John went back to Brottlund's and was impressed by the collection near the front door. He asked if it was for sale, and Brottlund answered, "No, this is a collection that I put together over the years, and I want to keep it." So the matter seemed to be dead. However, in the fall of 1979, Brottlund called

John and asked him if he wanted to come up and look at his collection again. Barlow asked if it was just to look or whether Brottlund thought they could make a deal. Brottlund replied, "Ya, I think we might make a deal." So Barlow drove up to Ironwood and looked over the collection, negotiated a price, and bought it. Don Olson selected two fine pieces including an aesthetic blue barite crystal on matrix, as his fee for finding the collection.

Barlow made another trip to go through the large assortment of Montreal Mine minerals in Brottlund's basement and selected several dozen more pieces for his collection. Thus was acquired the fine "living room" collection selected from what Brottlund had found in the mine plus those from Brottlund's "basement stock." While there are fine Montreal Mine pieces in other collections, Barlow clearly has the best overall collection from the mine. He was able to select the best of the best.

Not surprisingly, the Montreal Mine yielded a number of outstanding specimens of iron ore, including hematite and goethite. It also produced a number of other species, including barite, calcite (mainly manganocalcite), goethite, hematite, manganite, psilomelane, pyrolusite, and rhodochrosite. A selection of specimens of these species is included here.

### MONTREAL MINE, MONTREAL, WISCONSIN

### BARITE (BARIUM SULFATE)

**#1965 11.0×11.0×7.0cm Plate #16-16**

A superb blue "spear-point" barite crystal to 6.0cm on an edge, with associated slightly divergent barites, perched attractively on a matrix of brecciated hematite coated with drusy greenish brown ankerite crystals to about 2mm. This is the finest blue barite from the Montreal Mine and is from the #4 shaft, 37th level. It was pictured in *Rocks and Minerals*, Volume 59, Number 2, 1984, page 71.

***Plate #16-16, Specimen #1965*** **BARITE – Montreal Mine, Montreal, Wisconsin. (MH)**

**#1966 4.0×4.0×2.5cm Plate #16-17**

An exquisite miniature with outstanding aesthetics. The main crystals are an orange "spear-point" 2.0×1.2cm×2.0mm barite crystal and a 1.2cm manganocalcite ball perched on a colorful matrix of white and orange-tipped barites with scattered fragments of yellow-ochre limonite. From the #5 shaft, 48th level. This specimen was pictured on the cover of *Rocks and Minerals*, Volume 59, Number 2, 1984.

***Plate #16-17, Specimen #1966*** **BARITE – Montreal Mine, Montreal, Wisconsin. (JL)**

## MONTREAL MINE, MONTREAL, WISCONSIN
## MANGANOCALCITE

**#2008 19.0×14.0×11.0cm**

A superb, large 16.0cm dome of lustrous pale pink manganocalcite surrounded by a number of 1.0–2.0cm spheres on a matrix of brecciated and botryoidal goethite. The large globe of manganocalcite is composed of lustrous intergrown rhombohedral crystals to about 3.0mm with curved faces. This is the most aesthetic large manganocalcite "ball" from the mine. It is from the #5 shaft, 38th level.

**#2015 12.5×12.6×6.5cm Plate #16-18**

A superb specimen with eight undamaged partially intergrown lustrous pink 2.5cm to 3.0cm manganocalcite balls scattered on a matrix of brown, iron-stained calcite. This is one of the more aesthetic specimens to come from the Montreal Mine. The coarse brown calcite matrix, with curved rhombohedral faces, cements a matrix of brecciated earthy hematite and goethite. The manganocalcite balls, composed of small intergrown rhombs with curved faces, formed upon the earlier carbonate. From the #5 shaft, 38th level. Pictured in *Rocks and Minerals*, Volume 59, Number 2, 1984, page 70.

***Plate #16-18, Specimen #2015*** **MANGANOCALCITE – Montreal Mine, Montreal, Wisconsin. (JL)**

## GOETHITE (IRON OXIDE HYDROXIDE)

**#1992 9.0×5.5×6.0cm Plate #16-19**

An outstanding example of lustrous, botryoidal, golden-yellow goethite. Radiating acicular crystals cross the curved color bands of the colloform mounds. The acicular goethite crystals form curved 5.0–10.0cm mounds with a rich golden-yellow silky luster. It is from the #4 shaft, 38th level. This is perhaps the best sample of "golden" goethite surviving from the mine. Pictured in *Rocks and Minerals*, Volume 59, Number 2, 1984, page 70.

***Plate #16-19, Specimen #1992*** **GOETHITE – Montreal Mine, Montreal, Wisconsin. (JL)**

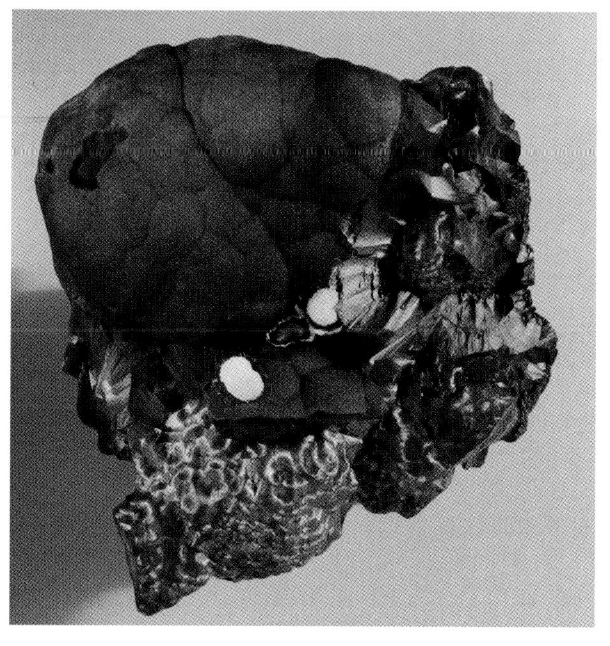

**#2007 14.0×12.0×5.0cm Plate #16-20**

An outstanding example of botryoidal goethite from the #5 shaft, 12th level. The natural exterior of the specimen has a brownish black felted surface. The remainder of the specimen has a lustrous golden-yellow-brown surface showing the fibrous cross-section of the larger mounds as well as the small chatoyant 3.0–8.0mm mounds on the broken surface. Two 1.0cm balls of pale pink manganocalcite are perched on the felted natural goethite surface. The botryoidal goethite is developed on a matrix of botryoidal hematite. This superb specimen aesthetically illustrates better than any specimen known many of the major features of the goethite ore from the Montreal Mine.

***Plate #16-20, Specimen #2007*** **GOETHITE – Montreal Mine, Montreal, Wisconsin. (MH)**

## MONTREAL MINE, MONTREAL, WISCONSIN

### HEMATITE (IRON OXIDE)

**#1998 8.0×6.0×4.0cm Plate #16-21**

A superb specimen of undamaged botryoidal hematite with a spectacular, green, purple, blue, and red iridescent surface. The sample consists of a 5.0cm rounded mass of hematite composed of numerous 7-10mm mounds and several smaller globular masses. It is from the #5 shaft, 12th level. This is perhaps the finest example of iridescent botryoidal "grape ore" from the Montreal Mine. Pictured in *Rocks and Minerals*, Volume 59, Number 2, 1984, page 69.

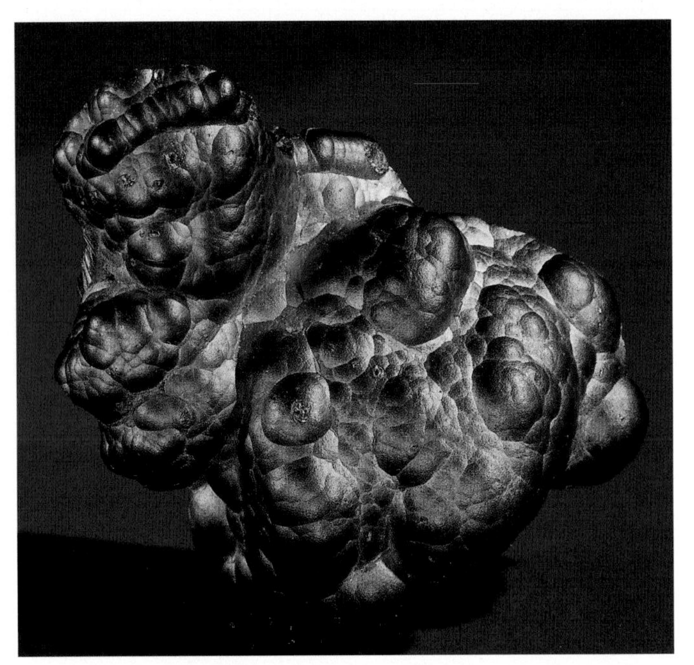

***Plate #16-21, Specimen #1998*** **HEMATITE – Montreal Mine, Montreal, Wisconsin. (JL)**

***Plate #16-22, Specimen #2000*** **HEMATITE – Montreal Mine, Montreal, Wisconsin. (MH)**

**#2000 9.0×7.0×6.0cm Plate #16-22**

An outstanding specimen of a lustrous dark reddish brown 4.0×5.0×6.0cm mound of botryoidal "grape ore" on a base of larger mounds. The lustrous mound of "grape ore" is perched aesthetically on the base of larger "kidney ore." From the #5 shaft, 12th level.

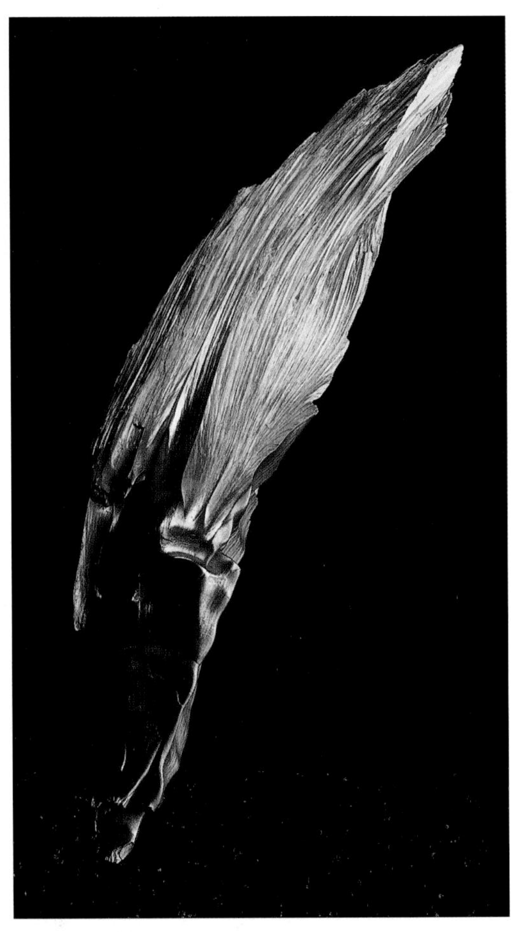

***Plate #16-23, Specimen #2002*** **HEMATITE – Montreal Mine, Montreal, Wisconsin. (JL)**

**#2002 26.0×6.0×3.0cm Plate #16-23**

A spectacular example of the deep reddish brown needle ore with long, curved hematite needles more than 20.0cm long from the #5 shaft, 12th level. One end of the specimen shows excellent development of smooth water-worn channels cutting the needle ore. Samples of water-worn needle ore are especially prized by collectors. This is a very aesthetic piece that combines long curved crystals and a water-worn portion. Pictured in *Rocks and Minerals*, Volume 59, Number 2, 1984, page 69.

## MANGANITE (MANGANESE OXIDE HYDROXIDE) WITH BARITE

**#1990 12.5×10.0×4.0cm Plate #16-24**

An exceptional specimen composed of an aesthetic combination of lustrous black manganite and white bladed barite. The lustrous, randomly oriented, tabular manganites to about 1.0cm formed in and around the several 2.0-5.0cm hemispherical mounds of radiating blades of creamy-white barite. The manganite crystals are prominently striated parallel to the long axis of the crystals. This superb undamaged specimen is probably the best combination of these minerals from the Montreal Mine; it came from the #6 shaft, 38th level. Pictured in *Rocks and Minerals*, Volume 59, Number 2, 1984, page 70.

***Plate #16-24, Specimen #1990*** **MANGANITE with barite – Montreal Mine, Montreal, Wisconsin. (MH)**

## PSILOMELANE (A GENERAL TERM FOR MASSIVE, HARD MANGANESE OXIDES) WITH HEMATITE

**#1999 17.0×12.0×11.0cm Plate #16-25**

A superb specimen of black botryoidal manganese oxides (psilomelane) with the 1.0-2.0cm balls of psilomelane attractively intergrown to produce a spectacular example of "grape ore." Tiny 1mm mounds decorate some of the 1.0-2.0cm balls. In places there are radiating stalactitic branches of manganese oxides that coalesce to form 2-5mm botryoidal surfaces. This is one of the finest examples of botryoidal manganese oxides from the Montreal Mine. It came from the #5 shaft, 12th level. Pictured in *Rocks and Minerals*, Volume 59, Number 2, 1984, page 70.

***Plate #16-25, Specimen #1999*** **PSILOMELANE with hematite – Montreal Mine, Montreal, Wisconsin. (JL)**

## RHODOCHROSITE (MANGANESE CARBONATE) WITH CALCITE

**#1988 9.0×7.0×5.0cm Plate #16-26**

A superb example of one of the several crystal habits of rhodochrosite from the Montreal Mine. This rich pink rhodochrosite is in radiating diverging sprays of long slender prismatic crystals about 1.0mm across and 12.0mm long covering and emerging from a matrix of crystalline manganite. The long prismatic crystals are uncommon for rhodochrosite. This is a very fine representative of the species from this locality. From the #5 shaft, 38th level. Pictured in *Rocks and Minerals*, Volume 59, Number 2, 1984, page 72.

***Plate #16-26, Specimen #1988*** **RHODOCHROSITE with calcite – Montreal Mine, Montreal, Wisconsin. (JL)**

## Other Localities

The southwestern corner of Wisconsin has historically been a major producer of lead and zinc. In fact, the area played an important role in the early settlement of the state.

From 1840 to 1860 the area was the major source of lead for the growing pioneer nation. Lead was mined from hundreds of small adits driven into the hillsides, some extensions of older Indian mines. Some of the early miners lived in these openings. The miners were referred to as "badgers" because of their primitive dwellings. The catchy name caught on and, to this day, Wisconsin is known as the "Badger State." Much of the lead used for bullets by pioneers and by the Union army during the Civil War came from southwestern Wisconsin.

Because of the vast area covered by the district, there were more than 400 mines that produced ore over the history of the district.

Minerals from the district in the Barlow Collection were obtained from William Figi, Sr., an avid collector in the area during the latter stages of mining operations. Because all the mines are now closed and the waste piles are being used for road materials, it is unlikely that new minerals from the area will come onto the market.

Major crystallized mineral specimens from the district include galena, sphalerite, marcasite, pyrite, calcite, and dolomite. Fine examples of these minerals are present in the collection, but only a few will be described here.

### CALCITE (CALCIUM CARBONATE)

**#5500 16.5×9.5×7.0cm Plate #16-27**

**Blackstone Mine, Shullsburg, Lafayette County, Wisconsin**

A superb milky-white 11.0cm scalenohedral crystal abundantly overgrown with lustrous 1-1.5mm rhombohedral calcites in parallel growth. The main scalenohedral crystal is capped with a lustrous 6.0cm rhombohedral calcite to produce a spectacular sceptered crystal. Four rather large flat rhombohedral crystals in parallel growth occur near the top of the crystal. The specimen illustrates well the multiple generations of calcite, with early scalenohedral crystals overgrown by later rhombohedral forms.

***Plate #16-27, Specimen #5500*** **CALCITE – Blackstone Mine, Shullsburg, Lafayette County, Wisconsin. (MH)**

## MARCASITE (IRON SULFIDE)

**#1567 23.0×4.0×2.0cm Plate #16-28**

**Blackstone Mine, Shullsburg, Lafayette County, Wisconsin**

This specimen is an especially aesthetic column of intergrown bladed marcasite crystals to about 5.0mm in fanlike sprays along the column. Several doubly terminated brownish gray scalenohedral calcites are scattered as accents along the column. The base of the column is a 6.0×8.0cm grouping of intergrown brownish gray scalenohedral crystals of calcite that range from several millimeters to several centimeters long. Discontinuous, thin, curved plates of calcite are present within the marcasite along the length of the column. This is a very fine marcasite from the Blackstone Mine.

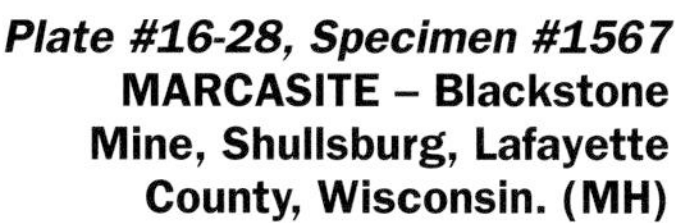

***Plate #16-28, Specimen #1567*** **MARCASITE – Blackstone Mine, Shullsburg, Lafayette County, Wisconsin. (MH)**

## The Vulcan Quarry

The midcontinent area contains hundreds of limestone and dolomite quarries that provide a wide array of materials used in our economy. The quarries also may provide fine specimens of many different mineral species including calcite, pyrite, marcasite, galena, sphalerite, and occasionally fluorite, barite, celestite, chalcopyrite, millerite, and quartz. One such locality is the Vulcan Quarry in Racine, Wisconsin, where fine calcite, marcasite, pyrite, and sphalerite specimens were exposed during quarry operations in 1993 and 1994.

## CALCITE (CALCIUM CARBONATE)

**#5470 15.0×12.0×11.0cm Plate #16-29**

**Vulcan Quarry, Racine, Racine County, Wisconsin**

A superb specimen, one of the best from the quarry, consisting of an 11.0cm high mound of lustrous, sharp, intergrown 1.5–3.0cm yellowish gray scalenohedral calcites. The translucent to transparent calcites contain multiple dark gray phantoms formed by coating of early stage calcites with fine pyrite. The calcites cover a 9.0×5.5cm triangular fragment of gray dolomite. Minor black petroleum and white chalky "clay" are included in some of the crystals.

***Plate #16-29, Specimen #5470*** **CALCITE – Vulcan Quarry, Racine, Racine County, Wisconsin. (MH)**

## Estabrook Park

Estabrook Park is a scenic area along the Milwaukee River in the northern Milwaukee suburb of Shorewood. Limestone outcrops of the Devonian age Milwaukee Formation have yielded occasional geodes with aesthetic sprays of millerite crystals. The park is an area that John visited as a youngster. It is not surprising, therefore, that John was pleasantly surprised to find an aesthetic specimen from his old "stomping grounds" for sale in a rock shop in Wisconsin Rapids, Wisconsin early in his collecting days. John had been in Wisconsin Rapids on business and was caught in a snowstorm. He decided to stay overnight rather than risk driving home. It so happened that a rock shop, "The House of Suzanne," was right across the street from the hotel where John was staying. To pass some time in the evening, he went over to see what the shop had to offer. The most interesting thing to John was on a shelf in the back room where the millerite stood in the dim light waiting to be discovered. He took the sample to the front of the shop to pay for it where he was told "Oh, that isn't for sale!", to which John replied, "What do you mean? It has a price tag on it and I'm willing to pay for it." After some negotiation, John finally bought the specimen for twice the price listed on the tag and it became part of his collection.

***Plate #16-30, Specimen #312*** **MILLERITE – Estabrook Park, Milwaukee, Milwaukee County, Wisconsin. (MH)**

### MILLERITE (NICKEL SULFIDE)

**#312  4.4×4.4×3.0cm  Plate #16-30**

**Estabrook Park, Milwaukee, Milwaukee County, Wisconsin**

A superb spray of radiating, lustrous yellow-green acicular millerite with needles to 3.0cm in a 3.0×3.5cm vug lined with light gray prismatic calcite crystals to about 8.0mm. The calcites form an attractive halo for the millerite spray. A thin coating of dark gray dolomite forms the exterior of the specimen. The specimen has been pictured on page 63 in *Wisconsin Through 5 Billion Years*, 1976, by Byron Crowns, the former owner of the piece. It was also pictured in *Rock and Minerals*, Volume 59, Number 2, 1984, page 72.

Refer to Chapter 17 for the whole story on this specimen.

# PART VI

## *CONCLUSION*

### *Introduction*

As a collector moves from beginner to advanced, he encounters a variety of stories and experiences, some funny, some tragic, some informative. So it is with the Barlow experience. The result is an interesting chapter that sheds light on his collecting activities and the philosophy John developed as he advanced through the hobby.

This section also contains useful references offered by the authors, either in the text or as supplemental reading. Again, these are not intended to be definitive, but rather a source of furthering the reader's interests.

> *Collecting is a world habit.*
> *Collectors practice it consciously and*
> *with a definite, recognized aim.*
> *The rest of us practice it more or less unconsciously.*
>
> *-Arnold Bennett*
> *"Collecting"*
> Woman's Home Companion
> *New York, N.Y., 1924*

Chapter 17

# Philosophy, Experiences, and Deaccession

*by F. John Barlow*

## Philosophy

Putting one's philosophy into words is difficult because there are so many different situations that arise in which one's philosophy plays a part. However, with respect to mineral collecting, and in business, I have tried to be fair with all those I have dealt with. If I feel that the price for a specimen is appropriate, I will buy it without haggling. If, in my opinion, the price is too high, I will do my best to get it at a lower price. And, on occasion, if the price is much too low I may offer more than the asking price (as happened with the priest who had the Japanese stibnite). The price of any mineral is rather arbitrary. It depends on how much the seller thinks he or she can get, and how much the buyer feels it is worth to his or her collection. Obviously, there are times when these two figures are very far apart, and this may lead to some hard bargaining or just walking away.

A second part of my philosophy is, and has been, the challenge of the hunt and the conquest. In the realm of mineral collecting it is essential that one hears about the fine specimens that are for sale. Nearly always, other collectors also hear about fine pieces, so it is necessary to be decisive and move quickly, lest your competitors get the piece before you do. If you develop a reputation of fairness and honesty and can be relied upon, good specimens will come to you. It has been a sense of accomplishment and pleasure for me that I have managed to "beat the crowd" on a number of occasions. I am well aware that I have not won all the matches, but I think I have done quite well to assemble the collection I have. The sense of accomplishment is greater, knowing that I have some very tough competition. Quite possibly this competition including Federation judges has driven me to succeed well above my early expectations. Mineral collecting is, indeed, like big game hunting; the greater the challenge, the greater the sense of accomplishment when you win. So I thank all of those who have competed with me over the years, because they have given me a greater feeling of pride in the contests I have won. There have been many challenges over the years — far too numerous to

*Mineral collecting is, indeed, like big game hunting; the greater the challenge, the greater the sense of accomplishment when you win. So I thank all of those who have competed with me over the years, because they have given me a greater feeling of pride in the contests I have won.*

recount them all — so I will provide just a few examples of some victories and some losses.

## Experiences

As I mentioned earlier, Bill Larson has been a friend of mine for many years, and he has helped me improve my collection on many occasions. He is not only a superb collector and top dealer but is also very competitive and wise. To beat Bill when he is after something is a real challenge. I know of several occasions that this has happened, but fortunately I won out. Two of them I will mention at this time. One is the number one cinnabar from Austria. While traveling in Vienna, Austria I met with Mr. Knobloch, a dealer, who took me to his home to show me his collection. I noted the cinnabar and asked if it was for sale. He said, "None of my specimens are for sale." I said, "I certainly like that cinnabar." He said, "You are not the only American; Bill Larson has been here twice. The first time he tried to get it and I wouldn't let him have it. The second time he was so pushy that I decided no, never will he ever get this specimen." I said, "Would you be interested in a super trade?" He said, "What do you mean?" I had carried with me a Michigan "Christmas tree" of superb silver crystals.

As I recall, the silver was approximately 6″ × 2½″. I brought it out of my pocket, and I could see his jaw drop. I said, "I might be willing to trade this top-quality Michigan silver if you would throw in a couple of pieces with the cinnabar." He said, "John, I can't do that." I said, "Well, OK." I started to put the piece back in my pocket, and he said, "Maybe we do have a deal under one condition." I said, "What is that?" He said, "You must promise me that when you get back to America you do not tell Bill Larson that you have this specimen. Of course, he will see it at sometime, but don't rush the situation." I said, "That's a deal."

One other time a very, very important specimen was being sought after for me by my Indian family friends who had contacts in the middle east mostly in gemstones. You may recall reading, I was good friends with the family and with P.C. Lunia (Raj). One evening his brother Malik called saying he had a specimen for me. I said, "Tell me about it." He explained it, and I said, "A pink topaz?" "No, Mr. Barlow, very, very red." I said, "Oh, what's the price?" He quoted a very high five-figure price. I said, "Malik, that is ridiculous. No way can any specimen be worth that. I will offer you one-tenth of that price." He seemed very upset, but said he would talk with his Parisian partner.

Incidentally, he was calling from Paris. He said, "I will talk to you tomorrow." The following day he called me and said, "My partner feels that the price was right and that he could probably get the same price in Europe." I said, "How can I accept that offer without seeing the specimen. I must see it." Because I was friends with the family and they trusted me, he mailed the red topaz on matrix from Pakistan (catalog #1420 illustrated as Plate #3-90 in Chapter 3).

Having the piece in hand gave me a better opportunity of negotiating. After lengthy letter writing and telephone conversations I was able to secure the specimen at my original offering price.

Several months later I learned that Bill Larson was trying to secure the piece and that Malik had sent him a colored photograph and asked if he would be interested in it. Unknowingly, and to my great pleasure, I had beaten Bill Larson again.

Collecting special and superb mineral specimens particularly with a plan is very much like big game hunting. The real fun, the real sport is working out your plan, tracking, following the arduous trail, and then deciding whether the specimen is worthy of taking. If not, another tracking and trailing takes place until the sought-after trophy is acquired.

I have a great many tracking stories that I do believe are of interest. Space will not permit telling many of them, but I will relate several for your pleasure.

First, I would like to start out with a fun sort of tracking and trailing.

I had been in contact with Bryan Lees, who was mining rhodochrosite from the Home Sweet Mine. He indicated to me that I would not have to worry, that he would be sure to have a fine piece for me. He did say that at Denver he would be showing a lot of material and would suggest that I have a look.

Bryan Lees knew that the buying frenzy would be hectic, and he did very much want to be fair to everyone. As most of you know, Brian rented a large enclosed room to display all sizes and prices of the new rhodochrosite find. To be fair, he had several people on the opening day with tickets that were given out to those interested in seeing the specimens. I realized that there would be a line to the entrance of the Denver Show early in the morning. I was not about to wait in line with a blanket from 3 or 4 AM in the morning and thought that I wouldn't even go to see them. However, at breakfast the urge was too great. I thought I would go down and see who was in the lineup. I drove over, parked my car, saw the line of people and thought to myself, "as long as I am here I might as well get in line and play the game." When the doors opened, a number of Bryan's friends were passing out the tickets. Each ticket had a number at one end. We were told that the doors would not be open to that room for about one hour. I peeked into the room and saw all the beautiful red things and decided, "Well I will get in line when to doors open." There were at least 50–75 people crowding around the front door, and I noticed several buyers inside of the room. This bothered me because I couldn't figure out why they were in there and all the rest of us were outside. Becoming very impatient and agitated, I made a remark to those around me to the effect: "Why are those people in there and we are out here?" Several of my friends chuckled. And the young man standing to my left said, "Well, you know when you are going to get in." I said, "I don't know." He said, "Yes, you do; look at your ticket." I said, "What does that mean?" He said, "Well, look up on the wall above those doors. You will see numbers, and if you look at your ticket, you can tell when you can go in." I looked at my ticket, and I said, "I am going to have to wait 40 minutes to get in there. This is ridiculous." The young man said to me, "I would like your ticket; would you like to sell it?" I said, "You bet. What will you give me?" He said, "How is 20 bucks?" I said, "Let me see your money." He pulled out a $20 bill. I grabbed it and I left. My satisfaction that morning was that I was the only one who made any money on the rhodochrosites except Bryan.

Of course, as you can see from the plates in this book, some of the very finest rhodochrosites from the Sweet Home Mine are represented.

In my mind, one of my great acquisition experiences was acquiring the green millerite #312 shown as Plate #16-30 in Chapter 16.

This story started in April of 1973 in Wisconsin Rapids, Wisconsin. It was a cold, wintry, raining and freezing day. I was calling on two of my customers, namely the Nekoosa Edwards Paper Company and Consolidated Paper Company. When I finished my business meetings at approximately 5 PM, it was obvious that the streets were freezing badly and the roads to Appleton would be treacherous. Therefore, I decided to check into the Mead Hotel and drive to Appleton the following morning. After checking into the Mead I was not in the mood to go to the bar, but I did remember that across the street from the Mead Hotel was a rock and gift shop called the House of Suzanne.

When I walked into the shop, it appeared they were closing. I asked the clerk if they had any minerals. She said, "Yes, we have minerals in the back room." I said, "May I go back there?" She said, "There is no heat on in there, but I could turn the lights on and you could go into that room." The room was an uninsulated steel building. It was poorly lighted, but I could see across the room. It was a typical lapidary shop with bins of cutting rocks of all types. I was about to turn around and leave when I noticed an unlighted case very close to me. I walked up to the case and, to my tremendous surprise, I saw a beautiful green millerite in a calcite vug. It said "Estabrook Park, Milwaukee, Wisconsin." There was also a pretty geode sliced and polished sitting next to it. I froze but soon came to and returned to the clerk. I said, "May I see the millerite specimen? I would like very much to look at it." Before she could say anything, a man's voice from behind a curtain said, "That mineral is not for sale." I said to the girl, "But it has a price on it of $125 and it must be for sale." The man came out and nicely told me it was not for sale at this time because he had only a 50% interest in it with a partner. I was determined that I was not going to leave the store without that millerite. So I said to him, "Call your partner." He said, "He is probably gone from his office. He is a lawyer and I can't reach him." I said, "Will you try, please call him." He tried and there was no answer. He said, "I am sorry, sir, but I have to talk to my partner. Perhaps you could come back next week. I said, "Sir, you have a price on that mineral. It should be for sale and I would like to buy it." He said, "I am sorry, sir, I can't do it." I said, "You have a price of $125. I'll tell you what I will do. I will double that price if you throw in the geode." He hesitated before saying, "Well, I really couldn't do that." I said, "You could give your partner more that his 50% interest in his piece and he should be very happy." I said, "Can I have the millerite?" "Well" he said, "I probably am going to get into trouble but for $250 I will sell it to you." I went out of that shop a happy man.

Why was I happy? Estabrook Park was a park on the east side of Milwaukee and had a little creek running through it. When I was nine or ten years old, a big hiking trip would be from my home on the west side of Milwaukee to Estabrook Park. I made this trip on several occasions, kicking stones down that creek. I never knew what I was kicking in those days, but I found out many years later that what lies under your feet could be of great value.

*When I was nine or ten years old, a big hiking trip would be from my home on the west side of Milwaukee to Estabrook Park which I did on several occasions kicking stones down that creek. I never knew what I was kicking in those days but I found out many years later what lays under your feet could be of great value.*

The rest of the story is very interesting. About 1978 the Wisconsin Rapids Rock and Mineral Club called me and asked if they could come down and see my collection on a Sunday afternoon in the near future. I said, "Yes, how many will there be?" They said there would probably be 16 or 17. I said, "OK, I will be waiting for you." They arrived about 2 PM on that Sunday afternoon. It was fun to chat with them and have them see twelve cases of fine minerals in my museum. John Almquist, who has been a friend of mine for years and a member of that club, looked in one case and saw the Estabrook Park millerite. "Oh, there is my millerite. I have always wanted that; Barlow, how did you get it?" I then told them the story with great pleasure, and I said "You know, I have never met the partner, the lawyer from Wisconsin Rapids." At that moment, a gentleman standing in the back of the group said, "Would you like to meet him?" I looked at him and said, "I sure would." He said, "You are looking at him. I am Bryon Crowns." He then said, "Did you know that your millerite was pictured in my book *Wisconsin Through 5 Billion Years of Change*? It was photographed by Ralph Boyer." I said, "No, I really didn't know that." He said, "Do you have a copy of the book?" and I said, "I must have." I found the book, turned to page 63, and there in color was my great acquisition from my home state, Wisconsin.

Tracking down and acquiring a superb specimen is not always as easy as finding the millerite. In 1975 Don Olson, a mineral and gem dealer today, was living in Milwaukee selling cardboard boxes, traveling from Milwaukee to Michigan. He and I were often able to get together and collaborate on specimens as well as stories. One Friday evening in April of 1975 Don and I were talking on the phone, and he said, "John, there is a superb, very special Michigan silver for sale." I said, "Sounds great, Don, where is it?" He said, "It is in Magdalena, New Mexico." I said, "Who's got it?" He said, "Tony Otero." I said, "What do you think the price might be?" Don said, "I feel it is about $2,500."

Another big game hunt, and I wasted no time. Saturday morning (the following day) I called Tony Otero on the telephone and questioned him about what I had heard of his possessing a Michigan silver. He said, "Si, I have silver." I said, "What's the price, Tony?" Tony hesitated and said, "In Mexican on up to $4-5,000," I realized immediately this would be a great challenge. I said, "Tony, are you open Sunday, tomorrow?" He responded, "Si, I here." I said, "Tony, I will come to see you." He said, "OK." I called my pilot, Don Hoyman, Saturday morning. I said, "Don, we are leaving for Magdalena, New Mexico tomorrow morning." He said, "John, I know the way."

The following morning we flew to this desolate little town with a very short gravel runway. As we were coming in, I noticed the little city about half mile from the airport. After landing there were no cars, no taxis, no one, no anything, so I said to Don, "Let's hike to town." We walked to the main street. We saw no cars, no movement, no action. It was about noon on Sunday. I saw one restaurant that seemed to be open. I said, "Don, let's go up there and see if we can locate Tony Otero." The cafe was small, a counter with a half dozen swivel chairs, a curtain behind the serving area. There was a man talking to a Mexican friend. I asked the man if he knew Tony Otero. He made believe he didn't understand me or was really not interested in answering my question or being friendly. The man at the counter spoke to him in Spanish and a woman spoke from behind the curtain in Spanish. I wasn't sure what was really going on. I said, "Is there a taxi in town?" They said, "No." I said, "I am a mineral collector from Wisconsin I would like to see Tony." The whole climate changed. All three were talking rapidly in Spanish. Then the man behind the counter said in English, "Oh! Tony, Tony Otero. He is down there," pointing with his hand. I said, "Will you call a taxi?" He said, "No taxi, come I take you." I had a feeling they thought Don and I were FBI or IRS agents.

We got into the cafe owner's 20-year-old station wagon that smelled like a barn. Riding to Tony's shop at 25 miles per hour, I felt like I was riding a bucking bronco. The road was so rough it should have been posted at 10 miles per hour. However, we did get to Tony's shop.

As we entered Tony's shop, he greeted us in Spanish. I introduced myself as the one who called him Friday evening interested in the Michigan silver.

Tony brought the silver out from the back room and placed it on his counter. I said, "What is the price, Tony?" Again, he mumbled that he was getting up to $4,000 to $5,000. I said, "Tony, too much money, not interested, I will see what else you have." I walked around his shop, picked out a couple of $35 Kelly Mine specimens, and walked back to the counter.

Tony then pushed the silver in front on me and said, "You like, you want?" I said, "No." He said, "I have a Batopilas silver for you." He reached down under the counter and brought out a large 17 cm × 13 cm plate of many small crystals. I said, "How much, Tony?" He said "$2,000." I said "Tony, that is not even silver." This angered him a little bit, and he went in the back brought a little tub with some TARN-X and, of course, it was silver. Tony said, "See, I told you." I said, "Well, I am not interested." He then said, "How about Michigan silver?" I said, "Too much money, no I don't think so." I started to walk toward the door, and he said, "Señor, señor, come back." I came back, and he said, "I make you a proposition, you buy the Batopilas silver for $1,000 and I sell you Michigan silver for $3,000." I walked to the door, and then I turned around, and I said, "Tony, too bad." He said, "Come, come, I show you." I went back to the counter, and at that point I said, "I will make you my last offer before I leave town. I will give you $3,300 for both the silvers." He hemmed and hawed and indicated that it was a bad deal for him. I started to leave, and he said, "Señor, we make deal."

I left Tony's very happy because the Michigan silver (catalog #1154 pictured in Chapter 3 as Plate #3-82) is a top Kearsarge Lode silver, Houghton County, Michigan specimen.

I was very elated; of course, Don was happy the mission was completed.

The very best part of the story really came later.

John Whitmire, a long-time wholesale dealer and miner, was a friend of mine for a number of years. I always looked him up at one of the wholesale shows in the area to chat with him. I even got into a poker game with him and his crowd one night when they thought they were going to really clean me. The story can be verified: to their shock I was the big winner of the evening. At the Tucson Show following the Tony Otero acquisition I looked for John and found him in the wholesale area. He was sitting in a chair in his usual posture. He usually reacts in his very dry humorous unfriendly manner (none of which is real). John saw me coming and looked away. I walked over and stood in front of him. He didn't get up from his chair. I said, "Hi, John, how are things?" He gave me a very ugly terrible look, "You blankety blankety blankety SOB!" which stopped me in my tracks. I said, "John, what's the matter, I don't understand?" He said, "The heck you don't, you so and so." "I said "John, you don't know what you are talking about." He said, "You blank stole my Michigan silver." I said, "I don't know what you are talking about." He said, "For your information, that Tony Otero SOB was holding that specimen for me to pick up on Tuesday and you were there and got it the Sunday before. You are a blankety blank." I was aghast, didn't know what to say, and did my best to tell him what happened and how it happened. Eventually the air was cleared, and we were again friends.

I would like to tell another story about Bill Larson. When Cal Graeber was working for Bill Larson back in the 1970s, he and Bill were driving their van with their stock to the Detroit Gem and Mineral Show. On their way across the country they stopped by to visit with me. We had a great afternoon talking minerals and sharing stories, and I think we did do some trading. However, they wanted to get on to Detroit and indicated they wanted to get going. I said, "Bill, I know you like mai tais from our experiences at Trader Vic's in San Francisco. How would you like a mai tai before you go?" Bill said, "It sounds like a great idea." The three of us went into my barroom, and I started making some mighty mai tais. To cut the story short, we had perhaps six or seven mai tais, and then they wanted to get on the road. I suggested that maybe they should stay overnight, but they indicated, no, they had to leave.

*We started looking through the flats, and it turned out they had acquired all of Lazard Cahn's Collection with very fine, rare old-classic pieces. I was in heaven, and truthfully, many of the old German silver-bearing minerals and many of the silvers in my collection were acquired that evening.*

To get to my mineral room, you had to drive to a lower level down a relatively narrow concrete walled drive. They had their van at the bottom close to the mineral room. When they were leaving, I said to them, "Be very careful when you drive out because you will be backing out and you will scrape your van on the concrete walls if you are not careful." They indicated it was no problem, and I watched. Cal Graeber stood in front of the van giving Bill directions left or right as he went up. Bill was about half way up when Cal felt he had it made and gave him the signal to go full speed. Bill hit the throttle and went out the drive and hit my limestone post. The post is approximately 30″ × 30″, three to four feet high. He hit it with a crash, and I thought they would not leave. The van would be a mess. It turned out that he had hit the post directly on the frame length and did little damage to the van but my mortared stone post was all over the street and the sidewalk. I saw them in Detroit, and it certainly made a good story to talk about.

This is an acquisition story that always gives me pleasure to tell. As a matter of fact, when I do, it tickles my funny bone. In the spring of 1978 I received word talking with Don Olson that the Northwestern University major collection was going to be sold. In an attempt to find out how to get in on it, I found that several dealers were making bids and offers. One of them was Gene Schlepp. I talked with Gene. He said that he was not at liberty to tell me very much but felt he had a good chance of getting the collection. I told Gene I would stay out of his way and not interfere, but he should let me know if I could be of any help. In April of 1978 Gene Schlepp and Dr. Richard Kelly completed their arrangements with Northwestern University and were making arrangements to transport the total collection to Dr. Kelly's home in Rochester, New York. Gene called me and said, "When we leave Chicago for Rochester, I will call and let you know." About a week later Gene called me and said they were leaving Chicago and they would probably be in Rochester in two days and I would have first opportunity. He pointed out there was a ticklish situation because Bill Pinch, who also lived in Rochester, wanted the first opportunity. I said to Gene, when you arrive in Rochester, call me.

The following week Gene called me, saying, we just came in with all the flats and they are in Dick's basement. We will be going into them tonight, tomorrow, and the following week. I said, Gene, I will be there tonight. I talked to Don Hoyman. We took off in a miserable April storm and flew to Rochester under difficult conditions, arriving about 5 PM. Dick Kelly was a superb host. He had several other friends for dinner and Don and I were both invited to dinner, too. I will say that Dr. Kelly was a real connoisseur of fine white wine.

However, during the meal I was itchy to get into the basement; eating and drinking was not what I came for. Finally, Gene said, "Shall we go downstairs?" I said, "Let's go." Well, we went into the basement. It was a typical midwestern basement. Concrete floor, block tile walls. The difficulty was the poor lighting. The boxes were piled high, and I told him I was interested in just silver-bearing minerals because I would not have time that evening to go through all the boxes. Fortunately, they had them segregated in certain species and classes, and the metals and silver-bearing minerals were easily available. We started looking through the flats, and it turned out they had acquired all of Lazard Cahn's Collection with very fine, rare old-classic pieces. I was in heaven, and, truthfully, many of the old German silver-bearing minerals and many of the silvers in my collection were acquired that evening.

You may recall reading in Chapter 10 about specimen #1690, Plate #10-17, an argentopyrite from St. Andreasberg, Germany. There is a short story of its acquisition that I want to enlarge upon in this chapter. Looking through the boxes, I noticed a specimen with no label as many of the boxes had no labels and were mixed up. In any case, I was looking at a dark sample. Having recently come back from the British Museum in London where I had attempted a trade with Peter Embrey on an argentopyrite, I made a wild guess now, for the fun of it. Looking at the mineral specimen through my loop, I said, "Dick, this is an argentopyrite." Dick looked at it, and he said, "That is not an argentopyrite, I am sure that is not an argentopyrite." This was fun, so I said, "Yes, Dick it is." Dick said, "I'll tell you what I am going to do, you may have that specimen for $300. You have it checked and if it is an argentopyrite, it its yours as a gift." I said, "OK, Dick it's a deal." On returning home, I sent it to the Smithsonian. John White reported several weeks later that, yes, tests indicated that it was argentopyrite. I gleefully advised Dick Kelly, showing him the tests, and he was amazed. This was one of those lucky, good fortune situations that occurs once in a while.

So many great acquisition stories come to mind such as beating Bill Larson out of the world-famous red topaz from Pakistan. Also beating Bill Larson out of the number one cinnabar from Austria. The red topaz is Catalog #1420 and pictured in Chapter 3 as Plate #3-90. The cinnabar is catalog #2347 and pictured in Chapter 3 as Plate #3-21.

Bill Larson is not only a collector and a dealer; he is very competitive and enjoys the hunt as I do. I have known Bill since 1972 when I bought my first two blue cap tourmalines from him. That relation has traveled a rough road ever since. However, I gratefully speak of him as a friend and as one who has helped me. His acquiring the number one fabulous gem tanzanite named the "Sleeping Beauty of Tanzania" was one of those moves. Also, the world-famous hydroxylherderite came from Bill, as well as the very large doubly terminated sapphire from Sri Lanka. I will tell that story as it is not a long one.

Bill had promised to get me a superb large sapphire crystal for a couple of years, and I kept pressing him. Before the 1986 Tucson show Bill called and said he had the sapphire; he would have it the opening day of AGTA Show. He indicated that there were several others interested and that I should be there when he opened. I arrived at his booth at the time he was opening, and he said, "Good thing you came, John, there are several people standing by." I said, "Let me see the sapphire." He showed it to me. It really is a superb sapphire, one that I had been waiting for for years. He said to me, "John, you must make up your mind immediately because, as I said, it will be sold. I promised several other people they would have a first look right after you examine it." I said, "How much, Bill?" He indicated the price with a low five figures. I said, "I will take it." He said, "Good thing, John, look around." I stepped aside and saw Paul Desautels, Barry Yampol, and Peter Keller. As I walked away, I recall Peter Keller saying to me in jest, "John, you are a pig." (Pigs are supposed to be very smart.) I learned later that number two was Paul Desautels, number three was Barry Yampol, and number four was Peter Keller. Shortly after I purchased the crystal, Paul Desautels came to me and said, "John, I will give you $3,000 more than you paid for it immediately." I said, "Thanks, Paul, but I really want it for my collection."

## Deaccession

Deaccession was always difficult for me. When I was a young boy I was looking for competitive challenges, whether it was marbles or bottle caps. I filled boxes with my winnings and became a pack rat. When material things (e.g., marbles, bottle caps) available to me were no longer there to win, I lost interest. Fortunately, the next step came easily, that of joining or leading a team effort to win in the juvenile fun of sports: grade school and junior high school softball pitcher; Tiger A.C., Crimson A.C., athletic clubs at age ten. West Side Boosters, hard ball in the sandlot leagues of Milwaukee 1927, 1928, and 1929. In high school it became boxing, wrestling, track, and football with the top prize being the all-city fullback on the 1931 city championship team. There was a short stint in racing hydroplane boats on Wisconsin lakes. In the military academy with a scholarship it was football, hockey, baseball, and discipline. I was lucky to be able to attend the University of Wisconsin during the 1929–1932 Depression. But the road was the same — drive, challenge, make the varsity football team. In my sophomore year the mission was accomplished.

This was one of the great turning points in my life. My interest in Dorothy plus the strong challenge dormant within me to be a great engineer in the business world and have a noteworthy family were maturing. Thinking of where I was, and where I was going, it was obvious, I was not a great athletic talent and, I was wasting my time. The next challenge was Dorothy and the world. 1935 was a great year; Dorothy and I were married. I was appointed business manager of the Sigma Chi fraternity, and my goal was to get my Bachelor of Science degree in Mechanical Engineering. Mission accomplished in 1937 with 146 credits in four years. I am sure you are wondering what this list of events has to do with deaccession.

Having read the biography, Chapter 2, you know we raised eight daughters and no sons. About the time our fourth daughter, Jacqueline, was born, my friends' sons were getting marbles and baseball, football, and hockey equipment. Slowly, all my collectibles being saved for my son were deaccessed.

I cannot say this deaccession was fun; but I can tell you deaccession can be great fun, sometimes scary, and sometimes difficult. I would like to tell a few real stories that will illustrate each situation.

*Another pleasure is to see one of your old friends (mineral specimen) proudly displayed in another major private collection.*

One of my very pleasant deaccessions started in Tucson in 1981. My daughter Alice and I were attending Marion Stuart's private dinner party. I was seated to the right of Dr. Richard Gaines, a long-time friend. After a very jovial cocktail hour, the laughter and conversation were loud in the dining room. Dr. Gaines was oblivious to the spirit and was falling asleep. I tried my best to engage him in conversation, as did others. I was about to leave him in his slumbering peace when I thought of his mineral collecting interest in Te, Se, Ta, Nd, Be, and Li rare minerals. I turned to Dick and said a bit loudly, "Dick, how is your rare mineral collecting coming?" He perked up and said, "What did you say?" When I repeated my question, he said "OK, fine," and went to sleep. Speaking loudly again, I said, "Dick, do you have a rickardite specimen?" This did bring him up, and he said, "No, I do not." It made me proud to say, "Dick, I have one." He looked at me in such a way that I was sure he thought I didn't know what I was talking about. Studying me very alertly, he slowly said, "John, describe it for me." I said, "Dick I can see it. It's about 2″ wide × 1½″ high and about 1″ deep. It's orelike, no crystals, colored violet-pink." His reply was, "Very interesting, you have a very fine rare mineral specimen, but where did you get it?" I felt this was his clincher question to be sure I knew what I was talking about. I said, "Dick, it came from the Lazard Cahn Collection, which Dr. Kelly and Gene Schlepp acquired from Northwestern University." I saw his face light up, and I said, "When I get home, if I can find it, I will send it to you." On my return to Appleton, the rickardite was where I thought it was. I mailed it to Dr. Richard Gaines as a gift for spending a pleasant evening with him at Marion Stuart's party. Quoting Richard's letter of March 11, 1981, "It was really kind of you to have remembered about the rickardite, and also to have thought about it in the first place and to have realized that I would want it. You know I have been searching for such a specimen for perhaps a dozen years. Although I have seen minute specks of the mineral from Moctezuma in Mexico, this one is much more desirable, being from the type locality and rich."

I have derived great pleasure and have wonderful memories of deaccession that are really "parts of me."

Flashing before me are deaccessions, some minor and some major, to the University of Wisconsin Madison; University of Wisconsin Oshkosh; National Museum of Natural History, Smithsonian; British Museum of Natural History; Fersman Museum of Moscow; Field Museum of Chicago; Harvard Mineralogical Museum; Carnegie Museum of Natural History; and Bergstrom Mahler Museum of Neenah, Wisconsin.

Another pleasure is to see one of your old friends (mineral specimen) proudly displayed in another major private collection.

Before I tell of a few unhappy deaccessions, as well as a scary near miss, I want you to feel the pleasure of a particular deaccession.

I have had in my collection a Wisconsin diamond. I was very sensitive to the lack of Wisconsin minerals, having been rubbed sore by collectors and dealers. If and when asked if I had a diamond from Wisconsin, I could always say, "Yes, of course." I had acquired it from John Almquist in a one-sided trade, tilted, as always, in John's favor. It was found on John's and my birthday, July 12, by Al Falster in the Rib Mountain area and is well documented.

The diamond was small (micro) but a perfect octahedron. It was part of me.

But it was my great pleasure to deaccess my only diamond from Wisconsin when I learned that a good friend of mine, Richard Thomssen, the author of Chapter 7, was a micromounter. I felt this specimen would be more important to him than to me. My pleasure, Dick.

I can think of many deaccessions that gave me great satisfaction and did me more good than they did the recipient. As I mentioned earlier, deaccession can be fun, scary, and sometimes unpleasant.

A scary — temporary — deaccession occurred in Tampa, Florida in 1979 with a very interesting outcome.

## Cops and Robbers

The Tampa Gem and Mineral Club was hosting the National Federation Show. Planning to attend, I selected a number of top-quality mineral specimens for display. It would also give me an opportunity to visit with a collector friend, Dr. Dale Dubin.

My pilot, Don Hoyman, grandson, Greg Helein, and I had a great time in Tampa. The show was a success, and I was fortunate to have Paul Desautels, curator of the Smithsonian, acquire a very fine Graves Mountain rutile for me.

Checking out of the Holiday Inn after lunch on Monday, our system was in force. I would check out and account for the bags, Greg would be responsible for the minerals that were in several flats under his bed, and Don would check the weather and file our IRF flight plan.

We had an excellent flight home, landing at the Outagamie Airport in Appleton at about 5 PM.

On the way to get the car while Don and Greg were unloading the plane, I had a horrible thought. I did not remember seeing the mineral flats when the plane was being loaded. When I got back to the plane, all our luggage was on the apron waiting to be loaded into the car. My heart stopped beating. I did not see the minerals. I said to Greg, "Where are the minerals?" He said, "Gee, I don't know, I thought you had them." I said, "Let's get going, I have to call the Holiday Inn!" It was 6:30 PM there. On the way home I was thinking if the room was not rented, the minerals would still be under the bed. If the room was rented, they might have been turned over to lost and found or they may have been stolen. I was sick.

Without taking my coat off, I called the Holiday Inn and asked the desk clerk if they had rented Room 230 for that night. The answer was no, and I was elated, hoping that everything was still OK. I asked the night clerk if he would send a bellhop up to look into the room under the bed and I described the boxes and how they were wrapped and that I would wait. He said he was the only person on duty at that time. The day shift had left and there were no bellhops. I then told him that I had to know whether several boxes we had left in the room were there. They were very important, and I would hold the phone while he or someone was able to confirm that the boxes were still there. He said he would do his best, and I should hold on. I must have waited for 15 minutes, anxiously wondering what I was going to do next. He finally came back and said, "There were no boxes under the bed." My heart again stopped. He said, "Housekeeping is closed, there will be no one there until 6 in the morning." I said, "Thank you very much."

I hung up and went through a painful decision process of what in the world was I going to do. The boxes represented low six figures in dollar value, and I was almost in panic. It was now 8:30 PM or 9:30 PM in Tampa. My only thought and hope was to call Dr. Dale Dubin to explain my predicament and wonder what could be done. Very fortunately, I reached him, and he too was in a quandary. All of a sudden, he said, "I have an idea! I have a very good friend who is a private detective. He runs a very successful agency and he certainly would know what to do." He said, "I will try to reach him by phone and will call you back." I waited, and in about a half hour Dr. Dubin called back. He said, "John, this is a difficult situation, but I think we may be in luck." He said, "Sam Ferrara was very ill; he had recently been through an operation, and I found him in bed. Sam heard the story and said, "Doctor, I will do it. Let me take hold," I told Dale, "Let him go, and hopefully they can be recovered." Dr. Dubin called me back and said that he had had another call from Sam, who said he had already been in contact with the night clerk and had the whole unit very much shook up because he wanted to talk to the night clerk's boss to advise him that there were very dangerous radioactive materials left in Room 230 and if anyone came in contact with them they would be in danger of not only their health but perhaps their life. He said the desk clerk told him that Sandy Brown, the head housekeeper, came in at 5:45 AM.

> *Sam said he had already been in contact with the night clerk and had the whole unit very much shook up because he wanted to talk to the night clerk's boss to advise him that there were very dangerous radioactive materials left in Room 230 and if anyone came in contact with them they would be in danger of not only their health but perhaps their life.*

Sam told Dale he would be there with his associates at 5:40 AM. I went to bed feeling I had done everything I could and hoped the plan would work. Dr. Dubin said he would call me in the morning, that he had surgery early but should get through to me about noon.

On the following day, Dr. Dubin called me and said, "John, you will not believe what has happened! But let me tell you. Sam arrived the following morning at 5:40 AM waiting for Sandy at the housekeeping door. He had three of his other associates posted at other locations. They were carrying Geiger counters, meters, and cameras over their shoulders. When Sandy came in, he explained to her that the items left were very dangerous and would probably be hazardous to the health of anyone who came in contact with them. He said Sandy shook and took him into the housekeeping room. She let Sam look around. He found one box in a corner with four minerals in it. Sandy was so scared she could hardly talk. She said, 'You may look all around.' Sam looked around and opened the housekeepers' lockers. In five lockers they had already split up the loot." One locker had the diamond on matrix and the topaz, another locker had the beautiful fine gem tourmaline, and on and on.

Sam called Dr. Dubin to advise him that he had recovered eighteen beautiful large minerals and that was all he could find. He was able to take them with him because of the danger that all parties on site were concerned about. This was a very happy ending to an scary, almost deaccession. Every mineral was recovered, and none of them had any damage.

I have a detailed report from Sam Ferrara Investigative Agency, Inc. with an invoice for $500. Paying that invoice was a great relief and pleasure.

Although I have had several unpleasant deaccessions, I have chosen to relate only one, which is perhaps the most unhappy deaccession that I have ever been a party to. The reason I would like to relate this to you is that it was a very excellent learning experience. I do believe any collector could learn a lesson from this true story, which happened about 1990.

Many of us in the hobby, in the world of collecting minerals, have had the experience dealing with a vacuum cleaner salesman from Colorado. He is a superb salesman, a very hard working, knowledgeable hobbyist who knows what he wants and where he wants to go. I've had many pleasant and some not-so-pleasant experiences trading minerals with him. I understand most top collectors have had similar experiences.

I'm sure most readers will recognize the man I am describing. He is a very prominent mineral collector and known to all of us as Keith. He has been to my home on numerous occasions and, as always, has his bags filled with his minerals to sell and trade. My problem with Keith has been the fact that he has so much drive that he never knows when to stop. Every time he has been to my home he has had his own room. The problem has been getting him to go to bed at a reasonable hour. On three separate occasions he has set off my security system, and one time I really thought we were under attack. The police department were there fast and found it was Keith who had set off the alarm. Because of these unhappy experiences, Keith is sort of a *persona non grata* in my home. His last visit came about in 1990. He called me from Chicago saying that he was traveling from his home in Colorado to the east coast visiting his numerous friends and customers and that he would like to stop by. At that time I was not in a buying or selling mood but thought that it would be fun to see Keith again. He arrived about 8 PM, and after an hour it became apparent that he had his eye on securing some of my fine minerals. He brought out nothing to trade and I became a little suspicious of his approach. After several hours of exchanging stories and talking minerals, he began praising my collection, in particular, certain individual specimens. It became obvious to me that he had a very strong interest in four special pieces. I finally agreed that he could take the four specimens out of the mineral cases to look more closely at them. As many of you know, this is rather hazardous; however, I did let him take out the Climax rhodochrosite, the Brazilian morganite with the phantom crystal, the superb Michigan copper, and a very fine Chivor emerald. I tried for several hours to convince Keith that absolutely under no circumstances was I going to be selling any of the specimens from my collection. Furthermore, I had an 8 AM board of directors meeting the following morning, and we were going to go to bed.

Keith realized that he had to move quickly. He said, "John, I want these four specimens; give me a price. I am willing to pay you many times what you paid for them." To end this session and with the feeling I had full control of the situation and that I certainly could stop him immediately, I said, "OK, Keith, give me $44,000." Knowing that the cost for my four pieces was $16,000, I was sure that Keith would not accept but would start to negotiate. I was shocked when he said, "John, you have a deal, $44,000, and I am going to take them with me." I said, "Keith, wait a minute, I said $44,000 in cash, that does not mean a check, any trading, part payment later; I meant cash hard dollars all at one time at this moment. Period." Knowing of Keith's past performance with regard to cash flow, I was *absolutely positive* that we would be able to go to bed. Keith said, "We got a deal, cash right now." As I held my breath, he stood up, reached into his inner coat pockets, and produced two thick envelopes. I couldn't believe my eyes. He started counting out hundred dollar bills and he stopped at 440. I was absolutely flabbergasted. I realized that I had set a trap for myself and made a terrible mistake.

I did not sleep all night; I was very, very upset. As a matter of fact, it took me several weeks to recover, because I did not want to part with those four specimens. I had been overconfident, feeling that I had full control of the situation.

The moral of this story and the lesson to be learned by any collector is, never feel you have absolute control of a buying, selling, or trading situation.

### The "Burma Beauty"

*Bill Larson, Pala International*

I and Dr. Gueblin were shown several pieces from Da Taw Mine, Mogok, Burma in 1992. Most were not good crystals, but one was. We offered a very high price — but no sale.

On my next trip to Mogok, I was again shown the "Burma Beauty" — and, again, not for sale. But the discussed black market price was 100 lakhs of kyat (~$100,000 US).

In 1996, MGE (Myanmar Gemstone Enterprises), which controls exports, reduced the tax from 33% to 10%; owners could now accept lower offers and still make a nice profit. The Burma Beauty was worth chasing again.

I visited Mogok again in March 1996 and traced down the Da Taw Mine owners only to find that they had sold a group of rough ruby to the famous Yaw Sett, the "King of Rubies," in Mogok (known to Westerners as Joseph). Fortunately, Joseph is a good supplier to me, and I immediately visited him. He had obtained two crystals of good form but not showing clear areas. For each he had paid 70 lkhs (approximately $70,000 US). His cutter had ground up one looking for gem quality areas. Yaw Sett's manager said $70,000 worth of ruby dust was now in Yaw Sett's garden! The Burma Beauty was spared a similar fate. But Joseph had sold it to his sister. I left empty-handed but returned in late April as part of a cutting project. Again, the one-day, 120-mile drive in a flat-tire van from Mandalay, only to be told that the sister wanted 100 lakhs. So no large crystal. But I promised to return in July for other business.

In July, a bit of good news: the US dollar was up 50% against the kyat. A chance. The king himself (Yaw Sett) intervened on my behalf. His only words were to take care of the "Beauty." This time I came home excited — even after the flights and airport waiting time added up to an incredible 47 hours!

Barlow, in accordance with my agreement with Yaw Sett, was the first collector offered the "Burma Beauty."

# CHAPTER 18

# An Important Dream

*by Knarf Wolrab*

This chapter was not intended. After completing Chapter 17, John had a dream which was long and intense. In his dream he made an agreement. It was so real he had to keep his agreement.

John was working in his mineral room and he was hearing certain sounds, mumbles, hums, so to speak, coming from several of his cabinets. One would quiet down, then another one would start the noise. He thought that it was an electrical noise and couldn't understand it. Then he saw a couple of mineral specimens moving about, and he thought he was over the San Andreas fault.

The following day (in his dream) John was back in his mineral room sorting the over 600 color transparencies for this book when he again heard the rumblings. They became louder. John was greatly concerned. He closed the mineral room and left for an hour.

When he returned, on his worktable in his mineral room was a sealed envelope. He opened the envelope to find a grievance notice. The letterhead indicated it was from Local 54912 Appleton, Wisconsin, affiliated with the International Union of Mineral Specimens. The demand letter was signed by Red Crocoite, business manager for Local 54912, and was addressed to F. John Barlow representing SANCO LTD., owner of the F. John Barlow mineral collection.

The body of the letter said:

*I, as business manager of Local 54912 of the Mineral Specimens Union, have been requested to advise you that there is great turmoil in our Local. The members are unhappy with the manner with which you have handled their positions. I am instructed to advise you that unless this matter is settled quickly, it will be necessary to go into the underground courts or seek binding arbitration.*

*Furthermore, if this matter is not settled quickly, the turmoil will probably destroy this Local and the membership.*

*Our demands are as follows:*

1. *You will stop deaccessing our members by transferring them to other locals or making trades with other nonunion groups.*

2. *Having heard rumors of your selling your collection to nonunion organizations out of the United States, our membership is in great turmoil. Therefore, we demand that you will do no further deaccessing without the permission of our executive board and under no circumstances will you split up our union group by selling our membership out.*

3. *Because our membership is in great turmoil and we are very unhappy with your photographing certain members and leaving other members out, I have been requested to demand that you add to your forthcoming book 25 additional color photographs of our members.*

4. *After all, there are minerals among us that have been judged the best in the world and they deserve to be pictured in the book also.*

*We request that you respond within 24 hours.*

*(Signed)*

*Red Crocoite*

John quickly met with their executive board and made a counter offer of ten additional photographs, but under no circumstances would he agree to their deaccession request. They immediately turned him down. They agreed to ten additional photographs but held firm to the deaccession demand.

John then made his final offer and stated, "I will agree to include seven additional photographs in the book but under no circumstances will I agree not to sell the entire membership to another owner. That is my last offer, take it or leave it." They came back accepting his offer of seven additional photographs and asked that he be fair to their entire membership if the entire body was sold.

John's dream having been completed, he felt he had to abide by this agreement. When John makes an agreement he always keeps his agreements, and he intended to follow through on this one.

Therefore, seven members' photos have been added to the book as listed below, and he agreed not to break up the collection but that it would go entirely to one major organization.

***Plate #18-1, Specimen #4785*** **BENITOITE – The Gem Mine, San Benito County, California. (MH)**

**BENITOITE (BARIUM TITANIUM SILICATE)**
**#4785 8.5×8.0×7.5 cm Plate #18-1**

**The Gem Mine, San Benito County, California**

This plate has a grouping of about twelve deep blue benitoite crystals on a white natrolite matrix.

There are several exceptionally large crystals of this unusual trigonal mineral that are present. The largest crystal measures 3.2 cm on an edge. Most of the crystals are deeper blue on the girdle than on the apex of the trigonal pyramids. A very aesthetic specimen.

## CROCOITE (LEAD CHROMATE)

**#4999 9.0×1.8×1.2cm Plate #18-2**

**Zeehan-Dundas Region, Tasmania, Australia**

This is an exceptionally large brilliant red-orange, crudely terminated crocoite crystal with minor attached goethite matrix. The brownish goethite provides interesting highlights to this fine crystal from the classic Dundas locality. From the Glen Bolick Collection.

***Plate #18-2, Specimen #4999***
**CROCOITE – Zeehan-Dundas Region, Tasmania, Australia. (VP)**

## FERBERITE (IRON TUNGSTATE)

**#5649 9.0×6.0×7.0cm**
**Plate #18-3**

**Potosi Department, Bolivia**

This is a spectacular grouping of lustrous black striated, tabular intergrown crystals from the famous Potosi, Bolivia area.The larger crystals are 6.0-7.0cm long, with numerous smaller crystals in the "valleys" formed by the intersection of the major crystals. Secured from Rock Currier.

***Plate #18-3, Specimen #5649*** **FERBERITE – Potosi Department, Bolivia. (MH)**

## GALENOBISMUTITE (LEAD BISMUTH SULFIDE)

**#5466 14.0×12.0×8.5 cm Plate #18-4**

**Hunan Province, China**

This is a very showy grouping of minerals consisting a jackstraw cluster of quartz crystals to 10cm long. A column of dolomite crystals to 1cm stands along the larger quartz crystals. The thin, elongated, striated crystals, metallic, lead-gray galenobismutite, occur in small sprays on the quartz. Several colorless 7×8mm fluorite cubes occur on the dolomite. A lustrous partial cube of pyrite forms the base of this fine specimen from the Chinese locality known better for its stibnite crystals. Traded with Scott Thieben of Boone, Iowa for Flambeau Mine chalcocites.

***Plate #18-4, Specimen #5466*** **GALENOBISMUTITE – Hunan Province, China. (MH)**

***Plate #18-5, Specimen #2023*** **RUTILE – Graves Mountain, Lincoln County, Georgia. (MH)**

## RUTILE (TITANIUM OXIDE)

**#2023 8.0×3.0×5.3cm Plate #18-5**

**Graves Mountain, Lincoln County, Georgia**

Several intergrown exceptionally lustrous brownish black rutile crystals with reddish internal reflections are perched on a matrix of randomly oriented kyanite and pyrophyllite matrix. The major crystal is 7.5×5.5×3.0cm. It is intergrown with a 2.5×2.4cm crystal. This excellent sample is from the famous Graves Mountain, Georgia locality. It was mined by Jack Hanahan in 1979. Barlow acquired it from Paul Desautels at the Tampa Federation Show in 1979. Refer to Chapter 17, page 397 ("Cops and Robbers").

*Plate #18-6, Specimen #5871* **SKUTTERUDITE with serpentine – Arhbar Mine, 270 Level, Bou Azze, Quarzazate, Morocco. (MH)**

**SKUTTERUDITE WITH SERPENTINE**
**(COBALT NICKEL ARSENIDE)**
**#5871 12.0×11.0×8.0cm Plate #18-6**

**Arhbar Mine, 270 Level, Bou Azze, Quarzazate, Morocco**

This specimen is a group of shiny lustrous rhombo-dodecahedra with one 7.0cm crystal. It was found in the late 1960s and is the world's finest Moroccan skutterudite according to all knowledgeable experts. It is from the noted French dealer Christian Gobin.

***Plate #18-7, Specimen #5882*** **SPINEL – Mogok, Myanmar (formerly Burma). (MH)**

**SPINEL (MAGNESIUM, ALUMINUM OXIDE)**
**#5882 7.5×7.5×4.5cm Plate #18-7**
**Mogok, Myanmar (formerly Burma)**

An exceptionally showy vitreous red XL specimen, consisting of a superb spinel twinned octahedron, 3.5cm on an edge, perched aesthetically on a vivid white etched calcite. The macle (twin) is unusually well displayed on this crystal from the classic locality in Myanmar (formerly Burma).

# Index

OFFICIAL FIRST DAY COVER
National Gem and Mineral Show
HOSTED BY
LINCOLN GEM AND MINERAL CLUB
A LINCOLN STAMP CLUB CACHET
LINCOLN JUN 13 1974 68501
FIRST DAY OF ISSUE
10 cents
UNITED STATES mineral heritage
Petrified wood
Tourmaline
Rhodochrosite

TOURMALINE—FOUND IN CONNECTICUT, NEW YORK, CALIFORNIA, OF VARIED COLORS.
COMMEMORATING
American Mineral Heritage
Fleetwood
OFFICIAL FIRST DAY COVER
LINCOLN, NE JUNE 13 1974
FIRST DAY OF ISSUE
10 cents
Tourmaline
UNITED STATES mineral heritage